T0240087

# Fractional Derivative Modeling in Mechanics and Engineering

Wen Chen · HongGuang Sun · Xicheng Li

# Fractional Derivative Modeling in Mechanics and Engineering

Wen Chen
College of Mechanics and Materials
Hohai University
Nanjing, China

HongGuang Sun
College of Mechanics and Materials
Hohai University
Nanjing, China

Xicheng Li
School of Mathematical Sciences
University of Jinan
Jinan, China

*Edited by*
HongGuang Sun
College of Mechanics and Materials
Hohai University
Nanjing, China

Yingjie Liang
Hohai University
Nanjing, China

This book was funded by the Natural Science Foundation of Jiangsu Province, grant number: BK20190024 and the National Natural Science Foundation of China, grant numbers 11972148.

ISBN 978-981-16-8804-1          ISBN 978-981-16-8802-7   (eBook)
https://doi.org/10.1007/978-981-16-8802-7

Jointly published with Science Press
The print edition is not for sale in China (Mainland). Customers from China (Mainland) please order the print book from: Science Press.
ISBN of the Co-Publisher's edition: 978-7-03-026857-0

Translation from the Chinese language edition: *Fractional derivative modeling in mechanics and engineering* by Wen Chen, et al., © China Science and Technology Press 2010. Published by China Science and Technology Press. All Rights Reserved.

This Springer imprint is published by the registered company Springer Nature Singapore Pte Ltd.
The registered company address is: 152 Beach Road, #21-01/04 Gateway East, Singapore 189721, Singapore

*We dedicate this book to the honorable memory of Professor Wen Chen*

# Preface

Classic Newtonian mechanics assumes that space and time are continuous everywhere. The basic physical quantities (e.g. speed, acceleration and force) can be described by an integer-order differential operator; thus, the physical and mechanics evolutions can be accurately described by using integer-order differential equations, for example, the Fourier heat conduction equations and Hamilton equations of classical mechanics. The above scientific research methods and models have achieved huge success in classical mechanics, acoustics, electromagnetics, heat transfer, diffusion theory and even in modern quantum mechanics and relativity. However, physicists, dynamicists and engineers have found that more and more anomalous phenomena cannot be explained from this viewpoint. For example, Richardson pointed out that in 1926, a turbulent velocity field is non-differentiable, which may be a critical obstacle in solving turbulence problems and cannot be tackled by using the traditional Newtonian mechanics. Moreover, a large number of experiments have shown that the stress relaxation of the viscoelastic material (including viscoelastic and rheology materials) is non-exponential decay (non-Debye) and has memory properties. It causes the conventional integer-order viscoelastic constitutive model can not accurately describe their mechanics behavior. Anomalous diffusion has attracted widespread concern in recent years, and involves the property of history dependency, path dependency and global correlation of physical and mechanics processes. It has been confirmed that classical Darcy law, the Fourier heat conduction law, Newtonian viscosity and Fickian diffusion law cannot accurately describe the above anomalous physical and mechanics processes.

From the mechanics modeling viewpoint, standard integer-order time derivative by local limit definition is not suitable to describe the history-dependent process. On the contrary, the time-fractional derivative is actually a differential-integral convolution operator; its integral term can fully reflect the history dependency of the system function, and is a powerful mathematical tool for modeling strong memory-dependent processes. On the other hand, the fractional Laplace operator (fractional Laplacian) is a typical non-local space fractional derivative, which can accurately describe the anomalous mechanics behavior (such as path dependence and long-range character-istics) in complex fractal spatial structures. It has overcome the theory of classical mechanics on the basis of Euclidean geometry and absolute time and space.

Fractional calculus is an ancient and fresh concept. In the early stages of integer-order calculus history, there are some mathematicians who began to consider the meaning of fractional calculus, such as L'Hospital, Leibniz and so on. However, it did not attract more attention and has not been further studied, due to the lack of application background and many other reasons. With the development of the natural and social sciences, the demand for complex engineering applications, espe-cially the fractal study of a variety of complex systems since the 1970s and 1980s, the theory of fractional calculus and its applications began to receive extensive atten-tion. From the beginning of the twenty-first century, fractional calculus modeling methods and theory have achieved successful applications in high-energy physics, anomalous diffusion, complex viscoelastic material mechanical constitutive rela-tions, system control, rheology, geophysics, biomedical engineering, economics and many other fields. Its unique advantages and irreplaceability are highlighted, and related theoretical and applied research has become a hot spot worldwide.

Meanwhile, the non-local nature of fractional calculus, resulting in the numerical simulation computation and storage capacity of the fractional derivative governing equation increases with the size of the problem. Hence, some effective numer-ical methods designed for integer-order equations are no longer applicable for fractional-order equations. Moreover, a lot of fractional derivative equation models are phenomenological descriptions, their physical and mechanical mechanism is not clear, pending further research.

Until now, several English language monographs on the introduction of fractional calculus have been published around the world. For example, Oldham and Spanier "The Fractional Calculus" (Academic Press, Inc., San Diego, 1974), Samko, etc. "Fractional Intergrals and Derivatives: Theory and Applications" (1987, Russian version, 1993 published in English), Miller and Ross "An Introduction to the Frac-tional Calculus and Fractional Differential Equations" (John Wiley & Sons, Inc. and, New York, 1993), Podlubny "Fractional Differential Equations" (Academic Press, New York, 1999) and Kilbas et al. "Theory and Applications of a Fractional Differ-ential Equations" (Elsevier, Amsterdam, 2006). However, these monographs focus on the mathematical theory of fractional calculus or its application in a particular area and did not fully take into account fractional calculus, and its applications are still a new thing for most researchers. We believe most researchers need a primer on fractional calculus theory and its application.

This book aims to provide graduate students with a textbook and an introductory book on the theory and applications of fractional calculus. This book offers detailed knowledge of fractional calculus modeling and numerical simulation of complex mechanical behavior, with the combination of recent research works of authors. This book introduces fractional calculus theory and its applications from the aspects of mathematical foundations, fractal and fractional calculus relations, unconventional statistics and anomalous diffusion, typical applications of fractional calculus, numerical solutions of fractional differential equations, etc. Here, we also further explore the prospects for the development of fractional calculus modeling.

The book focuses on the applications of fractional calculus in mechanics and physical modeling, emphasizing the physical and mechanics background and concept of fractional calculus modeling. Meanwhile, this book avoids too much introductions of mathematical background and rigorous mathematical proof, and strives to introduce the basic knowledge of fractional calculus to our readers. Interested readers may refer to the above-mentioned monographs and listed references about detailed mathematical analysis and proof. This book also contains some latest research achievements on fractional calculus theory and its application, such as positive fractional derivative, fractal derivative, variable-order derivative, distributed-order derivative and their applications, and fractional derivative continuum mechanics model of multi-scale turbulence.

This book provides some numerical examples and simulation results to enhance the understanding of fractional calculus definitions and concepts. Each chapter also discusses the relevant aspects of the existing problems. The last chapter is the summary and prospects. The key problems which should be solved in the future are also presented. Those key issues are raised by some scholars in the international conference "The Third IFAC Workshop on Fractional Differentiation and its Applications", held in Turkey in 2008. We hope that these elements can give readers inspiration to deepen their understanding and knowledge of the theory and applications of fractional calculus. Here, we should point out that there are too many papers on the theory and applications of fractional calculus; herein, it is impossible to enumerate all of them; interested readers can refer to the reference list of this book for more information; we would be pleased to receive the feedback from readers.

This book is presided over by Wen Chen, HongGuang Sun and XiCheng Li. The framework of this book was firstly proposed by Wen Chen, and was finally determined after a full discussion of all the authors. Wen Chen makes the overall arrangements for the book writing; Chap. 1 is mainly written by Wen Chen; Chap. 2 is mainly written by HongGuang Sun and Linjuan Ye; Chap. 3 is mainly written by Shuai Hu and Xiaodi Zhang; Chap. 4 is mainly written by Wen Chen and HongGuang Sun; Chap. 5 is mainly written by Xiaodi Zhang, Wen Chen and HongGuang Sun; Chap. 6 is mainly written by Wen Chen, HongGuang Sun and Xiaodi Zhang; Chap. 7 is mainly written by Wen Chen and Xi-Cheng Li; Appendix is mainly written by HongGuang Sun and Linjuan Ye. Xicheng Li is responsible for the main work of modification and typesetting. The authors of this book thank Prof. Yangquan Chen, Prof. Keqin Zhu, Prof. Changpin Li, Dr. Deshun Yin, Prof. Ning Chen and Dr. Yan Li for their help in writing this book and its modification; their suggestions and comments have improved the quality of this book.

This book was supported by National Basic Research Program of China (973 Project No. 2010CB832702), R&D Special Fund for Public Welfare Industry (Hydrodynamics, Grant No. 201101014), National Science Funds for Distinguished Young Scholars (Grant No. 11125208), Natural Science Foundation of China (Grant No.

11202066) and Program of Introducing Talents of Discipline to Universities (111 project, Grant No. B12032).

Time and knowledge being limited, errors and inadequacies are inevitable; any suggestions and comments are welcome.

Nanjing, China                                                    Wen Chen
February 2013

# Contents

# Chapter 1
# Introduction

## 1.1 History of Fractional Calculus

A watershed between modern mathematics and classical mathematics is the invention of differential and integral calculus by Newton (1642–1727) and Leibniz (1646–1716). It had undergone a fundamental change in development and applications of mathematics, and mathematical analysis, geometry and algebra become three basic mathematics research directions and tools thereafter [1]. However, fractional calculus is perhaps a strange and novel concept of mathematical analysis tool, for most researchers and engineers, even though it was actually put forward as early as three hundred years ago.

Leibniz firstly introduced $d^n y / dx^n$ to denote derivative, and this symbol promoted L'Hospital's contemplation on fractional derivative. In September 1695, L'Hospital wrote to Leibniz in the famous letter: For a simple linear function $f(x) = x$, whether the derivative order of a function can be a fraction rather than an integer, "What if the order will be 1/2". Leibniz wrote in his reply: "It will lead to a paradox, from which one day useful consequences will be drawn" [2–5]. Later, this problem becomes that the derivative of a function is an arbitrary order (fractional, irrational number or plural), so the "fractional calculus" is an inaccurate name (misnomer). However, due to historical reasons, fractional calculus has become the customary terminology; nowadays, the vast majority of researchers continues to use this name.

In history, Leibniz, Euler, Laplace, Lacroix and Fourier have paid attention on fractional calculus. Among them, Euler has made the critical first step. He noted that, when $p$ is a non-integer, the derivative of a power function $\frac{d^p}{dx^p} x^a$ is meaningful from a mathematical viewpoint. In 1812, Laplace proposed the idea that the fractional derivative of the function with integral form $\int T(t) t^{-x} dt$. In 1819, Lacroix repeated Euler's idea, and gave the answer to the question: $d^{1/2} x / dx^{1/2} = 2x^{1/2} / \sqrt{\pi}$ for the first time. The first scientist who used the fractional operators should be Niels Henrik Abel. In 1823, Abel introduced fractional calculus to represent the solution of the integral equation of the curve problems. In 1832, Liouville successfully applied his own proposed definition of fractional derivative, to solve the problem of potential

© Science Press 2022
W. Chen et al., *Fractional Derivative Modeling in Mechanics and Engineering*,
https://doi.org/10.1007/978-981-16-8802-7_1

theory. Thereafter, 1832–1837, a series of articles published by Liouville has made him the actual founder of the theory of fractional calculus. Following Liouville, Riemann, Fourier, Willian Center, Augustus De Morgan, GF Bernhard Riemann, A. Cayley and Weyl also did some important works and promoted the development of fractional calculus.

The first monograph on the theory of fractional calculus was published by KB Oldham and I. Spanier in 1974 [6]. In the same year, the first fractional calculus of the international conference was held at the University of New Haven in Connecticut, under the support of the USA National Science Foundation; the meeting attracted many famous mathematicians. Springer-Verlag published the proceedings of this conference [7]. In 1982, B. B. Mandelbrot pointed out for the first time the fact that a large number of fractal dimensions exist in nature and engineering, and self-similarity exists between the whole and its parts. Since then, fractional calculus has become a powerful tool in the research of fractal geometry and fractal dimension dynamics [8]. In 1984, Strathclyde District convened the Second International Conference on Fractional Calculus in Scotland. The Third International Conference was held at the University of Tokyo, Japan. In 1987, S. Samko, A. Kilbas and O. Marichev published a monograph which gave a more comprehensive introduction of fractional calculus. The book is originally in Russian; the English version was published in 1993.

In addition to the previously mentioned mathematician in the history of the development of fractional calculus, mathematicians and physicists such as Grünwald, Hadamard, Letnikov, Hardy, Riesz, Marchaud, Littlewood, Erdelyi, Fox, Kobe, Love, McBride, Ross, Srivastava and Caputo also made important contributions. As for the detailed history of the mathematical theory of fractional calculus, the reader can refer to the monograph of Samko, Kilbas and Marichev [5].

Currently, the definitions of the fractional operator include Riemann–Liouville type, Caputo type, Grünwald–Letnikov type, Weyl type, Erdelyi–Kober type and Riesz Marchaud–Hadamard Fractional Calculus [9, 10]. It should be pointed out that an intrinsical link can be established between these definitions. After long-term continuous efforts of many scholars, the theory of fractional calculus to some extent was established [3]. Practical engineering applications of fractional calculus still encounter a number of obstacles, including the mathematical foundation of fractional calculus is still not perfect, different definition is needed for different cases, and at the same time, Fourier and Laplace transform of the fractional derivative is also problematic. On the other hand, in 300 years of history since fractional calculus was proposed, it did not get a wide range of applications and concerns except for in the disciplines of physics and mechanics in the last two or three decades. It is considered a purely theoretical topic which is the object for mathematicians' curiosity-driven research. Perhaps, a more important reason is that the inherent contradictions exist between the fractional operators and the classical physics system established on a smooth continuous spacetime concept base [2, 3].

At the end of the nineteenth century, physical scientist Heaviside published a series of articles, which shows that fractional calculus can be applied to solve specific integer-order differential equations. Though his method is not strict from a mathematical viewpoint, his approach has been proved to be very effective in solving engineering problems such as current transmission in cables.

Heaviside's results were later proved to be correct, but his mathematical process is not perfect, this work was further improved by Bromwich in 1919 [4]. This method is called the Heaviside Operational Calculus. The novel idea of Heaviside contributed significantly to the development of fractional operators. However, fractional calculus has not been applied to physical and mechanics modeling in dealing with scientific and engineering problems.

In the 1940s, the mechanics scientists Scott Blair [11] and Gerasimov [12] proposed a fractional derivative model to characterize the mechanic property of material which is between Newtonian fluid and idea solid characterized by Hooke's law. Geophysicists Caputo and Mainardi [13, 14] applied fractional calculus to complex viscoelastic and rheological media, which is a new development of the mechanics model. More importantly, Caputo [13] developed a new definition different from the traditional definition of the Riemann–Liouville fractional derivative (in the literature, it is called the Caputo definition), to overcome the strong singularity and the naturally containing initial conditions in the former definition. Caputo's definition has been used to solve practical problems in a very wide range of applications. The first doctoral dissertation on viscoelastic material modeling of fractional calculus was completed by Bagley–Torvik under the guidance of Bagley [15]. After that, fractional calculus in modeling viscoelastic materials and other complex mechanics processes has drawn more and more attention.

In 1965, Prof. Mandelbrot from Yale University introduced the fractal concept, and pointed out that large numbers of phenomena in nature and engineering own the fractal dimension; its nature is the self-similarity between the global and local [8]. He noted that the fractional Brownian motion and the Riemann–Liouville fractional calculus definition have a close intrinsical link. Since then, as the basis of fractal geometry and fractal dynamics, fractional calculus, especially fractional calculus and fractional differential equations received widespread attention [3]; the focus of fractional calculus research has gradually transferred from pure mathematics to other disciplines.

Since the end of the last century, benefited by the researches of anomalous diffusion in porous media mechanics, non-Newtonian fluid mechanics, viscoelasticity, soft matter physics and mechanics [16], theory and application research of fractional derivative again attracted wide attention and the related research dramatically increased [17–20]. Because global correlation and history-dependent features of anomalous diffusion can be well represented by a fractional derivative, fractional diffusion models received a wide range of applications in the anomalous diffusion. Moreover, the successful application of the continuous-time random walk (CTRW) model in describing the anomalous diffusion and its close ties with the fractional calculus [21] has made anomalous diffusion to be most exciting research field. According to statistics [10], the total number of articles about fractional calculus application in anomalous diffusion has reached 1791 until 2003. There are about

1047 articles in the time period 1997–2003. Nowadays, anomalous diffusion is still a research hotspot in several research fields.

A series of international conferences on "Fractional Differentiation and its Applications" is held every 2 years, the first time in France in 2004, and in 2006 in Portugal, in 2008 in Turkey, in 2010 in Spain and in 2012 in China (Hohai University). In addition, in a series of international conferences sponsored by the American Society of Mechanical Engineers (ASME), the International Conference on Multibody Systems, Nonlinear Dynamics, and Control, all include the symposium "Fractional Derivatives and Their Applications". In some other conferences, the number of small meetings of fractional calculus gradually has increased, such as the National Conference of the Chinese Society of Theoretical and Applied Mechanics (2009), The Third International Conference on Dynamics, Vibration and Control (2010), "Academic Day of Fractional Dynamics and Control" organized by the Shanghai University 2010, etc.

There are also many academic journals focusing on fractional calculus, such as *Fractional Calculus and Applied Analysis, Journal of Fractional Calculus* and *Fractional Dynamic Systems.* Other related journals include *Chaos, Solitons & Fractals, Physics Review Letters、Physica A、Physics Letters A、Physics Review E、Journal of Computational Physics、Computers and Mathematics with Applications、Nonlinear Dynamics,* etc.

In recent years, there are more than 500 research papers involving fractional calculus each year; its theoretical and application research has penetrated almost all disciplines and application areas. This book is mainly related to the introduction of the basic mathematical theory of fractional calculus. In the application fields, we mainly focus on the application of fractional derivative, fractional integral beyond the main scope of this book (Fig. 1.1).

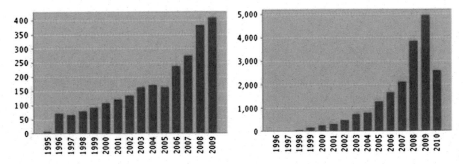

**Fig. 1.1** Statistical result of published papers on fractional calculus; the keywords are "fractional differential" or "fractional integral" or "fractional calculus" or "fractional order", from Web of Science (time interval: 1995–2009; Date: 2010-5-23). Notes: The retrieved result obtained by using the above keywords does not contain all of the literature related to fractional calculus, the right 2010 citation data until 2010-5-23

## 1.2  Geometric and Physical Interpretation of Fractional Derivative Equation

The modeling problems in physics, mechanics, biology and engineering are a major force in promoting the development of theoretical and application research of fractional calculus, and the order of fractional calculus in these models has a specific physical or geometric interpretation. From the application point of view, the nature of the difference between the fractional derivative model and integer-order derivative model lies in the time direction: integer-order derivative characterizes the property at a given moment in a physical or mechanics process, and the fractional derivative characterizes the property which is history-dependent. Meanwhile, an integer-order spatial derivative describes the local property of a given physical process, and the fractional derivative describes the property which is globally related.

Based on the assumption of absolute space and time, as well as Euclidean geometry, classical Newtonian mechanics considers space and time do not have start and end points, and everywhere continuous. The obtained physical quantities such as velocity, momentum, acceleration and force may be represented by an integer-order differential operator, and the evolution of the physical phenomena can thus be described using standard integer-order differential equations [3, 22]. The examples include the Euler–Lagrange equation and Hamilton equation in classical mechanics. These scientific research methods and models have received huge success in classical mechanics, acoustics, electromagnetics, heat transfer, diffusion theory and even modern quantum mechanics.

However, physical scientists have found more and more "anomalous" phenomena cannot be explained with this view and modeled by the classical physical and mechanics approaches. For example, Richardson in 1926 [23] pointed out that the turbulent velocity field is non-differentiable, it may be an important reason why the traditional Newtonian mechanics encounter long-term stagnancy in solving turbulence problems. For another example, a large number of experiments have shown that the stress relaxation of many viscoelastic materials (including viscoelastic solid and fluid substances) is non-exponential type (non-Debye). The stress relaxation process has memory, and the traditional viscoelastic differential constitutive model cannot accurately describe their mechanics behavior [24]. In recent years, anomalous diffusion has received widespread concern in the fields of high-temperature and high-pressure plasma motion [25], changes in the financial markets, the pollutant migration in the natural environment, turbulence [26, 27] as well as soft matter [28, 29] (also known as complex fluid, materials between the ideal solid and liquid states mostly are composed by macromolecules or groups, often multi-phase medium) heat conduction, diffusion and seepage [30], dissipation [31, 32] and electron transport [33], anomalous diffusion is the common features of a variety of complex phenomena. Note that "anomalous" diffusion is relative to the "normal" diffusion of ideal solid and fluid; its constitutive relation violates standard "gradient" laws (for example,

Darcy law, Fourier heat conduction law, Newtonian viscous fluid and Fickian diffusion law) [22, 29]; the corresponding empirical fitting formula obtained by experimental data is expressed as a function in the form of power law (power law). Note that "diffusion" should be broadly interpreted, also including the anomalous energy dissipation problem [31]. These "anomalous" phenomena involve the physical and mechanics processes of memory and hereditary [34–36], path dependence and global correlation.

Mingyu Xu and Wenchang Tan mentioned three challenges of the classic physical mechanics, when exploring the physical and mechanics backgrounds of fractional calculus: (1) Fluctuation of the turbulence velocity field, the randomness in dramatic amplitude and direction change of atmospheric turbulence velocity field cause the velocity is not differentiable in turbulence. (2) Brown motion. (3) Constitutive relation of complex viscoelastic material. Many complex viscoelastic materials have memory characteristics; Newton's law for viscous fluid and Hooke's law for elastic solid cannot accurately describe this feature; the complex viscoelastic material is actually the medium between the ideal elastic and viscous.

Next, we explore the physical and mechanics mechanisms of the fractional derivative in five specific areas.

### 1.2.1 Frequency-Dependent Energy Dissipation Process

This section gives an example using the definition of the Riemann–Liouville fractional derivative, to explain the difference between the classical model and the fractional model.

The definition of the Riemann–Liouville derivative can be written as

$$\frac{d^p f(t)}{dt^p}\Big|_a^t = \frac{1}{\Gamma(m+1-p)} \frac{d^{m+1}}{dt^{m+1}} \int_a^t (t-\tau)^{m-p} f(\tau) d\tau, (m \leq p < m+1),$$

$$(1.2.1)$$

where $p$ can be an arbitrary non-real number, and $m$ is the non-negative integer.

The above definition illustrates that the fractional derivative is in fact a differential integral operator; it means the current state is dependent on all previous history of the whole process. Hence, the fractional derivative can well represent the property of global correlation and history memory; $(t-\tau)^{m-p}$ is the kernel to characterize memory.

Applying the Fourier transform into fractional derivative, one can obtain

$$F\{d^p u(t)/dt^p, \omega\} = (i\omega)^p U(\omega). \qquad (1.2.2)$$

Thereby, the frequency domain expression of a response is obtained. Since $p$ can be an arbitrary non-real number, it is devoted to the arbitrary frequency-dependent property.

The actual physical or mechanics process is usually arbitrary frequency-dependent, while integer-order operators are only designed for integer-order frequency-dependent cases; hereby, it encountered significant problems in complex physical modeling of mechanical phenomena. Please read Sect. 5.5 for details. $(i\omega)^p$ can be decomposed as $ai + b$; the real part indicates the rate of dissipation, and the value of the imaginary part denotes the oscillation frequency.

## *1.2.2 Fractal Description and Power-Law Phenomena*

Mandelbrot [37] pointed out that there are a lot of fractal dimension phenomena that exist in nature and engineering fields, which relates to the complex physical and mechanics processes. The scale of time or space should be changed into $(x^\alpha, t^\beta)$ when investigating these kinds of processes [22]. In those fractal processes, the integer-order gradient does not exist and cannot be satisfied. Therefore, researchers began to introduce fractional calculus, to redefine the basic concepts of the physical and mechanics theory to overcome the limitations of integer-dimensional scale.

Most of the classical physics and mechanics theories were established on a conservative system using the definition of the integer-order calculus, which means there is no energy exchange between the inside and outside of the system, namely the conservation of energy. Contrarily, the fractional derivative-defined system is an open system, which has energy dissipation and is non-conservative. Traditional integer-order dynamic systems can be seen as a special case of the fractional-order dynamic system.

There is no determined definition of fractal, but it is expressed by using the following features. Fractal means 1. a fine structure at arbitrarily small scales; 2. self-similarity between part and whole, part and part, as well as a whole and the overall similarity (one is similar to the strict mathematical sense, another is similar to statistical sense); 3. scale invariance: there is no characteristic scale, amplification does not change characteristics; 4. fractal geometry is not Euclidean geometry but non-Euclidean geometry.

It means the fractal dimension cannot be simply defined by degrees of freedom dimension, but through the Hausdorff dimension [22]. The dimension value obtained thus is usually not an integer; the fractal dimension is greater than its topological dimension, less than the Euclidean dimension in the space.

Fractional-order differential operator is closely related to the fractal dimension. This relationship is particularly important in investigating the phenomenon of anomalous diffusion or solute transport problem in porous media (especially with a fractal porous media), but the exact relationship between the fractal dimension and fractional calculus is still not completely clear and needs further study.

It is also necessary to mention the concept of fractal functions. There are a lot of similarities between turbulence, Brownian motion in the mathematical sense. For example, the Weierstrass function raised by Weierstrass is a fractal function, which is everywhere continuous but not everywhere differentiable. A fractal function has no characteristic scale, is non-differentiable and scale invariant. Scale invariance is closely related to the renormalization group theory, and the memory integral is the feature of fractal. That is, a fractal function has fractal dimension; it is a bridge to exploring the intrinsic link between fractal and fractional calculus. Although the fractal function is non-differentiable from an integer-order derivative viewpoint, it has fractional differentiability; one can use the theory of fractional calculus to analyze it.

The power-law phenomenon is widespread in nature and is a hotspot in scientific research. In natural sciences, it was found in the field of complex networks, structure of the protein, process of decay of radioactive substances, turbulence and fractal phenomenon. In social sciences, the distribution of the population, the size distribution of the city and the appearing frequency of letters of the alphabet, all can be well described by using the power-law function.

In the fractional calculus study, researchers found the solution of fractional differential equations having the power-law function form. A fractional differential equation has a corresponding relationship with the power-law phenomenon, which can be used to describe the power-law phenomenon. Chapter 5 will give details about this topic.

### 1.2.3  Anomalous Diffusion

From a physical and mechanics modeling viewpoint, the standard integer-order time derivative is local-defined and not suitable to describe the history-dependent process. But the fractional differential operator is actually a differential integral operator in which the integral term fully reflects the history dependence of the system function, hereby is a powerful mathematical tool for strong memory process modeling [30, 35, 38]. On the other hand, fractional Laplacian is a typical non-local operator to describe an anomalous phenomenon which is global-related and path-dependent [22, 38], in fractal structure media [39]. Therefore, time and space fractional diffusion equations can well different types of anomalous diffusion processes, and has become the master equation for depicting power-law characteristics [22, 29, 40]. Below is the standard fractional diffusion equation:

$$\frac{\partial^\alpha s}{\partial t^\alpha} + \gamma\left(-\nabla^2\right)^\beta s = 0, \quad 0 < \alpha, \beta \le 1. \tag{1.2.3}$$

Here, $s$ denotes the considered physical quantity (such as concentration and temperature), $\gamma$ is the normalized physical coefficient (such as diffusion coefficient or thermal conductivity coefficient) and $(-\nabla^2)^\beta$ represents fractional Laplacian operator [39, 41]. Note that here $\alpha$ and $\beta$ can be real numbers. The fundamental solution of the above equation can be expressed as a time-dependent Lévy stable distribution form, and $2\beta$ is the Lévy stable distribution indicator [30, 35]. This equation also implies the fractional Brown motion with memory (because fractional Brown motion exhibits obvious long-range dependency feature, which can be analyzed by using fractional derivative), where $\alpha$ is the memory indicator of the system [22]. When $\beta = 1$, the smaller the $\alpha$ is, the strong the memory of the physical or mechanics process [35]. The fractional derivative model overcomes the backwards of integer-order model in bad data fitting result, and can give satisfied result with fewer parameters in describing some complex physical processes. However, it should be noted that the used fractional derivative model may be more complex than (1.2.3).

### 1.2.4 Constitutive Relation of Complex Viscoelastic Material

It is well known that the stress-strain relation of the ideal elastomer obeys the Hooke law:$\sigma(t)$ $\varepsilon(t)$; Newton fluid obeys the Newton law: $\sigma(t)$ $d^1\varepsilon(t)/dt^1$. If we replace the Hooke law with $\sigma(t)$ $d^0\varepsilon(t)/dt^0$., one can guess the stress–strain relation of material which is between ideal elastomer and Newton fluid, may follows $\sigma(t)$ $d^\beta\varepsilon(t)/dt^\beta$ $(0 \le \beta \le 1)$, this revolutionary view was proposed by Scott Blair [11] and Gerasimor [12], respectively. The new model can describe the ideal elastomer and Newton fluid, and can also characterize the material in between. This viewpoint can be applied to characterizing the constitutive relation of a series of viscoelastic materials (liquid crystal, rubber, polymers, sediments and proteins). In the fractional constitutive relation of viscoelastic materials, the derivative order is related to its physical property. Hence, we can define the type of viscoelastic material by using the fractional derivative order, and thus have a more in-depth understanding of the mechanics properties of the viscoelastic material. Section 5.4 will give details about this topic.

Classical mechanics, including the theory of mechanics, elasticity and fluid dynamics, and the basic mechanics constitutive relations are as follows:

**Hooke's law for elasticity**: $F = kx$ ($k$ denotes the elasticity coefficient; $x$ is the displacement);

**Newton's law for viscous fluid**: $F = \upsilon \frac{\partial u}{\partial y}$ ($\upsilon$ is the viscosity coefficient, $u$ represents the velocity and $y$ is spatial variable);

**Newton's second law for Rigid body**: $F = m \frac{d^2 x}{dt^2}$ ($m$ is the mass, $x$ is the displacement and $t$ is the time).

It can be observed that the above constitutive relations involve time derivative order 0, 1 and 2 (displacement, velocity and acceleration). When describing the mechanics behavior of multi-phase media (such as soft matter, complex fluid, mixture of air and solid), the classical model usually cannot work well. To overcome this problem, the often-used way is to couple several models into a complex model. However, from the fractional derivative modeling approach, Westerlund [42] has suggested a simple and universal way to describe the constitutive relations of complex materials using fractional derivative:

$$F = \rho \frac{d^\alpha x}{dt^\alpha}, \ 0 < \alpha \leq 2$$

### 1.2.5   Fractional Schrodinger Equation

The Schrodinger equation is a basic equation in non-relativistic quantum mechanics. Feynmann and Hibbs [43] described the non-relativistic quantum mechanics as path integral based on the Brownian motion. Their results are consistent with that described by the Schrödinger equation of quantum mechanics. When describing the motion of the particles via the traditional Schrödinger equation, it assumes that the motion does not have a memory, which means the current movement of particles is not dependent on its history; moreover, particle motion has the property of spatial locality. Laskin [44] obtained the operator represented with the space Riesz fractional Schrodinger equation using the path integral method, based on the Lévy flight path. Chen Wen [45] proposed a Fractional Plank quantum relationship, which is directly derived from the fractional derivative Schrödinger equation. Naber [46] gave the time-fractional Schrödinger equation employing the Caputo operator, but the obtained Hamilton function is time history-dependent. The main conclusion is the probability of the time-fractional Schrödinger equation described is not conserved. These results are very different from conventional quantum mechanics.

## 1.3   Application in Science and Engineering

In recent years, the fractional derivative has received widespread concern in the study of the "complexity" of the social and physical phenomena; a common characteristic of these complex phenomena is the power-law behavior. It is worth emphasizing that the traditional integer-order derivative model, including the nonlinear model cannot well characterize these power-law phenomena. In order to accurately characterize properties of power-law phenomena, such as the non-locality, frequency, path or history-dependent properties, researchers should use new mathematical modeling methods such as the fractional derivative modeling method.

Fractional derivative modeling research is interdisciplinary, widely used in a number of disciplines, such as physics, chemistry, biology, materials, medical science, mechanics, economics, social science, control theory and signal and image processing [4, 5, 9, 41]. The purpose of this book is to give readers a global understanding of fractional derivative modeling, which will help generate new research in terms of the theoretical basis, experiments, computational physics, etc. and further expands application potentials of the fractional derivative in different research areas.

Until now, although mathematicians have established a preliminary theoretical framework of fractional calculus, it still has not been widely used by scholars and engineers of other disciplines as an important modeling method. An important reason is that many engineering experts need adequate mathematical training. At the same time, most of the engineering experts believe that the integer-order differential equation theory has provided a good enough mathematical tool to solve engineering problems. Recently, however, in many areas such as viscoelastic material, medical testing, geophysical signal processing and control theory, the researchers have realized that fractional calculus is not only to provide a mathematical modeling tool, but it also brings dramatical change in engineering application and scientific research [3, 9].

## 1.4 Anomalous Diffusion Modeling in Environmental Mechanics

In the field of environmental mechanics, a lot of problems involve anomalous diffusion, such as geotechnical engineering seepage problem, oil recovery rate in reservoir engineering and transport of nuclear substances or pollutants in the formations. An important feature of these diffusion phenomena is that the diffusion process does not meet Fick's second law, and is a non-Markov process. Fractional calculus has served as a suitable mathematical tool to investigate such complex problems. Please read Chap. 4 in this book, to find more contents on modeling methods and applications of fractional calculus.

## 1.5 Constitutive Relation of Viscoelasticity

The mechanics properties of viscoelastic material used in a large number of practical engineering are essentially different from that of a single elastic material. Research concerns on viscoelasticity include creep, relaxation, flow, strain rate effect and the long-term strength effects, and research in this area is of great significance to the safety of the actual project structure. The presented methods such as the Maxwell model, Voigt model and Kelvin model cannot accurately describe the complex mechanics behavior of these materials. The fractional derivative has been widely applied to the modeling of the constitutive relationship of viscoelastic materials; please see Sect.5.4 for detailed applications.

## 1.6  Biomedical Science

In biomedical science, the wide range of applications of fractional calculus can promote biological engineers to improve the design of biomedical equipment, description and control ability. Fractional calculus is also used to simulate the cancer cell spread process in the human tissue. Fractional calculus can also be used as a modeling tool to simulate the diffusion process of the drug in human tissues. The theoretical study of fractional calculus applications in drug delivery system control has begun. Because of the frequency dependence property of ultrasonic dissipation, fractional calculus has been used for the numerical simulation of ultrasonic medical image detection and has achieved valuable results. The detailed content is stated in Sect. 5.5 of this book.

## 1.7  System Control

In 1960, Japanese scholar Manabe proposed the concept of fractional control. Subsequently, the French scholar Oustaloup [47] established CRONE control (Fractional robust control) in 1981, and it has been applied to many practical problems. CRONE control is currently the most widely used fractional control method, and Oustaloup has shown the CRONE controller is better than the conventional used PID controller. In addition, the Slovak Republic scholar Podlubny proposed the concept of fractional-order PID controller in 1999. Since then, the study of the fractional-order control has been rapidly developed, especially Chen Yangquan applied fractional control theory to the actual system control (such as Fractor). The practical application of the fractional-order PID controller has shown that it is better than the integer-order PID controller [48].

Nowadays, the fractional-order controller has been applied to various aspects of control theory: linear control, nonlinear control, adaptive control, optimal control, robust control, network control and so on. However, because the design and implementation of the fractional controller are more complex than that of the integer-order controller, it has not been widely applied in the real-world engineering field as a powerful tool. However, with the development of technology, computer science as well fast algorithms, we believe the fractional-order control will be widely applied in the actual engineering field.

## References

1. M. Klein, *Ancient and Modern Mathematical Thinking* [M] (Shanghai Science and Technology Press, 2009)
2. M.Y. Xu, W.C. Tan, Theory and method of fractional operator and its applications to the modern mechanics [J]. J. Taiyuan Univ. Technol. **36**(6), 752–756 (2005)

3. M.Y. Xu, W.C. Tan, Intermediate process, critical phenomena-fractional calculus theory, methods, progress, and its application in modern mechanics [J]. Sci. China Ser. G Phys. Mech. Astron. **36**(3), 225–238 (2006)

4. K.S. Miller, B. Ross, *An Introduction to the Fractional Calculus and Fractional Differential Equations* [M] (John Wiley and Sons, Inc., 1993)

5. S.G. Samko, A.A. Kilbas, O.I. Marichev, *Fractional Integrals and Derivatives: Theory and Applications* [M] (Gordon and Breach, 1993)

6. K.B. Oldham, J. Spanier, *The Fractional Calculus [M]* (Academic Press, New York-London, 1974)

7. B. Ross (Ed.), Fractional calculus and its applications, in Proceedings of the International Conference [C], New Haven, June 1974. Springer-Verlag New York, Inc., New York (1974)

8. B.B. Mandelbrot, *The Fractal Geometry of Nature [M]* (WH Freeman, New York, 1982)

9. I. Podlubny, *Fractional Differential Equations [M]* (Academic Press, San Diego, 1999)

10. A.A. Kilbas, H.M. Srivastava, J.J. Trujillo, *Theory and Applications of Fractional Differential Equations [M]* (Elsevier, Amsterdam, 2006)

11. G.W. Scott Blair, The role of psychophysics in rheology [J]. J. Colloid Sci. **2**, 21–32 (1947)

12. A.N. Gerasimov, A generalization of linear laws of deformation and its application to inner friction problems [J]. Prikladnaya Matematika i Mekhanika **12**, 251–259 (1948)

13. M. Caputo, *Elasticitàe Dissipazione [M]* (Zanichelli, Bologna, 1969)

14. M. Caputo, F. Mainardi, Linear models of dissipation in an elastic solids. Rivista del Nuovo Cimento (Ser. II) **1**(2), 161–198 (1971)

15. R.L. Bagley, Applications of generalized derivatives of viscoelasticity [D]. Ph. D. thesis, Air Force Institute of Technology (1979)

16. Q.S. Zheng, K.Z. Huang, ZP. Huang, J.X. Wang, QZ. Feng, L.L. Wang, Y.x. Gu, H.Y. Hu, T.Q. Yang, The mechanics at the turn of the century, in Thoughts on the 20th International Congress on Theoretical and Applied Mechanics [J]. Adv. Mech. **31**(1), 144–155 (2001)

17. R. Hilfer (ed.), *Applications of Fractional Calculus in Physics [M]* (World Scientific, Singapore, 2000)

18. W.C. Tan, W.Y. Wu, Z.Y. Yan, G.B. Wen, Moving boundary problem for diffusion release of drug from a cylinder polymeric matrix [J]. Appl. Math. Mech. **22**(4), 331–336 (2001)

19. J.A.T. Machado, R. Barbosa (Eds.), The proceeding of the 2nd IFAC workshop on fractional differentiation and its applications [C]. Porto, Portugal, 19–21 July 2006

20. J. Sabatier, O.P. Agrawal, J.A. Tenreiro Machado, *Advances in Fractional Calculus* [C]. Springer (2007)

21. R. Metzler, J. Klafter, The random walk's guide to anomalous diffusion: a fractional dynamics approach [J]. Phys. Rep. **339**, 1–77 (2000)

22. W. Chen, Time-space fabric underlying anomalous diffusion [J]. Soliton Fractal Chaos **28**(4), 923–929 (2006)

23. L.F. Richardson, Atmospheric diffusion shown on a distance-neighbour graph [J]. Proc. R. Soc. Lond. Ser. A. Containing Papers Math. Phys. Char. **110**, 709–737 (1926)

24. H. Schiessel, C. Friedrich, A. Blumen, Applications to problems in polymer physics and rheology, in *Applications of Fractional Calculus in Physics*, ed. by R. Hilfer [C] (Singapore: World Scientific Publishing Co Pte Ltd, 2000), pp. 331–376

25. D. del-Castillo-Negrete, B.A. Carreras, V.E. Lynch, Front dynamics in reaction-diffusion systems with Lévy flights: a fractional diffusion approach [J]. Phys. Rev.. Lett. **91**(1), 018301–4 (2003)

26. I.M. Sokolov, J. Klafter, A. Blumen, Ballistic vs. diffusive pair-dispersion in the Richardson regime [J]. Phys. Rev. E **61**, 2717–22 (2000)

27. W. Chen, A speculative study of 2/3-order fractional Laplacian modeling of turbulence: some thoughts and conjectures [J]. Chaos **16**, 023126 (2006)

28. T.A. Witten, Insights from soft condensed matter [J]. Rev. Mod. Phys. **71**(2), 367–373 (1999)

29. H.R. Ma, K.Q. Lu, The physics of soft condensed matter [J]. Physics **29**, 561 (2000)

30. R. Gorenflo, F. Mainardi, D. Moretti, G. Pagnini, P. Paradisi, Discrete random walk models for space-time fractional diffusion [J]. Chem. Phys. **284**(1/2), 521–541 (2002)

31. T.L. Szabo, J. Wu, A model for longitudinal and shear wave propagation in viscoelastic media [J]. J. Acoust. Soc. Am. **107**(5), 2437–2446 (2000)
32. W. Chen, S. Holm, Modified Szabo's wave equation models for lossy media obeying frequency power law [J]. J. Acoust. Soc. Am. **114**(5), 2570–2574 (2003)
33. H. Scher, E.W. Montroll, Anomalous transit-time dispersion in amorphous solids [J]. Phys. Rev. B **12**, 2455–2477 (1975)
34. M.F. Shlesinger, *Fractal Time and 1/f Noise in Complex Systems* [M] (New York: Annals of the New York Academy of Sciences, 1987)
35. Y.A. Rossikhin, M.V. Shitikova, Applications of fractional calculus to dynamic problems of linear and nonlinear hereditary mechanics of solids [J]. Appl. Mech. Rev. **50**, 15–67 (1997)
36. W. Chen, Lévy stable distribution and [0, 2] power law dependence of acoustic absorption on frequency in various lossy media [J]. Chin. Phys. Lett. **22**(10), 2601–2603 (2005)
37. B.B. Mandelbrot, *Fractals: Form, Chance and Dimension [M]* (W. H. Freeman & Co., San Francisco, 1979)
38. H.P. Xie, *Fractal—Introduction to Rock Mechanics* [M] (Sciences Press, 1996)
39. A.I. Nachman, J. Smith, R.C. Waag, An equation for acoustic propagation in inhomogeneous media with relaxation losses [J]. J. Acoust. Soc. Am. **88**(3), 1584–1595 (1990)
40. F.X. Chang, J. Chen, W. Huang, Anomalous diffusion and fractional advection-diffusion equation [M]. ACTA Phys. Sinica **54**(03), 1113–1117 (2005)
41. W. Chen, S. Holm, Fractional Laplacian time-space models for linear and nonlinear lossy media exhibiting arbitrary frequency dependency [J]. J. Acoust. Soc. Am. **115**(4), 1424–1430 (2004)
42. S. Westerlund, Causality, Report No. 940426, University of Kalmar (1994)
43. R.P. Feynman, A.R. Hibbs, *Quantum mechanics and path integrals [M]* (McGraw-Hill, New York, 1965)
44. N. Laskin, Fractional quantum mechanics[J]. Phys. Rev. E **62**, 3135–3145 (2000)
45. W. Chen, An intuitive study of fractional derivative modeling and fractional quantum in soft matter [J]. J. Vib. Control **14**, 1651–1657 (2008)
46. M. Naber, Time fractional Schrodinger equation [J]. J. Math. Phys. **45**, 3339–3352 (2004)
47. M. Moze, J. Sabatier, A. Oustaloup, LMI characterization of fractional systems stability, in *Advances in Fractional Calculus* [C], ed. by J. Sabatier, O.P Agrawal, J.A.T. (Machado, Springer, 2007)
48. Y.Q. Chen, K.L. Moore, Discretization schemes for fractional-order differentiators and integrators [J]. IEEE Trans. Circ. Syst. I: Fund. Theory Appl. **49**(3), 363–367 (2002)
49. J. Crank, *Free and Moving Boundary Problems [M]* (Clarendon Press, Oxford, 1987)

# Chapter 2
# Mathematical Foundation of Fractional Calculus

In the long history of the development of fractional calculus, a variety of definitions have been proposed by researchers from different perspectives, such as Riemann–Liouville, Grünwald–Letnikov, Weyl and Caputo typse definitions. Each definition has its expression and properties. Actually, the selection of fractional calculus definition depends on the problems of interest. In this chapter, four types of definitions commonly used are introduced.

Presently, there are four types of fractional calculus definitions commonly used in basic mathematics and application analysis: Grünwald–Letnikov, Riemann–Liouville, Caputo and Riesz type fractional calculus definitions. Compared to the Riemann–Liouville definition and the others, the Grünwald–Letnikov definition is defined as a limit of a fractional-order backward difference rarely applied for theoretical analysis, whereas it is commonly utilized for differential equation theory and numerical calculation. The Riemann–Liouville definition with an integro-differential expression which avoids limitation plays an important role in pure mathematics. Unfortunately, the Riemann–Liouville definition type leads to initial conditions containing the limit values of the Riemann–Liouville fractional derivative at the lower terminal of time or space. Their solutions are practically useless, because there is no known geometrical and physical interpretation for such types of initial conditions. For convenience in the real application, a modified definition, Caputo-type fractional derivative, is developed in viscoelasticity materials research. The Caputo definition type with initial conditions containing integer-order differential and integral expression is widely applied to practical problems.

This book is to introduce fractional calculus to more researchers from the application perspective. Therefore, in this chapter, we describe a few details of mathematical analysis and just briefly introduce four types of commonly used definitions and make a list of the other three types. This book focuses on the real domain; hence, the fractional orders are real values. For more details of the mathematical results and descriptions (for example, other types of definitions, complex domain and the differential order with complex values), readers can refer to the papers [1–3].

© Science Press 2022
W. Chen et al., *Fractional Derivative Modeling in Mechanics and Engineering*,
https://doi.org/10.1007/978-981-16-8802-7_2

## 2.1  Definition of Fractional Calculus

To distinguish fractional calculus from classical calculus, in this chapter, the symbol $_0D_t^\alpha$ is utilized, where $D$ represents fractional derivative and $I$ represents fractional integral; the script of the upper left represents the type of operator, as Riemann–Liouville is represented by "$RL$", "$C$" represents the Caputo operator; the script of the upper right $\alpha > 0$ denotes the fractional order; the scripts of the lower left and lower right represent the lower and upper terminals of the integral operator defined, respectively. To well understand, the fractional derivative is represented by the notation $d^\alpha/dt^\alpha$, the definition type and the lower and upper terminals of the integral will be illustrated additionally. If not specified, the default lower and upper terminals of integration can be considered as 0 and $t$, respectively.

### 2.1.1  Introduction of Fractional Calculus Definition

Scientific development makes more and more profound understanding of nature, for example, in mathematics, the first sight is to realize and use natural numbers, integers and fractions, and later negative numbers are involved, and then acknowledge is extended to real numbers, imaginary numbers, rational numbers and irrational numbers. Similarly, the emergence of fractal promotes the comprehension of the spatial dimension, which shows that only applied integer dimension to describe the space or substance geometric structure is inaccurate. The concept of fractal dimension also enhances the understanding of many physical processes. The integro-differential theory is one of the foundations for the development of modern mathematics and is also the cornerstone of the development of modern science. The emergence of the integro-differential approach greatly promotes the developments in natural science, social science and applied science, such as physics, mechanics, astronomy, biology, chemistry, engineering and economics. However, fractional calculus has a more open vision to recognize calculus.

For a specified function, the integrals and differentials can be expressed as follows:

Derivatives: $f, \frac{df}{dt}, \frac{d^2f}{dt^2}, \frac{d^3f}{dt^3}, \ldots\ldots$

Integrals: $f, \int f(t)dt, \int dt \int f(t)dt, \int dt \int dt \int f(t)dt, \ldots\ldots$

A sequence is composed of differentials and integrals together:

$$\ldots\ldots, \frac{d^{-3}f}{dt^{-3}}, \frac{d^{-2}f}{dt^{-2}}, \frac{d^{-1}f}{dt^{-1}}, f, \frac{df}{dt}, \frac{d^2f}{dt^2}, \frac{d^3f}{dt^3}, \ldots\ldots$$

Analogous to number set, it is obvious that these orders of differentials and integrals are integers and discrete. A greatly natural topic is proposed, whether the concept of calculus can be generalized or not, how to generalize it and what are the meanings of the generalized forms? Some scientists, especially mathematicians, thought deeply about this problem, put forward a variety of suppositions and theories, which opened up the history of the development of fractional calculus. This history of fractional calculus has been briefly introduced in the preface section. During more than two centuries of research, how to define a reasonable definition of fractional calculus is always one of the key topics. In the early stage, the major focus is the fractional calculus of power function.

From the knowledge of integer-order calculus, the following holds:

$$\frac{d^n x^m}{dt^n} = m(m-1)\dots(m-n+1)x^{m-n},$$

for $\Gamma(m+1) = m(m-1)\dots(m-n+1)\Gamma(m-n+1)$, hence, the above formula can be rewritten as

$$\frac{d^n x^m}{dt^n} = \frac{\Gamma(m+1)}{\Gamma(m-n+1)}x^{m-n}.$$

Based on the above expression, if $m$ and $n$ are non-integers, the fractional-order derivative can be obtained. A typical case is $m = 1$, $n = 0.5$, then it has $\frac{d^{0.5} x}{dt^{0.5}} = \frac{1}{\Gamma(3/2)}x^{0.5} = \frac{2}{\sqrt{\pi}}x^{0.5}$. The fractional calculus result of the power function reflects a basic comprehension of the definition form and properties of fractional calculus.

Two approaches related to establishing fractional calculus will be shown below. One approach starts with the limit definition (Grünwald–Letnikov definition) of the function derivative,

the first-order derivative: $f'(t) = \frac{df}{dt} = \lim\limits_{\Delta t \to 0} \frac{f(t)-f(t-\Delta t)}{\Delta t}$;

the second-order derivative:
$$f''(t) = \frac{d^2 f}{dt^2} = \lim\limits_{\Delta t \to 0} \frac{f'(t) - f'(t - \Delta t)}{\Delta t}$$
$$= \lim\limits_{\Delta t \to 0} \frac{f(t) - 2f(t - \Delta t) + f(t - 2\Delta t)}{(\Delta t)^2};$$

the third-order derivative:

$$f'''(t) = \frac{d^3 f}{dt^3} = \lim\limits_{\Delta t \to 0} \frac{f(t) - 3f(t - \Delta t) + 3f(t - 2\Delta t) - f(t - 3\Delta t)}{(\Delta t)^3};$$

and, by introduced, the $n$th derivative can be written as

$$f^{(n)}(t) = \frac{d^n f}{dt^n} = \lim\limits_{\Delta t \to 0} \frac{1}{(\Delta t)^n} \sum_{r=0}^{n}(-1)^r \binom{n}{r} f(t - r\,\Delta t).$$

If we generalize the fraction in the above formula, namely supposing $n$ as a non-integer number, then the Grünwald–Letnikov fractional derivative is derived. Simultaneously, for $n < 0$, it represents the Grünwald–Letnikov fractional integral.

This approach is a primary approach in early fractional calculus research, and the correlating works can be seen in the Refs. [1–3].

Another approach is based on the expression of the $n$-fold integral,

$$_aI_t^n f(t) = \int_a^t d\tau_1 \int_a^{\tau_1} d\tau_2 \ldots \int_a^{\tau_{n-1}} f(\tau_n)d\tau_n = \frac{1}{(n-1)!} \int_a^t (t-\tau)^{n-1} f(\tau)d\tau.$$

Suppose $n$ is a non-integer number; in view of the property of the gamma function, a type of left-side fractional integral is defined as

$$_aI_t^\beta f(t) = \frac{1}{\Gamma(\beta)} \int_a^t (t-\tau)^{\beta-1} f(\tau)d\tau.$$

Similarly, we also can have a right-side fractional integral,

$$_aI_t^\beta f(t) = \frac{1}{\Gamma(\beta)} \int_a^t (t-\tau)^{\beta-1} f(\tau)d\tau.$$

Based on the fractional integral definition, the fractional derivative is derived as below. Assume $n$ as a positive integer number, $\beta$ holds $n-1 < \beta < n$, then the $n$th derivative of the left-side fractional integral can be obtained as follows:

$$_aD_t^\beta f(t) = \frac{d^n}{dt^n} {}_aI_t^{n-\beta} f(t)$$

$$= \frac{1}{\Gamma(n-\beta)} \frac{d^n}{dt^n} \int_a^t (t-\tau)^{n-\beta-1} f(\tau)d\tau, \; n-1 < \beta < n,$$

which is the so-called left-side fractional derivative, where $_aD_t^\beta f(t)$ is a notation of the fractional derivative. Similarly, the right-side fractional derivative is defined.

There are three main expressions of fractional derivative: (1) $f^{(\beta)}(t)$, $\frac{d^\beta f}{dt^\beta}$; (2) $\frac{d^\beta f}{d(t-a)^\beta}$, $\frac{d^\beta f}{d(b-t)^\beta}$; (3) $_aD_t^\beta$, $_tD_b^\beta$; here, $\beta$ is the derivative order. For the convenience of reading, the first one is generally utilized, but the third one is considered to discuss different definitions of fractional derivative.

## 2.1.2  Riemann–Liouville Definition

Before introducing the Riemann–Liouville fractional derivative definition, we will introduce the Riemann–Liouville fractional integral definition firstly. For any arbitrary complex number $\alpha > 0$, the Riemann–Liouville fractional integral definition of the function $f(t)$ is defined as

$$I_{a+}^{\alpha} f(t) = \frac{1}{\Gamma(\alpha)} \int_{a}^{t} \frac{f(\tau)\mathrm{d}\tau}{(t-\tau)^{1-\alpha}}, \ (t > a; \alpha > 0), \tag{2.1.1}$$

where $\Gamma(\cdot)$ is the gamma function; the details of the definition can be seen in Appendix I. This definition is also called the left-side fractional integral; correspondingly, a right-side fractional integral can be defined as [4]

$$I_{b-}^{\alpha} f(t) = \frac{1}{\Gamma(\alpha)} \int_{t}^{b} \frac{f(\tau)\mathrm{d}\tau}{(\tau-t)^{1-\alpha}}, \ (t < b; \alpha > 0). \tag{2.1.2}$$

When $\alpha = n$ is an integer number, the two definitions are equivalent, i.e.

$$I_{a+}^{n} f(t) = \int_{a}^{t} \mathrm{d}\tau_1 \int_{a}^{\tau_1} \mathrm{d}\tau_2 \cdots \int_{a}^{\tau_{n-1}} f(\tau_n)\mathrm{d}\tau_n$$

$$= \frac{1}{(n-1)!} \int_{a}^{t} \frac{f(\tau)\mathrm{d}\tau}{(t-\tau)^{1-n}}, \ (n \in N). \tag{2.1.3}$$

The right-side integral has similar results; we will not repeat them again; the following is the same. For any arbitrary real number $\alpha$, the integer part of $\alpha$ is denoted by $[\alpha]$ (i.e. $[\alpha]$ is the largest integer number less than $\alpha$), then the Riemann–Liouville type of fractional derivative is defined as follows:

$$_{a}^{RL}D_{t}^{\alpha} f(t) = (\frac{\mathrm{d}}{\mathrm{d}t})^{n} I_{a+}^{n-\alpha} f(t)$$

$$= \frac{1}{\Gamma(n-\alpha)} (\frac{\mathrm{d}}{\mathrm{d}t})^{n} \int_{a}^{t} \frac{f(\tau)\mathrm{d}\tau}{(t-\tau)^{\alpha-n+1}},$$

$$(n = \lfloor \alpha \rfloor + 1; \ n-1 \le \alpha < n; \ t > a). \tag{2.1.4}$$

Furthermore, this definition can be rewritten as

$$\begin{aligned}
{}^{RL}_{a}D^{\alpha}_t f(t) &= \frac{1}{\Gamma(n-\alpha)} \left(\frac{\mathrm{d}}{\mathrm{d}t}\right)^n \int_a^t \frac{f(\tau)\mathrm{d}\tau}{(t-\tau)^{\alpha-n+1}}, \\
&= \sum_{k=0}^{n-1} \frac{f^{(k)}(0)t^{-\alpha+k}}{\Gamma(k+1-\alpha)} + \frac{1}{\Gamma(n-\alpha)} \int_0^t (t-\tau)^{n-\alpha-1} f^{(n)}(\tau)\mathrm{d}\tau. \quad (2.1.5)
\end{aligned}$$

Consider $\alpha \to n-1$ below (fractional derivative order trends to integer),

$$\begin{aligned}
\lim_{\alpha\to(n-1)_\alpha} {}^{RL}D^{\alpha}_t f(t) &= \lim_{\alpha\to(n-1)} \sum_{k=0}^{n-1} \frac{f^{(k)}(0)t^{-\alpha+k}}{\Gamma(n-\alpha)} \int_0^t (t-\tau)^{n-\alpha-1} f^{(n)}(\tau)\mathrm{d}\tau, \\
&= f^{(n-1)}(0) + \int_0^t f^{(n)}(\tau)\mathrm{d}\tau = \frac{d^{n-1}f(t)}{dt^{n-1}} \quad (2.1.6)
\end{aligned}$$

If $0 < \alpha < 1$, the fractional derivative in the sense of Riemann–Liouville can be defined as

$$ {}^{RL}_{a}D^{\alpha}_t f(t) = \frac{1}{\Gamma(1-\alpha)} \frac{\mathrm{d}}{\mathrm{d}t} \int_a^t \frac{f(\tau)\mathrm{d}\tau}{(t-\tau)^{\alpha}}, \quad (t > a). \quad (2.1.7)$$

Simultaneously, in view of the form (2.1.5), the formula (2.1.7) can be converted into

$$ {}^{RL}_{a}D^{\alpha}_t f(t) = \frac{f(0)t^{-\alpha}}{\Gamma(1-\alpha)} + \frac{1}{\Gamma(1-\alpha)} \int_0^t (t-\tau)^{-\alpha} f'(\tau)\mathrm{d}\tau. \quad (2.1.8)$$

### 2.1.3  Caputo's Definition

The Riemann–Liouville derivative definition is inconvenient for engineering and physical modeling because of its hyper-singularity. In the 1960s, Caputo, an Italian geophysicist, presented a weak-singular fractional differential definition. This kind of definition avoids the fractional derivative initial problems in the Riemann–Liouville-type definition. The fractional derivative in the sense of Caputo is defined as

$$\begin{aligned}
{}_{a}^{C}D_t^{\alpha} f(t) &= I_{a+}^{n-\alpha} f^{(n)}(t) \\[2mm]
&= \frac{1}{\Gamma(n-\alpha)} \int_a^t \frac{f^{(n)}(\tau)d\tau}{(t-\tau)^{\alpha-n+1}}, \quad (n = \lfloor \alpha \rfloor + 1; \; n-1 < \alpha \le n; \; t > a),
\end{aligned}$$

$$\tag{2.1.9}$$

where $n$ is the smallest positive integer number greater than $\alpha$; $f^{(n)}(\tau)$ is denoted as the $n$th derivative of the function $f(\tau)$. Compared with the Riemann–Liouville fractional derivative (2.1.4), the Caputo derivative definition becomes the integration of the $n$th derivative. Furthermore, we apply the principle of integration by parts, and then the formulation (2.1.8) can be converted into

$$\begin{aligned}
{}_{a}^{C}D_t^{\alpha} f(t) &= \frac{1}{\Gamma(n-\alpha)} \int_a^t \frac{f^{(n)}(\tau)d\tau}{(t-\tau)^{\alpha-n+1}}, \quad (n = \lfloor \alpha \rfloor + 1; \; n-1 < \alpha \le n; \; t > a), \\[2mm]
&= \frac{f^{(n)}(a)(t-a)^{n-\alpha}}{\Gamma(n-\alpha+1)} + \frac{1}{\Gamma(n-\alpha+1)} \int_a^t (t-\tau)^{n-\alpha} f^{(n+1)}(\tau)d\tau.
\end{aligned}$$

$$\tag{2.1.10}$$

The first term of the right-hand side in the above expression is initial conditions with integer-order derivative, which greatly increases the practicality of the definition. Therefore, Caputo's definition is widely chosen to use in physical and mechanics applications. For $\alpha \to n$, the derivative becomes a conventional $n$th derivative of the function $f(t)$.

$$\begin{aligned}
\lim_{\alpha \to (n-1)_\alpha}{}^{RL}D_t^{\alpha} f(t) &= \lim_{\alpha \to (n-1)} \sum_{k=0}^{n-1} \frac{f^{(k)}(0)t^{-\alpha+k}}{\Gamma(n-\alpha)} \int_0^t (t-\tau)^{n-\alpha-1} f^{(n)}(\tau)d\tau, \\[2mm]
&= f^{(n-1)}(0) + \int_0^t f^{(n)}(\tau)d\tau = \frac{d^{n-1}f(t)}{dt^{n-1}}
\end{aligned}$$

$$\tag{2.1.11}$$

From the above formula, it is noted that the integer-order differentiation is a special case of fractional differentiation and can be included in the fractional differentiation definition. Comparing formula (2.1.4) with (2.1.8), we note that the primary distinction between Caputo's definition and the Riemann–Liouville definition is the order of integral and differential, the latter integral first and the former converses. In view of pure mathematics, the conditions of the function $f(t)$ are different in the two definitions. Caputo's definition requires higher conditions, i.e. the function $f(t)$ is $n$-times differentiable.

If $0 < \alpha < 1, n = 1$, the definition above can be shorted for

$$_a^C D_t^\alpha f(t) = I_{a+}^{1-\alpha} f'(t) = \frac{1}{\Gamma(1-\alpha)} \int_a^t \frac{f'(\tau)d\tau}{(t-\tau)^\alpha}, \ (0 < \alpha \leq 1; \ t > a). \quad (2.1.12)$$

### 2.1.4  Grünwald–Letnikov Definition

The Grünwald–Letnikov definition is frequently applied for numerical evaluation which can be considered as a general form of the limit of differential definition of integer calculus. Let us consider a continuous function $f(t)$. According to the definition of an integer-order derivative, the first-order derivative of the function $f(t)$ is defined by

$$f'(t) = \frac{df}{dt} = \lim_{h \to 0} \frac{f(t) - f(t-h)}{h}. \quad (2.1.13)$$

Applying this definition twice gives the second-order derivative:

$$f''(t) = \lim_{h \to 0} \frac{f'(t) - f'(t-h)}{h} = \lim_{h \to 0} \frac{f(t) - 2f(t-h) + f(t-2h)}{h^2}. \quad (2.1.14)$$

More generally, the $n$th derivative can be expressed as

$$f^{(n)}(t) = \frac{d^n f}{dt^n} = \lim_{h \to 0} \frac{1}{h^n} \sum_{r=0}^{n} (-1)^r \binom{n}{r} f(t-rh), \quad (2.1.15)$$

where $\binom{n}{r}$ is the usual notation for the binomial coefficients,

$$\binom{n}{r} = \frac{n(n-1)(n-2)\cdots(n-r+1)}{r!}. \quad (2.1.16)$$

Generalizing the integer number $n$ of the above expression to a real number $\alpha$, a fractional derivative form can be defined as

$$f_h^{(\alpha)}(t) = \lim_{h \to 0+} \frac{1}{h^\alpha} \sum_{r=0}^{\left[\frac{t-a}{h}\right]} (-1)^r \binom{\alpha}{r} f(t-rh), \quad (2.1.17)$$

where $[c]$ is denoted as the integer part of $c$ ($[c]$ is the largest integer number less than $c$). For the function in the interval $[a, b]$, we have the left-hand and right-hand sides Grünwald–Letnikov fractional derivatives as follows:

$$\begin{aligned}
{}^{G}_{a}D^{\alpha}_{t} f(t) &= \lim_{h\to 0+} \frac{1}{h^{\alpha}} \sum_{r=0}^{\left[\frac{t-a}{h}\right]} (-1)^{r} \binom{\alpha}{r} f(t - rh), \\
{}^{G}_{t}D^{\alpha}_{b} f(t) &= \lim_{h\to 0+} \frac{1}{h^{\alpha}} \sum_{r=0}^{\left[\frac{b-t}{h}\right]} (-1)^{r} \binom{\alpha}{r} f(t + rh).
\end{aligned} \tag{2.1.18}$$

If the function $f(t)$ is $(m + 1)$-times continuously differentiable in the interval $[a, t]$, for $\alpha > 0, m = \lfloor \alpha \rfloor$, the limit of the above definition is

$$\begin{aligned}
{}^{G}_{a}D^{\alpha}_{t} f(t) &= \lim_{h\to 0+} \frac{1}{h^{\alpha}} \sum_{r=0}^{\left[\frac{t-a}{h}\right]} (-1)^{r} \binom{\alpha}{r} f(t - rh) \\
&= \sum_{k=0}^{m} \frac{f^{(k)}(a)(t - a)^{-\alpha+k}}{\Gamma(-\alpha + k + 1)}, \\
&+ \frac{1}{\Gamma(-\alpha + m + 1)} \int_{a}^{t} (t - \tau)^{m-a} f^{(m+1)}(\tau) d\tau,
\end{aligned} \tag{2.1.19}$$

where $f^{(k)}(a) = b_k$, $(k = 0, 1, 2, \ldots, m)$ are the known initial conditions.

The Grünwald–Letnikov fractional integral can be defined as

$$\sideset{^{G}_{a}}{^{\alpha}_{t}}{\mathop{I}} f(t) = \lim_{h\to 0+} \frac{1}{h^{\alpha}} \sum_{j=0}^{\left[\frac{t-a}{h}\right]} (-1)^{j} \binom{-\alpha}{j} f(t - jh), \tag{2.1.20}$$

The limitation of the expression above is ${}^{G}_{a}I^{\alpha}_{t} f(t) = \frac{1}{\Gamma(\alpha)} \int_{a}^{t} (t - \tau)^{\alpha-1} f(\tau) d\tau$. It is obvious that the Grünwald–Letnikov integral definition is equivalent to the Riemann–Liouville definition under certain conditions.

## 2.1.5 Riesz Definition of the Spatial Fractional Laplace Operator

Based on the conception of the Riesz fractional potential [5], Samko [1] proposed a definition of multi-dimensional spatial fractional Laplace operator:

$$(-\Delta)^{-\alpha/2} f = F^{-1} |X|^{\alpha} F f = \begin{cases} I^{-\alpha} f & \Re(\alpha) < 0 \\ D^{\alpha} f & \Re(\alpha) > 0 \end{cases}, \tag{2.1.21}$$

where $F$ and $F^{-1}$ represent the Fourier transform and inversion, respectively, $Ff$ denotes the Fourier transform of the function $f$, and $I^{-\alpha}f$ and $D^{\alpha}f$ represent the Riesz fractional integral and derivative, respectively. $D^{\alpha}f$ is the so-called fractional Laplacian operator. In the spacetime domain, the Riesz fractional integral can be expressed as the Riesz fractional potential in the following [5]:

$$I^{\alpha}f(x) = \int_{\mathbf{R}^n} k_{\alpha}(\mathbf{x}-t)f(t)dt \quad (\Re(\alpha)0), \tag{2.1.22}$$

where

$$k_{\alpha}(\mathbf{x}) = \frac{1}{\gamma_n(\alpha)} \begin{cases} |\mathbf{x}|^{\alpha-n} & \alpha - n \neq 0, 2, 4 \ldots, \\ |\mathbf{x}|^{\alpha-n} \log\left(\frac{1}{|\mathbf{x}|}\right) & \alpha - n = 0, 2, 4 \ldots, \end{cases}$$

$$\gamma_n(\alpha) = \begin{cases} 2^{\alpha}\pi^{n/2}\Gamma[\alpha/2][\Gamma((n-\alpha)/2)]^{-1} & \alpha - n \neq 0, 2, 4, \cdots, \\ (-1)^{(n-\alpha)/2}2^{\alpha-1}\pi^{n/2}\Gamma[1+(\alpha-n)/2]\Gamma[\alpha/2] & \alpha - n = 0, 2, 4, \cdots, \end{cases}$$

Samko [6] presented a definition of the spacetime Riesz fractional derivative as

$$D^{\alpha}f(x) = \frac{1}{d_n(l,\alpha)} \int_{R^n} \frac{\Delta_t^l f(x)}{|t|^{n+\alpha}}dt \quad (l > \alpha), \tag{2.1.23}$$

where

$$\Delta_t^l f(x) = \sum_{i=0}^{l} (-1)^i \binom{l}{i} f(\mathbf{x}-it),$$

$$d_n(l,\alpha) = \frac{2^{-\alpha}\pi^{1+n/2}A_t(\alpha)}{\Gamma(1+\alpha/2)\Gamma((n+\alpha)/2)\sin(\alpha\pi/2)}, \quad A_t(\alpha) = \sum_{i=0}^{l} (-1)^{i-1} \binom{l}{i} i^{\alpha}.$$

Note that the definition above contains a difference operator. Because of $n + \alpha > n$ ($n$ is the spatial dimension), this definition is a hyper-singular integral operator. The primary properties of the Riesz fractional integral and derivative are as follows:

$$I^{\alpha}I^{\beta}f = I^{\alpha+\beta}f, \ D^{\alpha}I^{\alpha}f = f, \tag{2.1.24}$$

which can be seen in Ref. [6]. More details of the definition of spatial fractional derivative can be found in Sect. 2.5.2.

These four definitions introduced above are frequently used in science and engineering applications, thus we make highlights on them. There are three other definitions of fractional integral and derivative can be listed briefly in the following [3].

**Hadamard fractional derivative:**

$$\frac{d^{\alpha}}{dx^{\alpha}}{}_{0+}\varphi(x) = \frac{1}{\Gamma(n-\alpha)}(x\frac{d}{dx})^{n}\int_{0}^{x}\frac{\varphi(t)}{t[\ln(x/t)]^{\alpha-1-n}}dt, \ x > 0, \ \alpha > 0, \quad (2.1.25)$$

**Hadamard fractional integral:**

$$J_{+}^{\alpha}\varphi(x) = \frac{1}{\Gamma(\alpha)}\int_{0}^{x}(\frac{t}{x})^{\mu}\frac{\varphi(t)}{t[\ln(x/t)]^{1-\alpha}}dt, \ x > 0, \ \alpha > 0, \quad (2.1.26)$$

where $\mu$ is a known parameter.

**Weyl–Marchaud fractional derivative:**

Assume $f \in C_{(-\infty,+\infty)}$, $0 < v < 1$, then the $v$th left and right Weyl–Marchaud derivatives can be defined as follows, respectively:

$$D_{l}^{v}f(x) = \frac{v}{\Gamma(1-v)}\int_{0}^{x}\frac{f(x)-f(t)}{t^{1+v}}dt, \quad (2.1.27)$$

$$D_{r}^{v}f(x) = \frac{(-1)^{v}v}{\Gamma(1-v)}\int_{0}^{x}\frac{f(x)-f(t)}{t^{1+v}}dt. \quad (2.1.28)$$

As we all know, so far, the three definitions listed above are primarily applied to pure mathematics, but rarely applied to physical and engineering modeling. And also, the relationships among them and the Riemann–Liouville definition and Caputo's definition are not yet clear.

## 2.2  Properties of Fractional Calculus

In this chapter, the variables are represented by $t$, $x$, $y$, $z$, etc.; the derivative or integral order are represented by $\alpha$, $\beta$, $\gamma$, $p$, $q$, etc.

### 2.2.1  Some Properties of Riemann–Liouville Operator

In this section, some common and simple properties of the Riemann–Liouville definition are introduced. The correlating properties of other types of definitions with similar properties but having minor differences will be introduced later. Here, we will no longer introduce one by one.

**Property 1  Linearity**

$$_aD_t^p(\lambda f(t) + \mu g(t)) = \lambda_a D_t^p f(t) + \mu_a D_t^p g(t). \tag{2.2.1}$$

This property can be proved by simple algebraic operations. Moreover, the other definitions also follow the property of linearity.

**Property 2** $\qquad _0D_t^{-\lambda}{}_0D_t^{-\beta}f(t) =_0 D_t^{-\lambda-\beta}f(t), \lambda > 0, \beta > 0. \tag{2.2.2}$

**Proof** In view of the Riemann–Liouville definition,

$$_0D_t^{-\lambda}{}_0D_t^{-\beta}f(t) = \int_0^t \frac{(t-t')^{\lambda-1}}{\Gamma(\lambda)}dt' \int_0^{t'} \frac{(t'-\tau)^{\beta-1}}{\Gamma(\beta)} f(\tau)d\tau, \tag{2.2.3}$$

exchange the integration order,

$$_0D_t^{-\lambda}{}_0D_t^{-\beta}f(t) = \frac{1}{\Gamma(\lambda)\Gamma(\beta)} \int_0^t f(\tau)d\tau \int_\tau^t (t-t')^{\lambda-1}(t'-\tau)^{\beta-1}dt' \tag{2.2.4}$$

holds. Using the following relation

$$\int_\tau^t (t-t')^{\lambda-1}(t'-\tau)^{\beta-1}dt' = (t-\tau)^{\lambda+\beta-1}B(\lambda,\beta), \lambda,\beta > 0, t > \tau, \tag{2.2.5}$$

we have

$$_0D_t^{-\lambda}{}_0D_t^{-\beta}f(t) = \frac{1}{\Gamma(\lambda+\beta)} \int_0^t (t-\tau)^{\lambda+\beta-1}f(\tau)d\tau =_0 D_t^{-\lambda-\beta}f(t). \tag{2.2.6}$$

**Property 3** *For the continuous function* $f(t)$ *(i.e. the functions in $L_1$ space), and the differential* $D_t^{\lambda-\beta}f(t)$ *that exists, the following relation holds:*

$$_0D_t^\lambda \, _0D_t^{-\beta} f(t) = _0D_t^{\lambda-\beta} f(t), \; \lambda > 0, \; \beta > 0. \tag{2.2.7}$$

In particular, $_0D_t^\lambda \, _0D_t^{-\lambda} f(t) = f(t) \; (\lambda > 0)$.

## Property 4  The Leibniz Rule

*Let us start with the known Leibniz rule with integer-order: for two functions, $\varphi(t)$ and $f(t)$, the nth derivative (n is an integer) of the product $\varphi(t) f(t)$ can be expressed as*

$$\frac{d^n}{dt^n}(\varphi(t) f(t)) = \sum_{k=0}^{n} \binom{n}{k} \varphi^{(k)}(t) f^{(n-k)}(t). \tag{2.2.8}$$

Suppose $\varphi(t)$ and $f(t)$ are differentiable and continuous in the interval $[a, t]$; the $p$th derivative of the product $\varphi(t) f(t)$ can be obtained in the following relation, namely the fractional derivative Leibniz rule,

$$_aD_t^p(\varphi(t) f(t)) = \sum_{k=0}^{\infty} \binom{p}{k} \varphi^{(k)}(t) \, _aD_t^{p-k} f(t). \tag{2.2.9}$$

## 2.2.2  Fractional Derivative of Some Functions

### 1  f(t) = Constant

The unit function $f(t) \equiv 1$, denoted by $[l]$, is very necessary for discussing the fractional integral and derivative of the unit step function $f(t) = \begin{cases} 0, t \in (-\infty, a) \\ 1, t \in (a, +\infty) \end{cases}$ (and is also called the Heaviside function $H(t - a)$).

The $p$th Riemann–Liouville fractional derivative of the unit function is

$$_a^{RL}D_t^p[1] = \frac{(t - a)^{-p}}{\Gamma(1 - p)}. \tag{2.2.10}$$

Applying the property of linearity, the $p$th Riemann–Liouville fractional derivative of a constant $c$ is

$$_a^{RL}D_t^p c = c\frac{(t - a)^{-p}}{\Gamma(1 - p)}. \tag{2.2.11}$$

Similarly, the $p$th Riemann–Liouville fractional derivative of the Heaviside function $H(t)$ is

$$^{RL}_{a}D^p_t H(t-a) = \frac{(t-a)^{-p}}{\Gamma(1-p)}, \; t > a. \tag{2.2.12}$$

## 2   Fractional Derivative of $(t-a)^v$

Consider the power function $f(t) = (t-a)^v$, where $v$ is a real number. Firstly, take the expression of fractional integral into account; according to the integral expression, the $p$th integral ($p > 0$) of the function $f(t)$ can be derived as

$$^{RL}_{a}I^p_t(t-a)^v = \frac{1}{\Gamma(p)} \int_a^t (t-\tau)^{p-1}(\tau-a)^v d\tau. \tag{2.2.13}$$

The integration is convergent if $v > -1$. Performing the substitution $\tau = a + \xi(t-a)$, we have

$$^{RL}_{a}I^p_t(t-a)^v = \frac{1}{\Gamma(p)}(t-a)^{v+p} \int_0^1 (1-\xi)^{p-1}\xi^v d\xi$$

$$= \frac{1}{\Gamma(p)}B(p, v+1)(t-a)^{v+p}$$

$$= \frac{\Gamma(v+1)}{\Gamma(p+v+1)}(t-a)^{v+p}. \tag{2.2.14}$$

For the $p$th fractional derivative of this function, if $p$ satisfies $0 \le m \le p < m+1$, by the definition of fractional derivative, we derive

$$^{RL}_{a}D^p_t(t-a)^v = \frac{1}{\Gamma(-p+m+1)}\frac{d^{m+1}}{dt^{m+1}} \int_a^t (t-\tau)^{m-p}(\tau-a)^v d\tau$$

$$= \frac{\Gamma(v+1)}{\Gamma(-p+v+1)}(t-a)^{v-p}. \tag{2.2.15}$$

From the relation (2.2.15), for $j = 1, 2, \cdots, [\Re(p)] + 1$, the following formula holds (Figs. 2.1 and 2.2):

$$^{RL}_{a}D^p_t(t-a)^{p-j} = \frac{\Gamma(p-j+1)}{\Gamma(-j+1)}(t-a)^{-j}. \tag{2.2.16}$$

**Fig. 2.1** Show the results of the fractional derivative of $x$ and $x^2$, respectively, with different orders p = 0.5, 0.75, 1 and 1.25

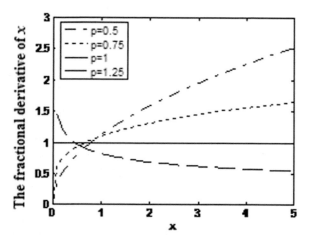

**Fig. 2.2** Show the results of the fractional derivative of $x$ and $x^2$, respectively, with different orders p = 0.5, 0.75, 1 and 1.25

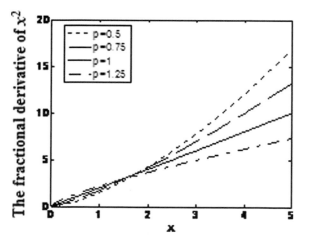

## 2.2.3 The Relationships of Different Definitions

1. **The relationship between the Grünwald–Letnikov definition and the Riemann–Liouville definition** [2, 7]

If the function $f(t)$ is $(m + 1)$-times continuously differentiable and $m \geq [\alpha] = n - 1$, the Grünwald–Letnikov definition is equivalent to the Riemann–Liouville definition, and both can be expressed as the formulation (2.1.13). However, if the conditions mentioned above are not established, the Riemann–Liouville definition may no longer be consistent with the Grünwald–Letnikov definition. Due to the expression of the integro-differential expression, the Riemann–Liouville definition is applied more widely.

2.    **Comparison of the Grünwald–Letnikov definition and Caputo's definition**

If the function $f(t)$ is $(m + 1)$-times continuously differentiable and $m \geq \lfloor \alpha \rfloor = n-1$, mightily suppose $m = n-1$, then $n = m+1$, and $f^{(k)}(a) = 0, k = 0, 1, 2, \cdots, n-1$; hence, Caputo's definition and the Grünwald–Letnikov definition are equivalent, otherwise, they are nonequivalent.

3.    **Comparison of the Riemann–Liouville definition and Caputo's definition**

(1)    Both of them are the modifications of the Grünwald–Letnikov definition;

(2)    If the lower terminal of the integration is negative infinite, for a sufficiently smooth function which has good properties at the terminal $t = -\infty$, both are equivalent;

(3)    If the function $f(t)$ is $(m + 1)$-times continuously differentiable and $m \geq \lfloor \alpha \rfloor = n - 1$, $f^{(k)}(a) = 0, k = 0, 1, 2, \cdots, n - 1$, both are equivalent;

(4)    For the Riemann–Liouville operator, we have the relation (2.2.16), however, for the Caputo operator, the following holds:

$$ {}_{a}^{C}D_{t}^{p}(t - a)^{k} = 0, \ k = 0, 1, 2, \cdots, \lfloor p \rfloor. \tag{2.2.17} $$

In particular, the Caputo derivative of a constant $A$ is zero, whereas the Riemann–Liouville derivative of a constant is not equal to 0, but

$$ {}_{0}^{RL}D_{t}^{\alpha}A = \frac{At^{-\alpha}}{\Gamma(1 - \alpha)}. \tag{2.2.18} $$

That is the reason the Caputo operator is more popularly applied in science and engineering applications;

(5)    Difference of the Laplace transform: the Laplace transform of the Riemann–Liouville fractional derivative is

$$ \int_{0}^{\infty} e^{-st}\{{}_{0}^{RL}D_{t}^{\alpha}f(t)\}\mathrm{d}t = s^{\alpha}F(s) - \sum_{k=0}^{n-1} s^{k}{}_{0}^{RL}D_{t}^{(\alpha-k-1)}f(0)|_{t=0}, $$

$$ (n - 1 \leq \alpha < n), \tag{2.2.19} $$

whereas the Laplace transform of Caputo's fractional derivative is

$$ \int_{0}^{\infty} e^{-st}\{{}_{0}^{C}D_{t}^{\alpha}f(t)\}\mathrm{d}t = s^{\alpha}F(s) - \sum_{k=0}^{n-1} s^{\alpha-k-1}f^{(k)}(0), \ (n - 1 < \alpha \leq n). $$

$$ \tag{2.2.20} $$

It seems that that the Laplace transform of the Riemann–Liouville fractional derivative requires initial conditions with fractional derivatives of the function $f(t)$. Unfortunately, the physical interpretation for such type of initial conditions is unclear

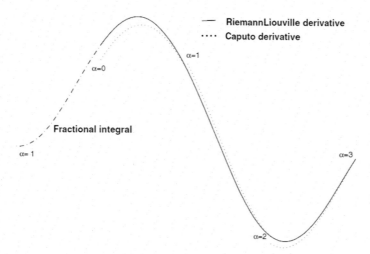

**Fig. 2.3** The sketch map to illustrate the consistency of fractional derivatives (integrals); it shows that between two adjacent integers, the Riemann–Liouville definition is left-continuous, whereas Caputo's definition is right-continuous

and uneasy to obtain. However, Caputo's fractional derivative allows utilization of initial values of classical integer-order derivatives. That is another reason for choosing the Caputo operator in real applications.

We will apply Li and Deng's [8] results to further illustrate the relationships among the Riemann–Liouville definition, Caputo's definition and the classical integer-order calculus, as shown in Fig. 2.3.

We can also distinguish the differences of different definitions by the following properties; suppose $m - 1 < \alpha < m \in Z^+$, the following holds:

(1)   $_0^C D_t^\alpha x(t) = {}_0^{RL} D_t^\alpha (x(t) - \sum_{k=0}^{m-1} \frac{t^k}{k!} x^{(k)}(0))$;

(2)   $_0^C D_t^\alpha I_{0,t}^\alpha x(t) = {}_0^{RL} D_t^\alpha I_{0,t}^\alpha x(t) = x(t),\ m = 1$;

(3)   $I_{0,t0}^\alpha \, {}^C D_t^\alpha x(t) = x(t) - \sum_{k=0}^{m-1} \frac{t^k}{k!} x^{(k)}(0)$;

(4)   $I_{0,t0}^\alpha \, {}^{RL} D_t^\alpha x(t) = x(t) - \sum_{k=0}^{m-1} [_0^{RL} D_t^{\alpha-k} x(t)]_{t=0} \frac{t^{\alpha-k}}{\Gamma(\alpha-k+1)}$.

## 4.   Conclusions

The Riemann–Liouville fractional definition is convenient for pure mathematics analysis of the fractional calculus, whereas Caputo's definition with weak-singularity, the Laplace transform of which is more concise, is more widely utilized in engineering applications. The Grünwald–Letnikov is frequently used for numerical analysis, which plays an important role to promote the practical application of fractional calculus. Riesz's definition can exactly describe the global characteristics of the spatial fractional derivative.

## 2.3   Fourier and Laplace Transforms of the Fractional Calculus

### 2.3.1   Fourier Transform of the Fractional Calculus

The Fourier transform is a powerful tool for frequency domain analysis and to solve differential equations. In this section, we primarily introduce the Fourier transform of the fractional calculus. We will firstly introduce the definition of the Fourier transform and some basic properties.

1.  **f(t) is a continuous function in the interval $(-\infty, +\infty)$; the Fourier transform and its inversion are**

$$G(\omega) = F\{g(t); \omega\} = \int_{-\infty}^{+\infty} e^{-i\omega t} g(t) dt, \tag{2.3.1}$$

$$g(t) = F^{-1}\{G(\omega)\} = \frac{1}{2\pi} \int_{-\infty}^{+\infty} e^{i\omega t} G(\omega) dt, \tag{2.3.2}$$

where $F$ denotes the Fourier transform and $F^{-1}$ represents the inversion. Moreover, $f(t)$ satisfies the Dirichlet conditions in finite domain and is absolutely integrable in $(-\infty, +\infty)$. Note that $\omega$ represents the frequency.

2.  **Properties of the Fourier Transform**

The linear additivity of the Fourier transform and inverse Fourier transform,

$$F\{g(t) + q(t); \omega\} = G(\omega) + Q(\omega), \tag{2.3.3}$$

$$F^{-1}\{G(\omega) + Q(\omega)\} = g(t) + q(t). \tag{2.3.4}$$

The Fourier transform of the convolution and the inversion of the product,

$$F\{g(t) * q(t); \omega\} = G(\omega)Q(\omega), \tag{2.3.5}$$

$$F^{-1}\{G(\omega)Q(\omega)\} = g(t) * q(t). \tag{2.3.6}$$

The Fourier transform of the product and the inversion of the convection,

$$F\{g(t)q(t)\} = \frac{1}{2\pi} G(\omega) * Q(\omega), \tag{2.3.7}$$

$$F^{-1}\{G(\omega) * Q(\omega)\} = 2\pi g(t)q(t).$$ (2.3.8)

The Fourier transform of the $n$th derivative,

$$F\{g^{(n)}(t); \omega\} = (i\omega)^n G(\omega).$$ (2.3.9)

3. **The Fourier Transform of Fractional Calculus**

Ensure the existence of the Fourier transform of the fractional calculus; the lower terminal of the integration should be defined as $-\infty$, then the definitions in the sense of Grünwald–Letnikov, Riemann–Liouville and Caputo have the same form as follows:

$$_{-\infty}D_t^{-\alpha}g(t) = \frac{1}{\Gamma(\alpha)} \int_{-\infty}^{t} (t-\tau)^{\alpha-1}g(\tau)d\tau.$$ (2.3.10)

The Fourier transform of the fractional integral can be written as

$$F\{_{-\infty}D_t^{-\alpha}g(t)\} = (i\omega)^{-\alpha}G(\omega).$$ (2.3.11)

The Fourier transform of the fractional derivative can be written as

$$F\{_{-\infty}D_t^{\alpha}g(t)\} = (i\omega)^{\alpha}G(\omega).$$ (2.3.12)

4. **The Fourier transforms of the space Riesz fractional integral and derivative can be expressed as follows, respectively** [3]:

$$F\{I^{\alpha}g; k\} = |k|^{-\alpha}G(k),$$ (2.3.13)

$$F\{D^{\alpha}g; k\} = |k|^{\alpha}G(k),$$ (2.3.14)

where $k$ represents a wave number, which is a variable in the Fourier field, corresponding to the space variable.

## 2.3.2 Laplace Transform of the Fractional Calculus

The Laplace transform is a powerful tool to solve differential equations. Applying the Laplace transform, a partial differential equation can be transformed into an ordinary differential equation, and also unifies differential order operators to the same algebraic field such that the solving process is simplified. In this section, the approach is introduced to solve fractional derivative equations. The Laplace transform of the function $f(t)$ is

$$L\{f(t); s\} = \int_0^\infty e^{-st} f(t)\mathrm{d}t = F(s), \qquad (2.3.15)$$

where the function $f(t)$ must be piecewise continuous in $t \in (0, +\infty)$, and the decreasing rate of $e^{-st}$ is faster than the increasing rate of $f(t)$ with respect to $t$, then the Laplace transform of the function $f(t)$ exists if $Re(s) > 0$, and the integration is absolutely uniformly convergent.

1. **The Characteristics of the Laplace Transform**

The linear additivity:

$$L\{f(t) + g(t); s\} = F(s) + G(s). \qquad (2.3.16)$$

The convolution can be defined by the following formula:

$$f(t) * g(t) = g(t) * f(t) = \int_0^t f(t - \tau)g(\tau)\mathrm{d}\tau. \qquad (2.3.17)$$

The Laplace transform of the convolution above is

$$L\{f(t) * g(t); s\} = F(s)G(s). \qquad (2.3.18)$$

The Laplace transform of the $n$th derivative can be obtained by

$$L\{f^{(n)}(t); s\} = s^n F(s) - \sum_{k=0}^{n-1} s^{n-k-1} f^{(k)}(0). \qquad (2.3.19)$$

The Laplace transform of the $n$-fold integral,

$$L\{\underbrace{\int_0^t \int_0^t \cdots \int_0^t f(t)\mathrm{d}t}_{n}; s\} = \frac{F(s)}{s^n}. \qquad (2.3.20)$$

2. **The Characteristics of the Inverse Laplace Transform**

The inverse Laplace transform is

$$f(t) = L^{-1}\{F(s); s\} = \frac{1}{2\pi i} \int_{\gamma-\infty}^{\gamma+\infty} e^{st} F(s)\mathrm{d}s, \ (t > 0, s = \gamma + i\omega). \quad (2.3.21)$$

The inverse Laplace transform of the product is

$$L^{-1}\{F(s)G(s); s\} = f(t) * g(t).  \qquad (2.3.22)$$

The inverse Laplace transform of the convolution is

$$L^{-1}\{F(s) * G(s); s\} = f(t)g(t).  \qquad (2.3.23)$$

3.  **Laplace Transform of the Mittag-Leffler Function**

$$\int_0^\infty e^{-st} t^{\alpha k+\beta-1} E_{\alpha,\beta}^{(k)}(\pm at^\alpha)dt = \frac{k!s^{\alpha-\beta}}{(s^\alpha \mp a)^{k+1}}.  \qquad (2.3.24)$$

The formula plays an important role to solve fractional derivative equations.
4.  **The Laplace Transform of the Fractional Derivative Operator**

The Riemann–Liouville fractional derivative operator is written as

$$_a^{RL} D_t^p f(t) = g^{(n)}(t),  \qquad (2.3.25)$$

$$g(t) = _a^{RL} D_t^{-(n-p)} f(t) = \frac{1}{\Gamma(n-p)} \int_0^t (t-\tau)^{n-p-1} f(\tau)d\tau, \ (n-1 \le p < n),$$

$$\qquad (2.3.26)$$

then the Laplace transform of the Riemann–Liouville fractional derivative operator is

$$L\{_a^{RL} D_t^p f(t); s\} = s^n G(s) - \sum_{k=0}^{n-1} s^k g^{(n-k-1)}(0), \ G(s) = s^{-(n-p)} F(s),$$

$$\qquad (2.3.27)$$

where

$$g^{(n-k-1)}(t) = \frac{d^{(n-k-1)}}{dt^{n-k-1}} {}_a^{RL} D_t^{-(n-p)} f(t) = {}_a^{RL} D_t^{p-k-1} f(t).  \qquad (2.3.28)$$

Finally, the Laplace transform of the Riemann–Liouville fractional derivative operator can be expressed as

$$L\{_a^{RL} D_t^p f(t); s\} = s^p F(s) - \sum_{k=0}^{n-1} s^k {}_a^{RL} D_t^{p-k-1} f(t)|_{t=0}.  \qquad (2.3.29)$$

If $0 \leq p < 1$, we have

$$L\{_a^{RL} D_t^p f(t); s\} = s^p F(s) - {_a}I_t^{1-p} f(t)|_{t=0}. \tag{2.3.30}$$

Now we rewrite the Caputo fractional derivative operator as

$$_a^C D_t^p f(t) = {_a^{RL}} D_t^{-(n-p)} g(t), \quad g(t) = f^{(n)}(t), \quad (n-1 < p \leq n), \tag{2.3.31}$$

so

$$L\{_a^C D_t^p f(t); s\} = s^{-(n-p)} G(s), \ G(s) = s^n F(s) - \sum_{k=0}^{n-1} s^k f^{(n-k-1)}(0), \tag{2.3.32}$$

then the Laplace transform of the Caputo fractional derivative operator is

$$L\{_a^C D_t^p f(t); s\} = s^p F(s) - \sum_{k=0}^{n-1} s^{p-k-1} f^{(k)}(0), \ (n-1 < p \leq n). \tag{2.3.33}$$

If $0 \leq p < 1$, the following holds:

$$L\{_a^C D_t^p f(t); s\} = s^p F(s) - s^{p-1} f(0). \tag{2.3.34}$$

5. **Examples of Applying the Laplace Transform to Solve Fractional Derivative Equations**

(1)  Consider an ordinary differential equation [2]

$$A_a^C D_t^p u(t) + B_a^C D_t^q u(t) = f(t), \ (0 < p < 1, \quad 0 < q < 1). \tag{2.3.35}$$

Because Caputo's fractional derivative operator is utilized, the same physical initial conditions as the conventional integer-order derivative equations, it is popular to apply to practical problems.

Assume $A = B = 1$; applying the Laplace transform on two sides of the equation ($s$ is the parameter of Laplace transform), we have

$$L\{_a^C D_t^p u(t); s\} = s^p U(s) - s^{p-1} u(0), \tag{2.3.36}$$

$$L\{_a^C D_t^q u(t); s\} = s^q U(s) - s^{q-1} u(0), \tag{2.3.37}$$

$$L\{f(t); s\} = F(s). \tag{2.3.38}$$

From the linear additivity of the Laplace transform, we have

$$U(s) = \frac{(s^{p-1} + s^{q-1})u(0) + F(s)}{s^p + s^q}, \qquad (2.3.39)$$

$$U(s) = \frac{s^{-1}u(0)}{s^{q-p} + 1} + \frac{s^{-1}u(0)}{s^{p-q} + 1} + \frac{s^{-p}F(s)}{1 + s^{q-p}}. \qquad (2.3.40)$$

The inverse Laplace transform of $U(s)$ can be obtained by the following processes:

Applying the formula of the Laplace transform of the Mittag-Leffler function, we have

$$L^{-1}\{\frac{s^{-1}u(0)}{s^{q-p} + 1}; s\} = t^{q-p}E_{q-p,\,q-p+1}(-t^{q-p})u(0), \qquad (2.3.41)$$

$$L^{-1}\{\frac{s^{-1}u(0)}{s^{p-q} + 1}; s\} = t^{p-q}E_{p-q,\,p-q+1}(-t^{p-q})u(0), \qquad (2.3.42)$$

$$L^{-1}\{\frac{s^{-p}F(s)}{s^{q-p} + 1}; s\} = \int_0^t f(t - \tau)t^{q-1}E_{q-p,\,q}(-t^{q-p})d\tau. \qquad (2.3.43)$$

Thus, the solution to this problem is written as

$$u(t) = t^{q-p}E_{q-p,\,q-p+1}(-t^{q-p})u(0) + t^{p-q}E_{p-q,\,p-q+1}(-t^{p-q})u(0)$$
$$+ f(t) * (t^{q-1}E_{q-p,\,q}(-t^{q-p})). \qquad (2.3.44)$$

The solution to the homogeneous equation ($f(t) = 0$) is

$$u(t) = t^{q-p}E_{q-p,\,q-p+1}(-t^{q-p})u(0) + t^{p-q}E_{p-q,\,p-q+1}(-t^{p-q})u(0). \qquad (2.3.45)$$

(2) Taking a partial differential equation, for example [9]:

$$_a^C D_t^p u(t) = \lambda^2 \frac{\partial^2 u(x, t)}{\partial x^2}, \quad (t > 0, \quad -\infty < x < \infty, \quad 0 < p < 1),$$

$$\lim_{x \to \pm\infty} u(x, t) = 0; \; u(x, 0) = f(x), \qquad (2.3.46)$$

where the fractional derivative operator is defined in the sense of the Caputo type. This equation is a time-fractional diffusion equation, where $u(x, t)$ is the solute concentration of the location $x$ at time $t$, $f(x)$ is the initial concentration of the location $x$ and the concentration vanishes at the infinite location.

The Fourier transform of the equation with respect to $x$ is

$$\substack{C \\ a}D_t^p\overline{u}(\beta, t) + \lambda^2\beta^2\overline{u}(\beta, t) = 0. \tag{2.3.47}$$

And then the Laplace transform of the equation above with respect to $t$ leads to

$$L\{\substack{C \\ a}D_t^p\overline{u}(\beta, t); s\} = s^p\overline{U}(\beta, s) - s^{p-1}f(\beta), \tag{2.3.48}$$

$$L\{\overline{u}(\beta, t); s\} = \overline{U}(\beta, s), \tag{2.3.49}$$

$$\overline{U}(\beta, s) = \frac{s^{p-1}f(\beta)}{s^p + \lambda^2\beta^2}. \tag{2.3.50}$$

The inverse Laplace transform of $\overline{U}(\beta, s)$ is

$$\overline{u}(\beta, t) = L^{-1}\{\overline{U}(\beta, s); s\} = f(\beta)E_{p,1}(-\lambda^2\beta^2t^p). \tag{2.3.51}$$

Applying the inverse Fourier transform to $\overline{u}(\beta, t)$, the solution to the problem is

$$u(x, t) = \int_{-\infty}^{\infty} G(x - \xi, t)f(\xi)d\xi,$$

$$G(x, t) = \frac{1}{2\lambda}t^{\rho-1}W(-z, \rho, \rho), \quad \left(z = \frac{|x|}{\lambda t^p}, \rho = p/2\right), \tag{2.3.52}$$

where $W$ represents the Wright function (see Appendix I). Figure 2.4 shows that the performance of $G(x, t)$ at different times and with different fractional orders $p$. From Fig. 2.4, we see that the nonzero values of $G(x, t)$ extend outwardly with the evolving of time. The larger the fractional derivative order $p$, the greater the values range $G(x, t)$. To a certain extent, the properties of the function $G(x, t)$ can illustrate the properties of the diffusion process. Therefore, furthermore, we can analyze the effect of the time and fractional order on the diffusion process. In Chap. 4, some details of fractional derivative anomalous diffusion models can be introduced.

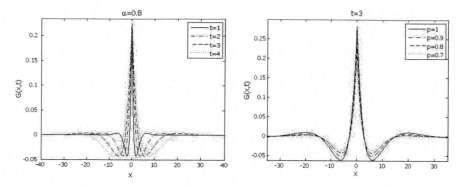

**Fig. 2.4** Left: $p$ fixed, with different times, Right: time fixed, $p$ is changed

## 2.4 Analytical Solution of Fractional-Order Equations [10]

### 2.4.1 Algorithms of Integral Transform

The integral transform methods can be carried out by the following steps[1]: (1) Applying the integral transform, partial differential equations can be transformed into ordinary differential equations or algebraic equations, and then the solution in the phase space is obtained; (2) Applying the inverse transform to the solution in the phase space, the solution to the original problem can be derived. The integral transform approach is one of the most important analytical approaches to solve differential equations (including integer-order equations). These algorithms contain the Laplace transform, Fourier transform, Hankel transform and Mellin transform. We already illustrated the Laplace and Fourier transforms by two examples in the two front sections, hence, we do not repeat them here.

The thought of the integral transform approach is very clear, and various transforms of basis functions have been listed, therefore, this approach is convenient to apply. However, it must be pointed out that the integral transform approach has many restrictions, i.e. the Fourier transform is no longer applicable for finite domain problems. Moreover, the inversion of some complex functions is difficult to obtain, which also restricts the applications of the integral transform approach. In addition, only a few nonlinear problems can apply the integral transform approach. Therefore, developing more types of analytical approaches to solve fractional-order equations is an important topic in the field of fractional calculus theory and application.

---

[1] Most of the contents of this section primarily refer to the literature [10].

## 2.4.2  Green's Function [2]

1.   Definition and Some Properties

Green's function plays a powerful role to solve nonlinear differential equations. In this section, we will introduce the fractional Green's functions by differential equations with the Riemann–Liouville-type definition.

Let us consider the following initial value problem:

$$_0L_t^{\sigma_n} y(t) = f(t, y); \quad \left[ _0D_t^{\sigma_k-1} y(t) \right]_{t=0} = b_k, \ k = 1, \cdots n, \qquad (2.4.1)$$

where $_0L_t^{\sigma_n}$ represents a linear combination of some fractional operators. And also assume that the equations above satisfy the homogenous initial conditions as follows:

$$b_k = 0, \ (k = 1, \ldots n). \qquad (2.4.2)$$

If the function $G(t, \tau)$ satisfies the following conditions:

(1)   $_\tau L_t G(t, \tau) = 0$, for any arbitrary $\tau \in (0, t)$;
(2)   $\lim_{\tau \to t-0}(_\tau D_t^{\sigma_k-1} G(t, \tau)) = \delta_{k,n}, k = 0, 1, \ldots n,$ ($\delta_{k,n}$ is Kronecker's delta function);
(3)   $\lim_{\tau, t \to +0, \tau < t}(_\tau D_t^{\sigma_k} G(t, \tau)) = 0, k = 0, 1, \ldots n - 1,$
       it is called Green's function of Eq. (2.4.1).
       The properties of Green's function:
   1.   If $G(t, \tau)$ is Green's function of Eq. (2.4.1), then $y(t) = \int_0^t G(t, \tau) f(\tau) d\tau$ is the solution to Eq. (2.4.1).
   2.   For linear fractional differential equations with constant coefficients, the following equation holds:

$$G(t, \tau) = G(t - \tau). \qquad (2.4.3)$$

2.   Examples of Green's Functions Method

   (1)   One-term fractional differential equation

$$A_0 D_t^\alpha y(t) = f(t). \qquad (2.4.4)$$

   Applying the Laplace transform, we have

$$g_1(p) = \frac{1}{A_0 p^\alpha}, \qquad (2.4.5)$$

and then the inverse Laplace transform gives

$$G_1(t) = \frac{1}{A_0} \frac{t^{\alpha-1}}{\Gamma(\alpha)}. \qquad (2.4.6)$$

Namely, $G_1$ is Green's function of the problem, thus, the solution of this equation is

$$y(t) = \frac{1}{A_0\Gamma(\alpha)} \int_0^t \frac{f(\tau)}{(t-\tau)^{1-\alpha}} d\tau. \qquad (2.4.7)$$

(2)  **Two-term fractional differential equation**

Let us consider the two-term fractional equation

$$A_0 D_t^\alpha y(t) + By(t) = f(t). \qquad (2.4.8)$$

Using the Laplace transform, we have

$$g_2(p) = \frac{1}{Ap^\alpha + B} = \frac{1}{A} \frac{1}{p^\alpha + \frac{B}{A}}, \qquad (2.4.9)$$

and then using the inverse Laplace transform, Green's function of Eq. (2.4.8) is given as

$$G_2(t) = \frac{1}{A} t^{\alpha-1} E_{\alpha,\alpha}(-\frac{B}{A} t^\alpha). \qquad (2.4.10)$$

Therefore, the solution of Eq. (2.4.8) can be written as

$$y(t) = \frac{1}{A} G_2(t) * f(t) = \frac{1}{A} \int_0^t G_2(t-\tau) f(\tau) d\tau. \qquad (2.4.11)$$

(3)  **$n$-term fractional differential equation**

More generally, $n$-term equation with real coefficients

$$A_{n0} D_t^{\beta_n} y(t) + A_{n-10} D_t^{\beta_{n-1}} y(t) + \ldots$$
$$+ A_{10} D_t^{\beta_1} y(t) + A_{00} D_t^{\beta_0} y(t) = f(t). \qquad (2.4.12)$$

Using the Laplace transform, we have

$$g_n(p) = \frac{1}{A_n p^{\beta_n} + A_{n-1} p^{\beta_{n-1}} + \ldots + A_1 p^{\beta_1} + A_0 p^{\beta_1}}, \qquad (2.4.13)$$

assume $\beta_n > \beta_{n-1} > \ldots > \beta_1 > \beta_0$, the inverse Laplace transform leads to

$$G_n(t) = \frac{1}{A_n} \sum_{m=0}^{\infty} \frac{(-1)^m}{m!} \sum_{\substack{k_0+k_1+\ldots+k_{n-2}=m}}^{k_0>0;\ldots k_{n-2}>0} (m; k_0, k_1, \ldots, k_{n-2})$$

$$\prod_{i=0}^{n-2} \left(\frac{A_i}{A_n}\right)^{k_i} t^{(\beta_n-\beta_{n-1})m+\beta_n+\sum_{j=0}^{n-2}(\beta_{n-1}-\beta_j)k_j-1}$$

$$\times E^m_{(\beta_n-\beta_{n-1}),\beta_n+\sum_{j=0}^{n-2}(\beta_{n-1}-\beta_j)k_j} \left(-\frac{A_{n-1}}{A_n} t^{\beta_n-\beta_{n-1}}\right). \qquad (2.4.14)$$

(4)   **Space fractional differential equation** [11, 12]

$$\begin{cases} D_t^{\eta} u(x,t) = D_{\theta}^{\gamma} u(x,t) \\ u(x,0) = \delta(x) \end{cases}, \qquad (2.4.15)$$

where $\delta(x)$ is the Dirac delta function, $D_{\theta}^{\gamma}$ denotes Riesz's operator. This equation is a time–space fractional derivative anomalous diffusion equation; for details, refer to Sect. 2.4.1. The derivation processes of the properties and expressions of Green's function and $K$ function are very complicated; for simplicity and for highlighting the purpose of this section, the details are omitted; readers who are interested can find the details in the Refs. [11, 12].

The relationship between Green's function and $K$ function of Eq. (2.4.15) can be expressed as

$$D_{\gamma,\eta}^{\theta}(x,t) = t^{-\lambda} K_{\gamma,\eta}^{\theta}(x/t^{\lambda}), \ \lambda = \eta/\gamma. \qquad (2.4.16)$$

The expression of the $K$ function with parameter $\theta$ is

$$K_{\gamma,\eta}^{\theta}(x) = \frac{1}{\gamma x} \frac{1}{2\pi i} \int_{\mu-i\infty}^{\mu+i\infty} \frac{\Gamma(s/\gamma)\Gamma(1-s/\gamma)\Gamma(1-s)}{\Gamma(1-\eta s/\gamma)\Gamma(\rho s)\Gamma(1-\rho s)} x^s ds \qquad (2.4.17)$$

where $0 < \mu < \min(\gamma, 1)$, $|\theta| \le 2 - \eta$.
If $\gamma = \eta$, the $K$ function can be rewritten as

$$K_{\gamma,\gamma}^{\theta}(x) = \frac{1}{\pi\gamma x}\frac{1}{2\pi i}\int\limits_{\mu-i\infty}^{\mu+i\infty} \Gamma(\frac{s}{\gamma})\Gamma(1-\frac{s}{\gamma})\sin[\frac{s}{\gamma}\frac{\pi}{2}(\gamma-\theta)]x^s ds. \qquad (2.4.18)$$

According to Green's function, the fundamental solution of this space fractional differential equation can be written as

$$G(x,t) = t^{-\eta/\gamma}K_{\gamma,\eta}^{\theta}\left(\frac{x}{t^{\eta/\gamma}}\right), \; 0 < \gamma \le 2, \; 0 < \eta \le 2. \qquad (2.4.19)$$

### 2.4.3 Adomian Decomposition Method (ADM)

The Adomian decomposition method has been widely applied to solve linear and nonlinear integer-order differential equations [6, 13]. Recently, this method has been introduced to get the solution of fractional derivative equations and obtains some good results [14–17]. Applying the Adomian decomposition method to solve equations, the equations are arranged into the following form [16]:

$$Lu + Ru + Nu = f, \qquad (2.4.20)$$

where $N$ is a nonlinear operator, $L$ and $R$ are linear operators and $L$ is easily or trivially invertible. The general solution of the given equation is decomposed into the sum

$$u = \sum_{n=0}^{\infty} u_n. \qquad (2.4.21)$$

Because $L$ is invertible, from the form (2.4.20), we have

$$u = \sum_{n=0}^{k} \frac{t^n u^{(n)}(0)}{n!} + L^{-1}f - L^{-1}Ru - L^{-1}\Phi u. \qquad (2.4.22)$$

$\Phi(u)$ represents the nonlinear term, $k = \lfloor \alpha \rfloor; \alpha$ is the order of the operator $L$. Substituting (2.4.21) into the linear term of (2.4.22), we have

$$\sum_{n=0}^{\infty} u_n = \sum_{n=0}^{k} \frac{t^n u^{(n)}(0)}{n!} + L^{-1}f - L^{-1}R\sum_{n=0}^{\infty} u_n - L^{-1}\Phi u, \qquad (2.4.23)$$

and the nonlinear term can be written as [17]

$$Nu = \sum_{n=0}^{\infty} A_n, \; A_n = \frac{1}{n!} \left[ \frac{d^n}{d\lambda^n} \Phi \left( \sum_{i=0}^{\infty} \lambda^i u_i \right) \right] \Bigg|_{\lambda=0}, \qquad (2.4.24)$$

where $A_n$ is called the Adomian polynomials, then the solution to the original problem can be written as

$$u_0 = \sum_{n=0}^{k} \frac{t^n u^{(n)}(0)}{n!} + L^{-1} f, \; u_{n+1} = -L^{-1} R u_n - L^{-1} A_n. \qquad (2.4.25)$$

Now, to illustrate this method, we take fractional KdV–Burgers equation [14], for example,

$$\begin{cases} \frac{\partial^\alpha u}{\partial t^\alpha} + \varepsilon u \frac{\partial^\beta u}{\partial x^\beta} + \eta \frac{\partial^2 u}{\partial x^2} + \upsilon \frac{\partial^3 u}{\partial x^3} = 0, \; t > 0, 0 < \alpha, \beta \le 1, \\ u(x, 0) = f(x) \end{cases} \qquad (2.4.26)$$

the nonlinear term $\Phi(x) = \varepsilon u \frac{\partial^\beta u}{\partial x^\beta}$, linear term $L(x) = \frac{\partial^\alpha u}{\partial t^\alpha}$, and according to the relationship between fractional derivative and integral, we get the following:

$$u(x, t) = f(x) - \varepsilon J^\alpha (\Phi(u)) - J^\alpha (\eta \frac{\partial^2 u}{\partial x^2} + \upsilon \frac{\partial^3 u}{\partial x^3}). \qquad (2.4.27)$$

For the decomposition method [16], we have

$$u(x, t) = \sum_{i=0}^{\infty} u_i(x, t),$$

$$\Phi(u) = \sum_{i=0}^{\infty} A_i(u_0, u_1, \dots, u_i), \qquad (2.4.28)$$

where [17]

$$A_i = \frac{1}{i!} \left[ \frac{d^i}{d\lambda^i} \Phi(\sum_{i=0}^{\infty} \lambda^i u_i) \right]_{\lambda=0} = \frac{1}{i!} \left[ \frac{d^i}{d\lambda^i} \left( \sum_{i=0}^{\infty} \lambda^i u_i ) D_x^\beta (\sum_{i=0}^{\infty} \lambda^i u_i) \right) \right], i \ge 0.$$

Substituting (2.4.28) into (2.4.27), all the components $u_1, u_2, u_3, u_n \cdots$ are determined by using the iteration form of (2.4.25), namely the approximate solution of the original equation is obtained.

The advantage of this method is involving the initial conditions to the iteration formula as iterative initial values, and is easy to program because of the regularity of the iteration formula. The accuracy of the method can be implemented by increasing the high-order term of the Adomian series, but this costs a huge computation expense.

### 2.4.4 Homotopy Function Method

Perturbation theory has been applied to solve nonlinear problems in science and engineering, and recently the theory has been used to solve approximate solutions of fractional calculus equations. Especially, in the recent 10 years, a number of researchers have obtained some valuable results. Based on these results, there are two perturbation algorithms introduced in this book: homotopy perturbation method and homotopy analysis method. The main contents of this section contain the main ideas of the two methods, and numerical examples.

1. **Homotopy perturbation method (HPM)**

The homotopy perturbation method was originally proposed to solve nonlinear equations [18–21]. With further research, researchers found that HPM has special merits to numerically solving the fractional differential equations. These merits include the following: (1) Different from the difference scheme and integral transform algorithm, it does not need to consider the truncation error and the stability of the difference form; (2) It does not cost a large amount of computation and computer memory usage; (3) Analytical solutions of some equations may be obtained by using this algorithm.

HPM becomes a hotspot to solve fractional differential equations, especially recently, this method has been widely applied to solve problems in fractional derivative vibration, anomalous diffusion, nonlinear wave and damping control, etc. [22–28]. To improve the convergent rate of this method, Nia proposed a modified HPM by modifying the iterative scheme, in which the initial iteration is an exponential function but not a polynomial function, such that the iterative scheme is stable and has a fast convergence rate [21]. The main idea of this method is that we construct a homotopy function according to the known equation. Take the following equation, for example.

Equation:

$$L(u) + N(u) = f(r), \ r \in \Omega, \tag{2.4.29}$$

establishing the homotopy form as follows:

$$H(v, p) = (1 - p)[L(v) - L(u_0)] + p[L(v) + N(v) - f(r)] = 0, \tag{2.4.30}$$

or simplifying as

$$H(v, p) = L(v) - L(u_0) + pL(u_0) + p[N(v) - f(r)] = 0 \tag{2.4.31}$$

where $p \in [0, 1]$ is an implicit parameter.

Assuming the solution has the following form:

$$v = v_0 + p v_1 + p^2 v_2 + \cdots, \qquad (2.4.32)$$

if $p$ tends to 1, we have

$$u = \lim_{p \to 1} v = v_0 + v_1 + v_2 + \cdots, \qquad (2.4.33)$$

this form is the solution of the original equation.

Here, take the fractional Riccati equation, for example [24],

$$\frac{\partial^\alpha u}{\partial t^\alpha} = -u^2 + 1, \ t > 0, \ (0 < \alpha < 1). \qquad (2.4.34)$$

The homotopy function can be constructed as follows:

$$H(v, p) = (u' - 1) - p[u' - u^2 - d^\alpha u / dt^\alpha] = 0. \qquad (2.4.35)$$

Note that $p$ is not necessarily a small parameter and also can be a sufficient large parameter [29]. Similar to the analysis above, if $p \to 1$, the form above is the solution of the original equation.

Here, suppose the solution of the equation above can be written as a power series in $p$,

$$v = v_0 + p v_1 + p^2 v_2 + p^3 v_3 + p^4 v_4 + \cdots \qquad (2.4.36)$$

Thus, the approximate solution to the original problem is

$$u = \lim_{p \to 1} v_0 + v_1 + v_2 + v_3 + v_4 + \cdots \qquad (2.4.37)$$

Substitute the formula (2.4.36) into (2.4.35); the values of $(I^\alpha f)(x) = \frac{1}{\gamma_n(\alpha)} \int_{R^n} \frac{f(t) dt}{|x-t|^{n-\alpha}}, \ (R(\alpha) > 0; \alpha - n \neq 0, 2, 4, \cdots)$, can be obtained by equating the terms with identical powers of $p$, namely we have the approximate solution to the original problem.

The drawbacks of this method are also obvious: (1) The initial and boundary conditions of the mechanics and the physical process should be good enough to get the solutions of $v_0, v_1, v_2, v_3, v_4 \ldots$; (2) We should ensure the series is convergent, which is the key condition to make the method feasible, but, in general, it cannot be satisfied.

**Example 1** Here, consider the fractional Riccati Eq. (2.4.34), the initial condition $u(0) = 0$. In view of the homotopy perturbation method, the expressions of the approximate solution are [24]

$$v_0 = t,$$

$$v_1 = t - \frac{t^3}{3} - \frac{t^{2-\alpha}}{\Gamma(3-\alpha)},$$

$$v_2 = t - t^3 + \frac{2t^5}{15} - \frac{2t^{2-\alpha}}{\Gamma(4-\alpha)} + \frac{t^{3-\alpha}}{\Gamma(4-2\alpha)},$$

$$v_3 = t - 2t^3 + \frac{2}{3}t^5 - \frac{17}{315}t^7 - \frac{3t^{2-\alpha}}{\Gamma(3-\alpha)} + \left[\frac{6}{\Gamma(3-\alpha)} + \frac{6}{\Gamma(4-\alpha)}\right.$$

$$+ \frac{2\Gamma(5-\alpha)}{\Gamma(4-\alpha)^2}\left]\frac{\Gamma(4-\alpha)}{\Gamma(5-\alpha)}t^{4-\alpha}\right.$$

$$- \left[\frac{2}{3\Gamma(3-\alpha)} + \frac{4}{\Gamma(4-\alpha)} + \frac{16}{\Gamma(6-\alpha)}\right]\frac{\Gamma(6-\alpha)}{\Gamma(7-\alpha)}t^{6-\alpha}$$

$$- \left[\frac{1}{\Gamma(3-\alpha)^2} + \frac{2}{\Gamma(4-2\alpha)} + \frac{2\Gamma(5-\alpha)}{\Gamma(4-\alpha)\Gamma(5-2\alpha)}\right]\frac{\Gamma(5-2\alpha)}{\Gamma(6-2\alpha)}t^{5-2\alpha}$$

$$+ 3\frac{t^{3-2\alpha}}{\Gamma(4-2\alpha)} - \frac{t^{4-3\alpha}}{\Gamma(5-3\alpha)}$$

$$\cdots$$

$$(2.4.38)$$

The results from the expressions above can be shown in Fig. 2.5.

To illustrate the HPM features more clearly, compare HPM with the explicit finite difference method (FDM) by numerical solution; the results are shown in Fig. 2.6.

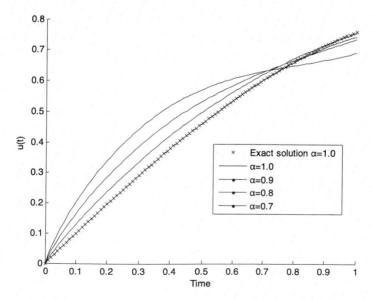

**Fig. 2.5** The solutions of the Riccati Eq. (2.4.34) with different fractional orders

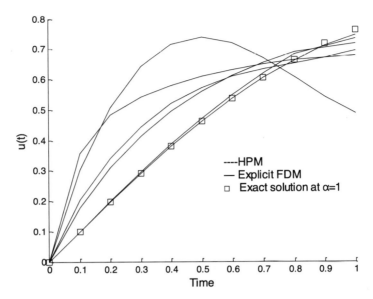

**Fig. 2.6** The results of comparing the homotopy perturbation method with the explicit finite difference method; the fractional derivative orders are $\alpha = 0.4,\ 0.7,\ 1.0$

From Fig. 2.6, we can observe that, although the numerical results of the HPM are better in the cases of $\alpha = 0.7,\ 1.0$, its numerical result is very bad when $\alpha = 0.4$. For the FDM, although there is a big numerical error, the general trend is correct. According to numerical experiments, for $t > 1$, the numerical result of HPM is far away from the analytical solution. The result shows that this method is only feasible with a small value of $t$.

***Example 2***

$$\frac{\partial^\alpha u(x,t)}{\partial t^\alpha} = \sum_{i=1}^{3} \frac{\partial^2 u(x,t)}{\partial x_i^2}, 0 < \alpha < 1. \tag{2.4.39}$$

Here, construct a homotopy function as follows:

$$H(v,p) = (1-p)(D_x^\alpha v) + p(D_x^\alpha v - \sum_{i=1}^{3} \frac{\partial^2 v}{\partial x_i^2}) = 0. \tag{2.4.40}$$

We also can use the modified homotopy function

$$H(v,p) = \frac{\partial v}{\partial t} - p(\frac{\partial v}{\partial t} + \sum_{i=1}^{3} \frac{\partial^2 v}{\partial x_i^2} - D_t^\alpha v) = 0. \tag{2.4.41}$$

The form (2.4.41) should be the solution to the original problem if $p \to 1$.

Assume the solution of the equation above can be written as a power series with $p$,

$$v = v_0 + pv_1 + p^2 v_2 + p^3 v_3 + p^4 v_4 + \ldots \qquad (2.4.42)$$

then the approximate solution of the original problem is

$$u = \lim_{p \to 1} v_0 + v_1 + v_2 + v_3 + v_4 + \ldots \qquad (2.4.43)$$

Substitute the formula (2.4.41) into (2.4.40), and the values of $v_0$, $v_1$, $v_2$, $v_3$, $v_4 \ldots$ can be obtained by equating the terms with identical powers of $p$, namely we have the approximate solution to the original problem.

To illustrate the details, we take a one-dimensional time-fractional anomalous diffusion equation with absorption (or reaction) term, for example,

$$D_t^\alpha u = \frac{\partial^2 u}{\partial x^2} + u, \ u(x, 0) = \cos(\pi x), \ \alpha \in (0, 1). \qquad (2.4.44)$$

Applying the modified homotopy perturbation method, we construct a homotopy function as follows:

$$\frac{\partial u}{\partial t} = p \left[ \frac{\partial u}{\partial t} + \frac{\partial^2 u}{\partial x^2} + u - D_t^\alpha u \right], \qquad (2.4.45)$$

and then assuming the expression of the approximate solution as a power series, substituting it into the homotopy function, we have

$$
\begin{aligned}
v_0 &= \cos(\pi x), \\
u_1 &= [1 - \pi^2] \cos(\pi x) \frac{t^\alpha}{\Gamma(\alpha+1)}, \\
u_2 &= [1 - \pi^2]^2 \cos(\pi x) \frac{t^{2\alpha}}{\Gamma(2\alpha+1)}, \\
&\cdots \\
v_{j+1} &= [1 - \pi^2]^{j+1} \cos(\pi x) \frac{t^{j\alpha}}{\Gamma(j\alpha+1)}, j = 1, 2, \ldots
\end{aligned}
\qquad (2.4.46)
$$

Combining the series above together, we can have the exact solution to the original equation

$$u(x, t) = \cos(\pi x) E_\alpha([1 - \pi^2]t), \qquad (2.4.47)$$

where $E_\alpha$ represents the one-parameter Mittag-Leffler function.

Fig. 2.7 shows the result of the exact solution (2.4.47).

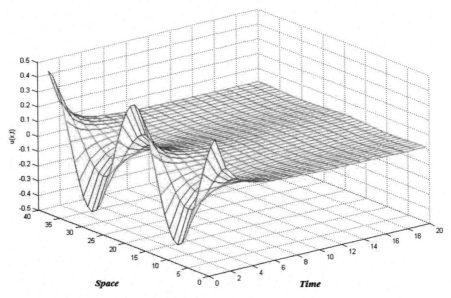

**Fig. 2.7** The exact solution of the equation $D_t^\alpha u = \frac{\partial^2 u}{\partial x^2} + u$, $u(x,0) = \cos(\pi x)$, $\alpha = 0.5$.

## 2.  Homotopy Analysis Method (HAM)

The homotopy analysis method is an approximate scheme, similar to the homotopy perturbation method, and is widely applied to solve nonlinear partial differential equations. Now it is also applied to solve fractional differential equations. We will introduce the main idea of this method below, and apply an example to specify the implementation process.

Let us consider a nonlinear differential equation

$$A(u(t)) = 0, \tag{2.4.48}$$

where $A$ represents a nonlinear operator, and $u(t)$ is an unknown function.

Then we will construct a homotopy function. According to different numbers of parameters, there are three kinds of homotopy functions: single parameter family; dual-parameter family; three-parameter family. Referring to the papers [30, 31], the HPM belongs to the single parameter family. Take the three-parameter family into account below.

Firstly, we introduce two auxiliary parameters $p, h$, and an auxiliary function $H(t)$, then the generalized zero-order deformation equation is obtained as

$$(1 - p)L(\phi - u_0(t)) = phH(t)A(\phi), \ p \in [0, 1], \tag{2.4.49}$$

where $L$ represents a linear auxiliary operator, and the function $\phi$ is a function with respect to $t$, $p$ and $h$. If $p = 0$, $p = 1$, the following holds:

$$\phi(t, h, 0) = u_0(t), \quad \phi(t, h, 1) = u(t). \tag{2.4.50}$$

The formula above shows that if $p$ changes from 0 to 1, $\phi$ is continuously changing from the guess initial solution $u_0(t)$ to the solution $u(t)$ of the original equation.

Next, we expand the function $\phi$ to the Taylor series of $p$

$$\phi(t, h, p) = u_0(t) + \sum_{k=1}^{+\infty} u_k(t) p^k, \tag{2.4.51}$$

where

$$u_k(t) = \frac{1}{k!} \frac{\partial^k \phi(t, h, p)}{\partial p^k}\Big|_{p=0}.$$

If an appropriate linear operator and a parameter $h$ can be selected, this series is convergent for $p = 1$, then the solution in series is obtained as

$$u(t) = u_0(t) + \sum_{k=1}^{+\infty} u_k(t). \tag{2.4.52}$$

Similarly, for the high-order deformation equation

$$L(u_k(t) - \chi_k u_{k-1}) = h H(t) R_k(t), \tag{2.4.53}$$

where

$$\chi_k = \begin{cases} 0, k \leq 1 \\ 1, k > 1 \end{cases}, \quad R_k(t) = \frac{1}{(k-1)!} \frac{\partial^{k-1} A[\phi(t, p)]}{\partial p^{k-1}}\Big|_{p=0},$$

according to the processes mentioned above, by choosing appropriate parameters $p$, $h$ and the auxiliary function $H$, the solution of the nonlinear equation can be derived.

**Example 3**  To illustrate the implementation of this method, we take a time-fractional differential equation, for example [32],

$$D_t^\alpha u(x, t) = \frac{1}{2} x^2 u_{xx}, \quad 0 < x < 1, \quad 0 < \alpha \leq 1, t > 0, \tag{2.4.54}$$

subject to the initial and boundary conditions

$$u(0, t) = 0, \ u(1, t) = e^t, \ u(x, 0) = x^2.$$

**Main scheme**:

Firstly, choose the trial initial function referring to the initial condition

$$u_0(x, t) = x^2. \tag{2.4.55}$$

Assume $H(t) = 0$ in view of the features of the equation, and construct a linear auxiliary operator

$$L[G(x, t, q)] = D_t^\alpha[G(x, t, q)], \tag{2.4.56}$$

which satisfies the following property:

$$L(\sum_{k=0}^{n-1} G(x, 0^+) \frac{t^k}{k!}) = 0. \tag{2.4.57}$$

From zero-order and high-order deformation equations, the following formula holds:

$$\begin{cases} u_0(x, t) = x^2 \\ u_m(x, t) = (h + \chi_m) u_{m-1}(x, t) - \frac{1}{2} x^2 h J^\alpha[u_{xx(m-1)}(x, t)] - (h - \chi_m) u(x, 0) \end{cases} \tag{2.4.58}$$

If $h = -1$, it leads to

$$u(x, t) = x^2[1 + \frac{t^\alpha}{\Gamma(\alpha + 1)} + \frac{t^{2\alpha}}{\Gamma(2\alpha + 1)} + \frac{t^{3\alpha}}{\Gamma(3\alpha + 1)} \cdots]$$
$$= x^2 E_\alpha(t^\alpha). \tag{2.4.59}$$

The result of the solution is shown in Fig. 2.8.

## 2.4.5  Other Iteration Methods

### 1.  Variational Iteration Method (VIM)

The variational iteration method was initially applied to solve linear or nonlinear differential equations, especially for partial differential equations. In recent years, some researchers have applied this method to solve fractional differential equations, and then extended it to solve nonlinear fractional differential equations [24–29, 33–37].

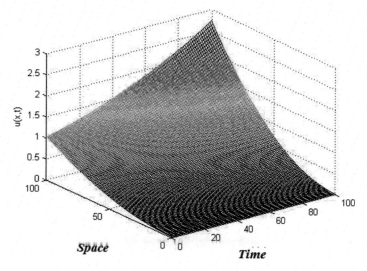

**Fig. 2.8** The solution of Eq. (2.4.54)

Here, take a linear fractional derivative equation, for example, to introduce the VIM.

$$\frac{\partial^\alpha u}{\partial t^\alpha} + a_0 u + a_1(x)\frac{\partial u}{\partial x} + a_2(x)\frac{\partial^2 u}{\partial x^2} + a_3(x)\frac{\partial^3 u}{\partial x^3} + \ldots + a_n(x)\frac{\partial^n u}{\partial x^n}$$
$$= q(x,t), t > 0, x \in R. \quad (2.4.60)$$

The discrete form of the equation above can be approximately written as

$$u_{k+1}(x,t) = u_k(x,t) + \int_0^t \lambda(\xi)(\frac{\partial^m}{\partial\xi^m}u_k(x,\xi) + a_0(x)\tilde{u}_k(k,\xi) + \ldots$$
$$a_n(x)\frac{\partial^n}{\partial x^n}\tilde{u}_k(k,\xi) - q(x,\xi))d\xi \quad (2.4.61)$$

where $\lambda$ represents the Lagrange multiplier, which can be obtained by the variation principle; $\tilde{u}_k$, $\frac{\partial\tilde{u}_k}{\partial x}$, ..., $\frac{\partial^n\tilde{u}_k}{\partial x^n}$ are restrict variables. Noting that $\delta\tilde{u}_k = 0$, we have

$$\delta u_{k+1}(x,t) = \delta u_k(x,t) + \delta \int_0^t \lambda(\xi)(\frac{\partial^m}{\partial\xi^m}u_k(x,\xi) - q(k,\xi))d\xi. \quad (2.4.62)$$

The Lagrange multipliers are obtained by computing the formula above,

$$\begin{cases} \lambda = 1, \, m = 1 \\ \lambda = \xi - 1, \, m = 2 \end{cases}. \quad (2.4.63)$$

When $m = 1$, the iterative expression is

$$u_{k+1}(x, t) = u_k(x, t) - \int_0^t (\frac{\partial^\alpha}{\partial \xi^\alpha} u_k(x, \xi) + a_0(x)u_k(x, \xi) + \ldots$$

$$+ a_n(x)\frac{\partial^n}{\partial x^n} u_k(x, \xi) - q(x, \xi))d\xi. \tag{2.4.64}$$

When $m = 2$, the iterative expression is

$$u_{k+1}(x, t) = u_k(x, t) + \int_0^t (\xi - t) \times (\frac{\partial^\alpha}{\partial \xi^\alpha} u_k(x, \xi) + a_0(x)u_k(x, \xi) + \ldots$$

$$+ a_n(x)\frac{\partial^n}{\partial x^n} u_k(x, \xi) - q(x, \xi))d\xi \tag{2.4.65}$$

Take $m = 1$, for example, subjecting to the initial conditions, the iteration scheme is

$$u_1(x, t) = f(x) - \int_0^t (a_0(x)f(x) + \ldots + a_n(x)\frac{\partial^n}{\partial x^n} f(x) - q(x, \xi))d\xi, \ n = 1,$$

$$\tag{2.4.66}$$

$$u_{k+1}(x, t) = u_k(x, t) - \int_0^t (\frac{\partial^\alpha}{\partial \xi^\alpha} u_k(x, \xi) + a_0(x)u_k(k, \xi) + \ldots$$

$$+ a_n(x)\frac{\partial^n}{\partial x^n} u_k(k, \xi) - q(x, \xi))d\xi \tag{2.4.67}$$

The final exact solution can be obtained by

$$u(x, t) = \lim_{n \to \infty} u_n(x, t). \tag{2.4.68}$$

This method is effective to find approximate solutions and exact solutions of a sequence of complex mechanics model equations. The limitation of this method is that the trial initial value of $u_0$ $u_0$ is restricted by the initial and boundary conditions, and also directly affects the accuracy and correctness of numerical solutions. Therefore, the iteration formula is generally complex and difficult to program.

**Numerical example**:

The time-fractional differential anomalous diffusion equation

$$\begin{cases} \dfrac{\partial u(x,t)}{\partial t} = \dfrac{\partial^{1-\alpha}}{\partial t^{1-\alpha}} \dfrac{\partial^2 u(x,t)}{\partial x^2} + \dfrac{\partial^{1-\alpha}}{\partial t^{1-\alpha}} \dfrac{\partial}{\partial x}(xu(t)) \\ u(x,0) = f(x) \end{cases}. \qquad (2.4.69)$$

Assuming $f(x) = 0$, the govern equation above can be converted into (the time-fractional derivative is defined in the sense of Caputo).

$$\frac{\partial^\alpha u(x,t)}{\partial t^\alpha} = \frac{\partial^2 u(x,t)}{\partial x^2} + \frac{\partial}{\partial x}(xu(t)). \qquad (2.4.70)$$

By using the variational iteration method, we can obtain the iteration formula of the time direction as

$$u_{n+1}(x,t) = u_n(x,t)$$

$$+ \int_0^t \lambda(\xi) \left[ \frac{\partial u_n(x,\xi)}{\partial \xi} - \frac{\partial^{1-\alpha}}{\partial \xi^{1-\alpha}} \frac{\partial^2 \bar{u}_n(x,\xi)}{\partial x^2} - \frac{\partial^{1-\alpha}}{\partial \xi^{1-\alpha}} \frac{\partial}{\partial x}(x\bar{u}_n(x,\xi)) \right] d\xi. \qquad (2.4.71)$$

To get an appropriate parameter $\lambda$, we have

$$\delta u_{n+1}(x,t) = \delta u_n(x,t) + \delta \int_0^t \lambda(\xi) \frac{\partial}{\partial \xi} u_n(x,\xi) d\xi = 0. \qquad (2.4.72)$$

From the formula above, the following conditions hold:

$$\lambda'(\xi) = 0, \ 1 + \lambda(\xi) = 0. \qquad (2.4.73)$$

Therefore, $\lambda = -1$.

Substituting the value into the iteration formula, we have

$$u_1(x,t) = f(x) + \int_0^t \left[ \frac{\partial^{1-\alpha}}{\partial \xi^{1-\alpha}} \frac{\partial^2 f(x)}{\partial x^2} + \frac{\partial^{1-\alpha}}{\partial \xi^{1-\alpha}} \frac{\partial}{\partial x}(xf(x)) \right] d\xi, \ n = 1, \qquad (2.4.74)$$

$$u_{n+1}(x,t) = u_n(x,t)$$

$$- \int_0^t \left[ \frac{\partial u_n(x,\xi)}{\partial \xi} - \frac{\partial^{1-\alpha}}{\partial \xi^{1-\alpha}} \frac{\partial^2 u_n(x,\xi)}{\partial x^2} \right.$$

$$\left. - \frac{\partial^{1-\alpha}}{\partial \xi^{1-\alpha}} \frac{\partial}{\partial x}(xu_n(x,\xi)) \right] d\xi, \ n > 1. \qquad (2.4.75)$$

The final exact solution can be expressed as

**Fig. 2.9** The curve of the
exact solution with
$\alpha = 0.5, x = 1.0$.

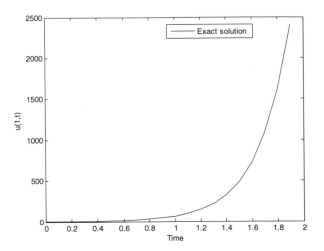

$$u(x, t) = \lim_{n \to \infty} u_n(x, t). \tag{2.4.76}$$

Assume $f(x) = x$; thus, we have

$$u_0(x, t) = x;\ u_1(x, t) = x + \frac{2xt^\alpha}{\Gamma(\alpha+1)};$$
$$\ldots,\ u_n(x, t) = x + \frac{2xt^\alpha}{\Gamma(\alpha+1)} + \frac{4xt^{2\alpha}}{\Gamma(2\alpha+1)} + \frac{8xt^{3\alpha}}{\Gamma(3\alpha+1)} + \cdots \tag{2.4.77}$$

The exact solution can be written as

$$u(x, t) = \lim_{n \to \infty} u_n(x, t) = x \sum_{k=0}^{\infty} \frac{2^k t^{k\alpha}}{\Gamma(k\alpha + 1)} = x E_\alpha(2t^\alpha). \tag{2.4.78}$$

The result is shown in Fig. 2.9.

2.   A New Iteration Method

Sachin and Varsha proposed and applied a new iteration method to solve fractional differential equations. We will briefly introduce the main idea of this new method below; for details, readers can refer to the Refs. [38, 39].

To introduce the new iteration method, we take the following fractional derivative equation, for example:

$$\begin{cases} D_t^\alpha u(x, t) = A(u, \partial u) + B(x, t),\ m - 1 < \alpha < m,\ m \in N \\ \frac{\partial^k u(x,0)}{\partial t^k} = h_k(x),\ k = 0, 1, \ldots, m - 1 \end{cases}, \tag{2.4.79}$$

where $A$ is a nonlinear term, and $B$ is a source function; the time-fractional derivative is defined in the sense of Caputo.

Firstly, integrating both sides of the equation with respect to $t$, we have

$$u(x, t) = \sum_{k=0}^{m-1} h_k(x) \frac{t^k}{k!} + I_t^\alpha B + I_t^\alpha A = f + N(u), \quad (2.4.80)$$

where $f = \sum_{k=0}^{m-1} h_k(x) \frac{t^k}{k!} + I_t^\alpha B$, $N(u) = I_t^\alpha A$.

And then, assume the form of the solution of this equation can be written as

$$u(x, t) = u_0(x, t) + \sum_{k=1}^{\infty} u_k(x, t). \quad (2.4.81)$$

By utilizing the following iterative formula:

$$u_0(x, t) = f(x, t), \ u_1(x, t) = N(u_0), \ u_2(x, t) = N(u_1) \\ \dots u_n(x, t) = N(u_{n-1}) \quad (2.4.82)$$

and the formula (2.4.82), we have the solution to the original problem.

## 2.5 Questions and Discussions

### 2.5.1 Fractal Derivative, Positive Fractional Derivative, Variable-Order Derivative and Random-Order Derivative

In recent years, some new definitions are proposed from theoretical study and engineering applications, besides the definitions introduced in Sect. 2.1. Combining the research work by the authors of this book, we focus on the following four new definitions.

1. **Fractal derivative**

The essential characteristics of the anomalous diffusion process can be described by the random walk model, namely the mean square displacement of a random walker, $\langle \Delta x^2 \rangle$, depends on time as follows:

$$\langle \Delta x^2 \rangle \propto \Delta t^\eta, \quad (2.5.1)$$

where $\Delta x$ represents distance, $\Delta t$ denotes time interval and $\eta$ is a real number. For $\eta = 1$, the motion is the Brownian diffusion, otherwise, for $\eta \neq 1$, anomalous diffusion. The corresponding phenomenological anomalous diffusion equation is

$$\partial^\alpha s / \partial t^\alpha + \gamma \left(-\nabla^2\right)^\beta s = 0 \tag{2.5.2}$$

where $s$ is the physical quantity of interest (e.g. temperature in heat conduction or concentration in diffusion), $\gamma$ is the corresponding physical coefficient, $\left(-\nabla^2\right)^\beta$ represents a symmetric fractional Laplacian, and $\alpha$ and $\beta$ can be real numbers, $0 < \alpha, \beta \leq 1$. The fundamental solution of Eq. (2.5.2) is the time-dependent Lévy probability density function. For $\alpha = 1$, $\beta < 1$, it is called the fat-tailed distribution. For $\alpha = 1$, $\beta = 1$, the equation is reduced to the classical diffusion equation. In view of the mesoscopic statistics, $\alpha$ is the memory strength index of the process (long time range correlation), and $\beta$ is the stability index of the Lévy distribution. The smaller the $\alpha$, the stronger the memory. The Lévy statistics and fractional Brownian motion are often considered the statistical mechanism leading to anomalous diffusion and accordingly $\eta = \alpha / \beta$ $\eta = \alpha / \beta$ can be derived. The second moment of the random system $\eta \neq 1$ $\eta \neq 1$ diverges, and the kinetic energy is infinite, i.e.

$$\left\langle (\Delta x)^n \right\rangle = \infty, \quad 2\beta < n. \tag{2.5.3}$$

It says that the potential energy cannot trap the particle, and the kinetic energy for a particle with finite mass diverges [40]. To solve this paradox, Chen [41], one of the authors of this book, proposed a spacetime transform:

$$\begin{cases} \Delta \hat{x} = \Delta x^\beta \\ \Delta \hat{t} = \Delta t^\alpha \end{cases}, \quad 0 < \alpha, \beta \leq 1. \tag{2.5.4}$$

Unlike the Lorentz transforms in the inertial reference system of special relativity and the spacetime transforms in the acceleration reference system of general relativity, the transforms above are nonlinear transforms which reflect the spatial–temporal distortions caused by complex fractal physical fields. The transforms coincide with the classical definition of the Hausdorff fractal dimension, thus, the transforms can be considered as a fractal metric spacetime transforms. To solve the counterintuitive paradox on the entropy production of the anomalous diffusion process, Hoffmann [42] and Li [43] also presented a time–scale transformation, referred to as "internal clock".

In terms of the spacetime transforms (2.5.4), the mean square displacement (2.5.1) is recast as a "normal" Brownian motion under the fractal metric spacetime

$$\left\langle \Delta \hat{x}^2 \right\rangle \propto \Delta \hat{t}. \tag{2.5.5}$$

The second moment of the anomalous diffusion process is finite and the corresponding kinetic energy exists. It is worth pointing out that the corresponding definition of velocity needs to be changed (see the formula 2.5.3). In terms of the transforms (2.5.4), Lévy statistics and fractional Brownian motion are considered as a consequence of the fractal metric spacetime, while the classical Gaussian distribution and

Brownian motion (white noise) correspond to the limiting $\alpha = 1$ and $\beta = 1$ spacetime fabric, respectively.

On the other hand, the restoration of the "normal" diffusion formalism in (2.5.5) implies the invariance and equivalence of physical law under scale transformation under the spacetime transforms. Corresponding to the equivalence principle and the general covariance principle in general relativity, Chen [41] gave two hypotheses on "anomalous" physical processes:

(1) The hypothesis of fractal invariance: the laws of physics are invariant regardless of the fractal metric spacetime.

(2) The hypothesis of fractal equivalence: the influence of anomalous environmental fluctuations on physical behaviors equal that of the fractal time–space transforms.

Anomalous diffusion equations can be derived from the two hypotheses above. Unlike existing phenomenological models to describe "anomalous behavior", Chen [41] proposed the fractal spacetime transforms above and two hypotheses to try to illustrate a variety of "anomalous" phenomena from the basic physical spacetime principal concept. The hypothesis of fractal invariance is very similar to the so-called scale relativity principle pioneered by Nottale [44]. Unlike the latter, Chen's work did not intend to incorporate Einstein's relativistic effects arising from the reference frame of motion transforms such as acceleration and velocity (inertial). Chen [41] also developed different spacetime transforms, and proposed and applied different differential and statistical methods to discuss different physical problems.

Fractals are self-similar phenomena widely existing in material spatial structure and physical evolving processes. Under fractal scale, the integer dimension spacetime $(x, t)$ is converted into the fractal spacetime $(x^\beta, t^\alpha)$ [41, 45]. As a local approximation of the fractional derivative method in fractal spacetime, based on the spacetime transforms above, Chen [41] defined a fractal Hausdorff derivative under fractal scale, which can be shortly called a fractal derivative

$$\frac{\mathrm{d}u(t)}{\mathrm{d}t^\alpha} = \lim_{t' \to t} \frac{u(t) - u(t')}{t^\alpha - t'^\alpha}. \tag{2.5.6}$$

The elementary physical concepts in fractal spacetime need to be redefined. For instance, the definition of velocity:

$$\hat{v} = \frac{\mathrm{d}\hat{x}}{\mathrm{d}\hat{t}} = \frac{\mathrm{d}x^\beta}{\mathrm{d}t^\alpha} \hat{t}, \hat{x} \, \forall S^{\alpha,\beta}, \tag{2.5.7}$$

where $S^{\alpha,\beta}$ represents time–space fabric having scaling indices $\alpha$ and $\beta$. Note that the time and space derivative orders of the definition of velocity (2.5.7) can be different. The traditional definition of velocity makes no sense in the non-differentiable fractal spacetime.

Time and space are the most basic physical and mathematical concepts, and the spacetime transforms have an impact on the basic physical concepts and mathematical methods. By using the fractal derivative, Chen [41] derived a new diffusion equation and obtained the corresponding stretched Gaussian fundamental solution (details will be introduced in Chap. 4) and other interesting results.

## 2. Positive Time-Fractional Derivative

The dissipation of acoustic wave propagation in a wide variety of loose soft mediums obeys the power frequency dependence [46, 47]

$$P(x + \Delta x) = P(x)e^{-\alpha(\omega)\Delta x}, \quad \alpha(\omega) = \alpha_0|\omega|^p, \tag{2.5.8}$$

where $0 < p < 2$, $P$ represents the pressure, $x$ is the wave line of propagation, $\omega$ denotes the angular frequency, $\alpha_0$ is the attenuation coefficient and $\Delta x$ is the wave propagation distance. The standard integer-order derivative model cannot describe the frequency-dependent attenuation of acoustic waves. Currently, fractional calculus has become one of the main modeling tools to describe complex mechanics behaviors with power-law features [48–53]. From the definition, the standard time-fractional derivative does not have the property of positivity and cannot accurately reflect the frequency dependency of acoustic attenuation of the dissipative medium.

Szabo [47] proposed an acoustic wave equation to accurately describe this acoustic frequency-dependent dissipation. This model can effectively describe the acoustic anomalous energy dissipation processes and satisfies the causality requirements of wave equations due to the positivity of the dissipative term. However, due to hyper-singular improper integral in the dissipative term of Szabo's model, its numerical solution is feasible. Noting the similarity of Szabo's model and fractional derivative model, Chen and Holm [46] introduced the positive fractional derivative and further proposed the modified Szabo's wave equation. Unlike the Fourier transform of general time-fractional derivatives, the positive time-fractional derivative has the expression as $|\omega|^p$ and agrees well with the power-law frequency-dependent dissipation. The definition of positive fractional derivative is stated as [46]

$$\frac{\partial^{|p|} f(t)}{\partial t^{|p|}} = -\frac{1}{q(p)} \int_a^t \frac{f(\tau)}{(t-\tau)^{p+1}} d\tau, \tag{2.5.9}$$

where $p$ is the order of positive fractional derivative, and the constant $q(p)$ is given by $q(p) = \frac{\pi}{2\Gamma(p+1)\cos[(p+1)\pi/2]}$; in terms of the Caputo fractional derivative, the positive fractional derivative can be defined as

$$\frac{\partial^{|p|} u(t)}{\partial t^{|p|}} = \begin{cases} \frac{-1}{pq(p)} \int_0^t \frac{u'(\tau)}{(t-\tau)^p} d\tau & 0 < p \le 1 \\ \\ \frac{1}{p(p-1)q(p)} \int_0^t \frac{u''(\tau)}{(t-\tau)^{p-1}} d\tau & 1 < p < 2 \end{cases}. \tag{2.5.10}$$

In Sect. 5.5, we will apply the positive fractional derivative to model the arbitrary order frequency-dependent dissipation in complex medium and also take some applications in medical ultrasound imaging, for example.

3. **Differences and Relations of the Fractal Derivative, Positive Fractional Derivative and Fractional Derivative**

From the definition of fractal derivative, we can see that the fractal derivative does not include the convolution integral and is a local operator [41]. However, the definitions of fractional derivative and positive fractional derivative both containing convolution integrals are non-local operators and can describe the history of dependence of the system evolution through the convolution integral [2, 46].

To compare three models established by the three derivatives, take the following relaxation vibration and damping vibration processes, for example, as below.

(1)   **Relaxation-Oscillation Equation**

$$\frac{d^p u(t)}{dt^p} + Bu(t) = f(t), (B = \omega^p). \tag{2.5.11}$$

The relaxation-oscillation equation above is the process of relaxation and oscillation control equation. For $0 < p \leq 1$, the equation is considered as a relaxation equation. For $p = 1$, this equation is reduced to a standard relaxation equation. Actually, a variety of substances exhibit viscoelastic characteristics and further microscopic fractal characteristics under certain conditions, but not the ideal elastic body or Newtonian fluid. The stress and strain responses of viscoelastic materials depend on self-loading and self-deformation history and exhibit memories of mechanics behaviors. The stress relaxation of complex viscoelastic materials often exhibits as a slow relaxation process with the memory of exponential index. Fractional derivative and positive fractional derivative both contain time convolution integrals which can describe the memory [2, 46, 48]. However, the fractal derivative can describe mechanics behaviors under the fractal scale [41]. Therefore, instead of the Newton dashpot which is based on the integer-order derivative standard models, the dashpot models based on fractal derivative, fractional derivative [48–54] and positive fractional derivative are applied to characterize the slow stress relaxation with memory.

For $1 < p \leq 2$, the equation is an oscillation equation. For $p = 2$, the equation is reduced to an undamped oscillation equation, and can be considered as a conservative system. For $p \neq 2$, the equation is a damped oscillation equation. Since non-integer order derivatives have been introduced, standard oscillation models become energy dissipation systems. Simultaneously, the damping effects are caused by the internal damping of viscoelastic materials [48, 54].

(2)   **Fractal Derivative Relaxation-Oscillation Equation**

For $0 < p < 1$, fractal derivative relaxation equation is

$$\frac{du(t)}{dt^p} + Bu(t) = f(t); \tag{2.5.12}$$

when $f(t) = 0$, we can have the analytic solution as

$$u(t) = C \exp(-Bt^p), \tag{2.5.13}$$

which reflects stretched exponential relaxation.

When $1 < p < 2$, the fractal derivative oscillation equation is given by

$$\frac{d^2 u}{d(t^{p/2})^2} + Bu = f(t); \tag{2.5.14}$$

when $f(t) = 0$, the analytic solution of the equation can be obtained as

$$u(t) = C \cos \sqrt{B} t^{p/2} + D \sin \sqrt{B} t^{p/2}. \tag{2.5.15}$$

For the formula (2.5.15), we can see that the fractal derivative oscillation model only manifests undamped oscillation and can be considered as an energy conservation system.

(3)   **Fractional Derivative Relaxation-Oscillation Equation**

$$\frac{d^p u(t)}{dt^p} + Bu(t) = f(t). \tag{2.5.16}$$

By using the Laplace transform, the analytic solution of Eq. (2.5.16) can be derived [2]. When $0 < p < 1$ (fractional derivative relaxation equation), the analytic solution is

$$u(t) = u(0) E_{p,1}(-Bt^p) + G(t) * f(t), \; G(t) = t^{p-1} E_{p,p}(-Bt^p); \tag{2.5.17}$$

when $1 < p < 2$ (fractional derivative oscillation equation), the analytic solution is

$$\begin{aligned} u(t) &= u(0) E_{p,1}(-Bt^p) + \dot{u}(0) E_{p,2}(-Bt^p) + G(t) * f(t), \; G(t) \\ &= t^{p-1} E_{p,p}(-Bt^p), \end{aligned} \tag{2.5.18}$$

where $E_{\alpha,\beta}(z)$ represents the two-parameter Mittag-Leffler function:

$$E_{\alpha,\beta}(z) = \sum_{k=0}^{\infty} z^k / \Gamma(\alpha k + \beta). \tag{2.5.19}$$

### (4)   **Positive Fractional Derivative Relaxation-Oscillation Equation**

When $0 < p < 1$, the positive fractional derivative relaxation equation is

$$\frac{d^{|p|}u}{dt^{|p|}} + Bu = f(t);$$   (2.5.20)

when $1 < p < 2$, the positive fractional derivative oscillation equation is

$$-\frac{d^{|p|}u}{dt^{|p|}} + Bu = f(t).$$   (2.5.21)

To describe the attenuation of the oscillation, the damping term of the positive fractional derivative is taken negative when $1 < p < 2$ to ensure being the energy dissipation term. The analytical solution of Eqs. (2.5.20) and (2.5.21) can be obtained by applying the Laplace transform [2]. When $0 < p < 1$ (the positive fractional derivative relaxation equation), the analytical solution is

$$u(t) = u(0)E_{p,1}\left(-\frac{B}{A}t^p\right) + G(t) * \frac{f(t)}{A},$$   (2.5.22)

where $G(t) = t^{p-1}E_{p,p}(-\frac{B}{A}t^p)$, $A = \frac{-\Gamma(1-p)}{pq(p)}$.

When $1 < p < 2$ (the positive fractional derivative oscillation equation), the analytic solution is

$$u(t) = u(0)E_{p,1}(-\frac{B}{A}t^p) + \dot{u}(0)E_{p,2}(-\frac{B}{A}t^p) + G(t) * \frac{f(t)}{A},$$   (2.5.23)

where $G(t) = t^{p-1}E_{p,p}(-\frac{B}{A}t^p)$, $A = \frac{-\Gamma(2-p)}{p(p-1)q(p)}$.

These three models mentioned above to describe stress relaxation and oscillation can be verified by numerical examples. For example,

$$B = 1, u(0) = 1, \dot{u}(0) = 0, f(t) = 0.$$

The numerical results shown in Figs. 2.10 and 2.11 suggest that the three models mentioned above decay slower than the exponential relaxation and exhibit stress relaxation with memory. The fractal derivative model decays the fastest, and the slowest one is the positive fractional derivative model with strong memory. Figures 2.12 and 2.13 show that the fractional derivative and positive fractional derivative model both reflect damped oscillation, whereas the fractional derivative model decays faster. However, the fractal derivative model is undamped, which coincides with the analytical solution (2.5.9) of the fractal derivative equation.

In addition, the numerical results show that, in the stress relaxation and damping oscillation processes, the positive fractional derivative model always decays more

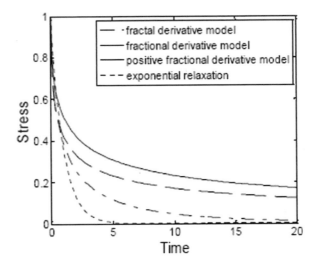

**Fig. 2.10** Numerical solution of relaxation equation with the order 0.5

**Fig. 2.11** Numerical
solution of relaxation
equation with the order 0.8

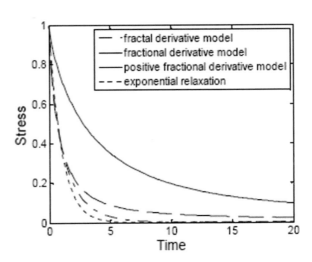

slowly than the fractional derivative model. By analyzing the analytical solution, we
can have the corresponding illustrations. The analytical solutions of the fractional
derivative and positive fractional equations both contain the Mittag-Leffler function
[2], which is the generalization of the exponential function and represents the attenuation of the strong memory effect [2]. The variable in the function is considered as
the attenuation coefficient. When the absolute attenuation coefficient is larger, the
corresponding mechanical process decays faster. From formulas (2.5.11), (2.5.12),
(2.5.16) and (2.5.17), the attenuation coefficient of the fractional derivative model is
$-Bt^p$, and the absolute value is greater than the absolute value of $-(B/A)t^p(A>1)$
of the positive fractional derivative model. Thus, the fractional derivative model

**Fig. 2.12** Numerical solution of oscillation equation with the order of 1.5

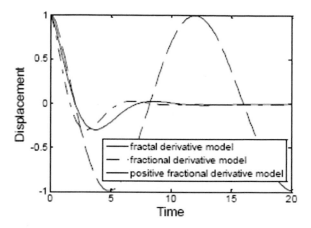

**Fig. 2.13** Numerical solution of oscillation equation with the order of 1.8

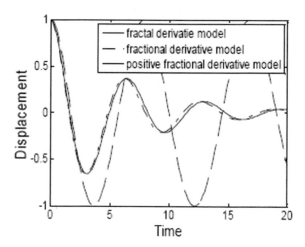

decays faster than the positive fractional derivative model. In terms of the analytical solution, we can have the same results as the results of the numerical solution.

(2) Damped Vibration Equation

The damped vibration equation is a vibration control equation of an oscillator under the external damping. In general, we apply the viscous damping model to describe this vibration system. However, in view of the viscous damping model, the damping medium is considered as Newtonian viscous fluid, and the model only considers the viscous shear effects. The mechanism of viscoelasticity lies between the ideal elasticity and Newtonian fluid, and the damping has a strong memory [55] and the fractional power-law frequency-dependent property [56]. To more accurately describe the time and frequency dependence of viscoelastic damping vibration, the integer-order derivative damping term (viscoelastic damping) of the standard damped vibration equation is replaced by the fractal derivative, fractional derivative and positive fractional derivative, respectively; thus, we have the three following damped vibration equations.

(1)  **Fractal Derivative Damped Vibration Equation**

$$mu''(t) + c\frac{du(t)}{dt^{\alpha}} + ku(t) = f(t), \ (t > 0). \tag{2.5.24}$$

(2)  **Fractional Derivative Vibration Equation**

$$mu''(t) + c\frac{d^{\alpha}u(t)}{dt^{\alpha}} + ku(t) = f(t), \ (t > 0, \ \ 1 < \alpha < 2). \tag{2.5.25}$$

By using Green's function method, the analytical solution of Eq. (2.5.25) [2] is

$$u(t) = \sum_{i=0}^{j} \frac{d^{p-i-1}G_3(t)}{dt^{p-i-1}}u^{(i)}(0) + G_3(t) * f(t), \ j < p < j + 1, \tag{2.5.26}$$

where $j$ is a non-negative integer number, $G_3(t) = \frac{1}{m}\sum_{n=0}^{\infty}\frac{(-1)^n}{n!}(\frac{k}{m})^n t^{2(n+1)-1}E_{2-p,2+pn}^{(n)}(-\frac{c}{m}t^{2-p})$, and $E$ represents the Mittag-Leffler function [2].

(3)  **Positive Fractional Derivative Damped Vibration Equation**

$$mu''(t) + (-1)^{\lceil\alpha\rceil-1}c\frac{d^{\lceil\alpha\rceil}u(t)}{dt^{\lceil\alpha\rceil}} + ku(t) = f(t), \tag{2.5.27}$$

where $0 < \alpha < 2$, and $[\alpha]$ is the smallest integer greater than $\alpha$. To describe the attenuation vibration process, the damping term of the positive fractional derivative is taken negative when $1 < \alpha < 2$ to ensure this term is the energy dissipation term. By using Green's function method, the analytical solution of Eq. (2.5.21) [2] is

$$u(t) = \sum_{i=0}^{j} \frac{d^{p-i-1}G_3(t)}{dt^{p-i-1}}u^{(i)}(0) + G_3(t) * f(t), \ j < p < j + 1, \tag{2.5.28}$$

where $j$ is a non-negative integer, $G_3(t) = \frac{1}{m}\sum_{n=0}^{\infty}\frac{(-1)^n}{n!}(\frac{k}{m})^n t^{2(n+1)-1}E_{2-p,2+pn}^{(n)}(-\frac{cA}{m}t^{2-p})$, $A = \frac{-\Gamma(1-p)}{pq(p)}$ $(0 < p < 1)$, $A = \frac{-\Gamma(2-p)}{p(p-1)q(p)}(1 < p < 2)$.

Because of the very complication of the analytic solution, comparing the three models by numerical examples is more intuitional. For example,

$$m = 4.5 \times 10^5, \quad c = 8.2 \times 10^5, \quad k = 1.5 \times 10^8, \quad u(0) = 0, \quad u'(0) = 0$$

$$f(t) = \begin{cases} 4 \times 10^8, & 0 \leq t \leq 0.1 \\ 0, & t > 0.1 \end{cases}, \quad \text{numerical solution lead to.}$$

From Figs. 2.14, 2.15, 2.16, 2.17, 2.18 and 2.19 , all of the three models can manifest damped vibration processes. When $p$ is smaller than about 0.75 or greater than about 1.9, the fractal derivative model decays fastest; when $p$ is greater than about 0.75 and smaller than 1, the positive fractional derivative model decays fastest, whereas the slowest one is the fractional derivative model; when $p$ is greater than 1 and smaller than about 1.9, the positive fractional derivative model decays fastest, whereas the slowest one is the fractal derivative model. Furthermore, from Figs. 2.14, 2.15, 2.16, 2.17, 2.18 and 2.19 , the positive fractional derivative model decays faster than the fractional derivative model, and the two models are comparable when $p$ is close to 2.

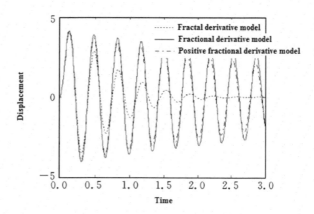

**Fig. 2.14** Numerical solution of damped vibration equation with the order of 0.5

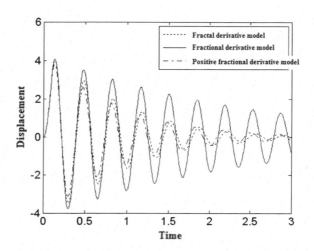

**Fig. 2.15** Numerical solution of damped vibration equation with the order of 0.75

**Fig. 2.16** Numerical
solution of damped vibration
equation with the order of 0.8

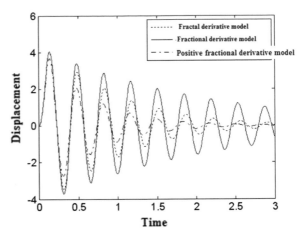

**Fig. 2.17** Numerical
solution of damped vibration
equation with the order of 1.2

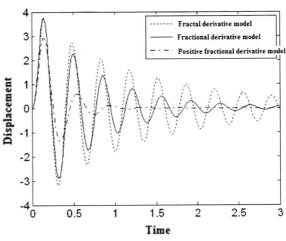

**Fig. 2.18** Numerical
solution of damped vibration
equation with the order of 1.5

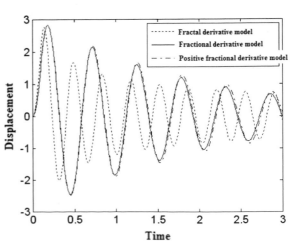

**Fig. 2.19** Numerical
solution of damped vibration
equation with the order of 1.8

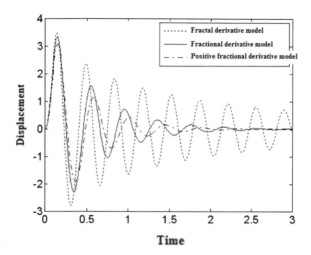

In the following, combined with the analytical solutions, the numerical results can be analyzed. Because it is difficult to obtain the analytical solution of the fractal derivative damped vibration equation, we only analyze analytic solutions of the fractional derivative damped vibration equation and positive fractional derivative damped vibration equation. The solutions of the fractional derivative equation and positive fractional equation both contain the Mittag-Leffler function [2]. The absolute attenuation coefficient is larger; the damped vibration process decays faster. From formulas (2.5.20)–(2.5.22), we can see that the attenuation coefficient of the fractional derivative model is $-(c/m)t^{2-p}$, the attenuation coefficient of the positive fractional derivative model is $-(B/A)\ t^{\ p}(A>1)$ and the absolute value is greater than that of the fractional derivative model. Therefore, according to the analytical solution of the equation, the same results are derived. In addition, when $p$ tends to 2, and $A$ tends to 1, then the fractional derivative model is close to the positive fractional derivative model.

### 4.  Fractional Integral and Differentiation of Variable Order

The concept of fractional integral and differentiation of variable order was firstly proposed by Samko in 1993. However, after a long period, this model was not widely concerned and developed. Until recent years, the variable-order derivative model has been applied to model the viscoelastic materials and viscous fluid. Recently, the variable-order derivative has become more and more popular, and a hot research issue in the field of fractional calculus.

1.  The first type definition

(1)  The definition of Riemann–Liouville [57]

The definition of the fractional variable-order integral in the sense of the Riemann–Liouville type can be stated as

$$I_{a+}^{\alpha(t)} f(t) = \frac{1}{\Gamma(\alpha(t))} \int_a^t (t - \tau)^{\alpha(t)-1} f(\tau) d\tau, \ Re(\alpha(t)) > 0. \quad (2.5.29)$$

The definition of fractional variable-order derivative in the sense of the Riemann–Liouville type can be given by

$$D_{a+}^{\alpha(t)} f(t) = \frac{1}{\Gamma(m - \alpha(t))} \frac{d^m}{dt^m} \int_a^t (t - \tau)^{m-1-\alpha(t)} f(\tau) d\tau, \ m - 1 \le \alpha(t) < m,$$

$$(2.5.30)$$

where $a \ge -\infty$.

If $\alpha(t) = c$ (constant), the definitions are reduced to the fractional Riemann–Liouville integral and derivative definitions.

**Property** $$D_{a+}^{\alpha(t)} I_{a+}^{\alpha(t)} f \ne f. \quad (2.5.31)$$

(2)   The definition of the fractional variable-order derivative in the sense of the Caputo type [58] is

$$D^{q(t)} f(t) = \frac{1}{\Gamma(1 - q(t))} \int_{0+}^t (t - \tau)^{-q(t)} f'(\tau) d\tau + \frac{(f(0+) - f(0-))t^{-q(t)}}{\Gamma(1 - q(t))},$$

$$(2.5.32)$$

where $0 < q(t) \le 1$. If $q(t) = C$ (constant), the definition is reduced to the fractional derivative in the Caputo sense.

Assume the initial condition is good enough, then we can have the following definition

$$D_{0+}^{q(t)} f(t) = \frac{1}{\Gamma(m - q(t))} \int_0^t (t - \tau)^{m-1-q(t)} f^{(m)}(\tau) d\tau, m - 1 < q(t) \le m.$$

2.   The Second Type Definition

(1)   Definition of the Riemann–Liouville type [59, 60]

The definition of the fractional variable-order integral in the sense of the Riemann–Liouville type can be defined as

$$I_0^{\alpha(t)} f(t) = \int_0^t \frac{(t - \tau)^{\alpha(t,\tau)-1}}{\Gamma(\alpha(t, \tau))} f(\tau) \, d\tau, \ \alpha(t) > 0. \quad (2.5.33)$$

The definition of the fractional variable-order derivative in the sense of the Riemann–Liouville type is

$$D_a^{\alpha(t)} f(t) = \frac{d^m}{dt^m} \int_a^t \frac{(t-\tau)^{m-1-\alpha(t,\tau)}}{\Gamma(m-\alpha(t,\tau))} f(\tau) d\tau, \ m-1 < \alpha(t) \le m,$$

(2.5.34)

where $\alpha(t, \tau)$ is a function with respect to $t$ and $\tau$.

From the expressions of the definitions defined above, the integral and derivative of time or spatial location manifest the memory property. This characteristic shows better results in the description of some complex systems.

5. **Fractional Integral and Differentiation of Random Order**

In this section, we only introduce the simplest of definitions, and other types of definitions can be similarly derived from the fractional integral and differentiation of variable order. In this book, the fractional integral of random order is defined as

$$I_0^{\alpha_0+\varepsilon_t} f(t) = \int_0^t \frac{(t-\tau)^{\alpha_0+\varepsilon_t-1}}{\Gamma(\alpha_0+\varepsilon_t)} f(\tau) d\tau, \ \alpha_0 + \varepsilon_t > 0, \ \forall t > 0.$$

(2.5.35)

The fractional derivative of random order is given by

$$D_{0+}^{\alpha_0+\varepsilon_t} f(t) = \frac{1}{\Gamma(m-\alpha_0-\varepsilon_t)} \int_{0+}^t (t-\tau)^{m-1-\alpha_0-\varepsilon_t} f^{(m)}(\tau) d\tau,$$

$$P(\varepsilon_t | m - 1 < \alpha_0 + \varepsilon_t \le m) = 1,$$

(2.5.36)

where $\alpha_0$ is a constant, and $\varepsilon_t$ is a random term. The physical background of the expression is that, in many practical systems, system parameters or force fields may fluctuate randomly, such that they are fluctuating in the entire system. To accurately describe some systems with random fluctuations, the concept of fractional derivative integral and differentiation of random order was proposed [61, 62].

The expression of the random term $\varepsilon_t$ can be divided into two types: the first type is a random term and also a function with respect to time or spatial variable, which is closer to the fractional integral or differentiation of random order, even it is more complex to analyze; The second type is that the random term does not change with other variables, at an arbitrary time or spatial location, which is a random number allowing the specified distribution. The second type is simpler for analysis, which is closer to the constant fractional calculus.

### 2.5.2 Discussions on the Spatial Fractional Derivative

Currently, for the theory and application of fractional derivatives, the previous papers mainly paid attention to time-fractional derivatives. Because of numerical solutions and mathematical definitions, applications of the spatial fractional derivative are seldom reported, which is widely applied to describe anomalous diffusion, acoustic wave dissipation and quantum mechanics. The fractional Laplace operator of the spatial fractional derivative is very important, also called Riesz's operator in some papers, and has an inherent relationship with fractional Riesz's potential. In Sect. 2.1.4, we introduce Riesz's definition of the space Laplace operator. From the definition, the acting domain of this operator is infinite. However, most of the practical modeling problems in science and engineering are considered in finite field, thus, researches on spatial fractional derivative in finite domain becomes one of the key issues of the fractional derivative equation.

In recent years, there are some meaningful results of the spatial fractional derivative in the finite domain, such as the description of the reflection barriers; Krepysheva [63] proposed the following operator:

$$\nabla^{\alpha}_{x,refl} = K \frac{-1}{2 \cos \alpha \pi / 2 \Gamma(2 - \varepsilon)} \partial^2_x \int\limits_{y=0}^{\infty} \left[ |x - y|^{1-\alpha} + (x + y)^{1-\alpha} \right] p(y, t) \mathrm{d}y.$$

$$(2.5.37)$$

Describing the Lévy distribution with absorbed boundary conditions in a bounded domain, Buldyrev et al. [64] proposed a definition as

$$D_{\alpha} f(y) = P \int\limits_{o}^{L} \frac{\mathrm{sgn}(x - y) f'(y) \mathrm{d}x}{2|y - x|^{\alpha}} - \frac{f(0)}{2y^{\alpha}} - \frac{f(L)}{2(y - x)^{\alpha}}.$$

$$(2.5.38)$$

Note that both the type definitions above are in one-dimensional space. For the mechanics modeling of the acoustic wave propagation with frequency-dependent dissipation, Chen [65] analyzed the definition of the fractional Laplace operator of the spatial fractional derivative in the multi-dimensional bounded domain. The details can be introduced further below.

It is worth pointing out that the fractional Laplace operator is a special space fractional derivative. It is defined by a singular convolution, however, it requires the positivity, the same as the classical integer-order Laplace operator [66]. The general Fourier transforms of the time and spatial fractional derivatives are [47, 67]

$$F_- \left( \frac{\partial^n \varphi}{\partial z^n} \right) = (ik)^n \Phi(k, \omega),$$

$$(2.5.39)$$

$$F_+\left(\frac{\partial^n \varphi}{\partial t^n}\right) = (-i\omega)^n \Phi(k, \omega), \tag{2.5.40}$$

where $\Phi(k,\omega)$ is the two-dimensional Fourier transform of the function $\phi\,(z,t)$ with respect to time and space

$$\Phi(k, \omega) = \int\limits_{-\infty}^{\infty} \int\limits_{-\infty}^{\infty} \varphi(z, t) e^{-i(kz-\omega t)} dz dt, \tag{2.5.41}$$

where $\omega$ represents frequency, and $k$ represents wave number; the spatial inverse Fourier transform is represented by $F_-^{-1}$, and the time inversion is represented by $F_+^{-1}$. Differing from the general spatial fractional derivative (2.5.33), the Fourier transform of the fractional Laplace operator [1, 68] is

$$F_-\left\{(-\nabla^2)_*^{s/2}\phi\right\} = k^s \Phi, \quad 0 < s < 2. \tag{2.5.42}$$

The formula above clearly portrays the positivity of the fractional Laplace operator. The fractional Laplace definition in the infinite domain is derived from the corresponding inverse Fourier transform

$$(-\nabla^2)_*^{s/2}\phi = F_-^{-1}\{k^s \Phi\} = \frac{1}{2\pi} \int \Phi k^s e^{ikx} dk. \tag{2.5.43}$$

The fractional Laplace operator is often called Riesz's fractional derivative in the current papers. The reason is that the fractional Laplace operator has an inherent relationship with Riesz's fractional potential (integral). The traditional fractional Laplace definition [1] is involved in the approximate finite difference scheme described in Sect. 2.1.4, which is not appropriate to deal with multi-dimensional irregular domain problems and the boundary conditions are not included in the definition. Chen et al. [65] noted that the strict fractional derivative definition should be derived from the fractional integral, thus, they gave a different definition of the fractional Laplace operator.

As we know, the strict analytic formula of the $s$th Riesz fractional potential (integral) in $d$-dimensional space is given by [1, 5, 69]

$$I_d^s\phi(x) = \frac{\Gamma[(d-s)/2]}{\pi^{s/2} 2^s \Gamma(s/2)} \int\limits_{\Omega} \frac{\phi(\xi)}{\|x-\xi\|^{d-s}} d\Omega(\xi), \quad 0 < s < 2, \tag{2.5.44}$$

where $\Gamma$ represents the Euler gamma function, and $\Omega$ represents the integral domain. When $d = 1$, $s = 1$, this function is singular; here, we do not discuss further. In view of the fractional integral definition, Chen et al. [65] presented an analytical definition of the fractional Laplace operator:

$$\left(-\nabla^2\right)_*^{s/2}\phi(x) = -\nabla^2\left[I_d^{2-s}\phi(x)\right]. \tag{2.5.45}$$

The distance function of the Laplace operator is

$$\nabla^2\phi(x) = \frac{d^2\phi}{dr^2} + \frac{d-1}{r}\frac{d\phi}{dr}, \tag{2.5.46}$$

where the Euclidean distance function $r = \|x - \xi\|$. From the formulas (2.5.44), (2.5.45) and (2.5.46), we can derive

$$
\begin{aligned}
\left(-\nabla^2\right)_*^{s/2}\phi(x) &= -\frac{\Gamma[(d-2+s)/2]}{\pi^{2-s/2}2^{2-s}\Gamma[(2-s)/2]}\nabla^2\int_\Omega \frac{\phi(\xi)}{\|x-\xi\|^{d-2+s}}d\Omega(\xi) \\
&= -\frac{(d-2+s)s\Gamma[(d-2+s)/2]}{\pi^{(2-s)/2}2^{2-s}\Gamma[(2-s)/2]}\int_\Omega \frac{\phi(\xi)}{\|x-\xi\|^{d+s}}d\Omega(\xi),
\end{aligned}
\tag{2.5.47}
$$

where $\Omega$ denotes integral domain, and $d$ represents spatial dimension. Note that the definition above is hyper-singular, i.e. its singular order $d + s$ is greater than the topological dimension $d$. It does not contain boundary conditions and is inconvenient to get a numerical solution and apply. Comparing the Riemann–Liouville time-fractional derivative definition with the Caputo definition, Chen et al. [65] presented a definition which differs from the definitions (2.5.46) and (2.5.47)

$$
\begin{aligned}
\left(-\nabla^2\right)^{s/2}\phi(x) &= -I_d^{2-s}\left[\nabla^2\phi(x)\right] \\
&= -\frac{\Gamma[(d-2+s)/2]}{\pi^{(2-s)/2}2^{2-s}\Gamma[(2-s)/2]}\int_\Omega \frac{\nabla^2\phi(\xi)}{\|x-\xi\|^{d-2+s}}d\Omega(\xi). \quad (2.5.48)
\end{aligned}
$$

Compared to the definition (2.5.47), the weak-singular order is $d - 2 + s$. The relationship of two definitions mentioned above can be established by the second Green's theorem,

$$\int_\Omega v\nabla^2\phi d\xi = \int_\Omega \phi\nabla^2 v d\Omega(\xi) - \int_S \left(\phi\frac{\partial v}{\partial n} - v\frac{\partial\phi}{\partial n}\right)dS(\xi), \tag{2.5.49}$$

where $S$ denotes boundary of the domain of interest, and $n$ denotes external normal direction of the boundary. Suppose

$$v = 1/\|x - \xi\|^{d-2+s}, \tag{2.5.50}$$

and

$$\phi(x)|_{x \in S} = D(x), \tag{2.5.51}$$

$$\frac{\partial \phi(x)}{\partial n}|_{x \in S} = N(x). \tag{2.5.52}$$

By applying the second Green's theorem (2.5.49), the fractional Laplace operator definition (2.5.48) can be converted into

$$
\begin{aligned}
\left(-\nabla^2\right)^{s/2} \phi(x) = {} & -\frac{(d-2+s)s\Gamma[(d-2+s)/2]}{\pi^{(2-s)/2}2^{2-s}\Gamma[(2-s)/2]} \int_{\Omega} \frac{\phi(\xi)}{\|x-\xi\|^{d+s}} d\Omega(\xi) \\
& + h \int_{S} \left[\phi(\xi)\frac{\partial}{\partial n}\left(\frac{1}{\|x-\xi\|^{d+s-2}}\right) - \frac{1}{\|x-\xi\|^{d+s-2}}\frac{\partial \phi(\xi)}{\partial n}\right] dS(\xi) \\
= {} & \left(-\nabla^2\right)^{s/2}_{*} \phi(x) \\
& + h \int_{S} \left[D(\xi)\frac{\partial}{\partial n}\left(\frac{1}{\|x-\xi\|^{d+s-2}}\right) - \frac{N(\xi)}{\|x-\xi\|^{d+s-2}}\right] dS(\xi),
\end{aligned}
\tag{2.5.53}
$$

where

$$h = \frac{\Gamma[(d-2+s)/2]}{\pi^{(2-s)/2}2^{2-s}\Gamma[(2-s)/2]}. \tag{2.5.54}$$

From the expression of (2.5.53), the definition of fractional Laplace operator $\left(-\nabla^2\right)^{s/2}$ is that the definition of fractional Laplace operator $\left(-\nabla^2\right)^{s/2}_{*}$ plus the integration of the boundary conditions, this link is very similar to the relation between the definition of the Caputo time-fractional derivative and the Riemann–Liouville definition.

These two definitions $\left(-\nabla^2\right)^{s/2}_{*}$ and $\left(-\nabla^2\right)^{s/2}$ are the definitions of the symmetric fractional Laplace operator only for the homogeneous medium [70, 71]. Based on these two definitions, we also can have new definitions of the anisotropic fractional Laplace operator. In Sect. 2.5, we will introduce the applications of the fractional Laplace operator to model the frequency-dependent acoustic wave dissipation in mathematics and mechanics.

In addition, in the mathematics field, researches on the fractional Laplace operator in the bounded domain have a meaningful promotion. The details can be found in the papers [72–74], which we do not discuss further in this book.

### 2.5.3 Discussions on the Geometric and Physical Interpretation of Fractional Calculus

What is the geometric and physical interpretation of fractional calculus? This is an ancient and unsolved problem until now [75]. Because of the name, perhaps, much effort has been devoted to relating fractional calculus and fractal geometry. However, it has been clearly shown by Rutman [76] that this approach is inconsistent. Besides these attempts, there are also other attempts [77, 78], however, they are only some special examples of fractional calculus. Obviously, there is still a lack of geometric and physical interpretation of fractional calculus. This section introduces the geometric and physical interpretation of the Riemann–Liouville fractional integral described by Podlubny [75].

Let us consider the left-sided Riemann–Liouville integral

$$_0I_t^\alpha f(t) = \int\limits_0^t f(\tau)dg_t(\tau), \qquad (2.5.55)$$

where $g_t(\tau) = \frac{1}{\Gamma(\alpha+1)}\{t^\alpha - (t - \tau)^\alpha\}$. Let us take the axes $\tau$, $g_t(\tau)$ and $f(\tau)$. In the plane $(\tau, g_t(\tau))$, we plot the function $g_t(\tau)$ for the interval $(0, t)$, and along the obtained curve we "build a fence" of the varying height $f(\tau)$, so the top edge of the "fence" is a three-dimensional line; see Fig. 2.20.

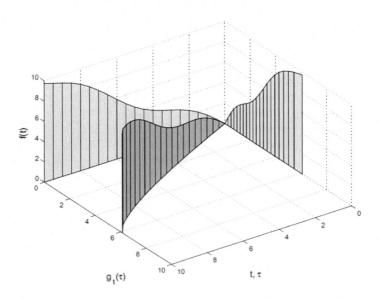

**Fig. 2.20** The "fence" and its shadows: $\alpha = 0.75$, $f(t) = t + 0.5\sin(t)$, $0 \le t \le 10$

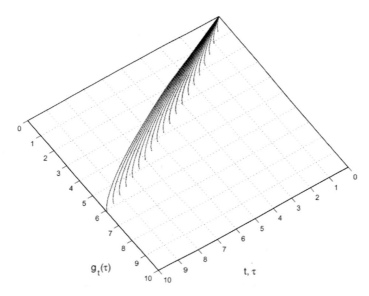

**Fig. 2.21**  The process of change of the fence basis shape with the time interval 0.5

From the figure, it is obvious that the area of the projection of this "fence" onto the plane $(\tau, f)$ corresponds to the value of the integral $_0I_t^1 f(t) = \int_0^t f(\tau)d\tau$, whereas the area of the projection of the same "fence" onto the plane $(g, f)$ corresponds to the value of the fractional integral. Therefore, this projection is a geometric interpretation of the fractional integral of the function $f(t)$. Obviously, for $\alpha = 1$, $g_t(\tau) = \tau$, both "shadows" are equal. This shows that integer-order integration is a particular case of the left-sided Riemann–Liouville fractional integration. As $t$ changes, the "fence" changes simultaneously; see Fig. 2.21. The projection onto the wall $(g, f)$ changes; see Fig. 2.22. Then we have a dynamical geometric interpretation of the fractional integral.

Suppose $f(\tau)$ is velocity, and then its fractional integration can be considered as the moving distance of a moving object. The new time $T = g_t(\tau)$ is not only relevant to the local time $\tau$ but also the total time $t$. When $t$ changes, the entire interval of the preceding time $T$ changes as well. This is in agreement with the current views in physics.

The interpretation introduced above is just one of several current interpretations. To get a clear physical interpretation of fractional calculus, we have a long way of research to go in the future.

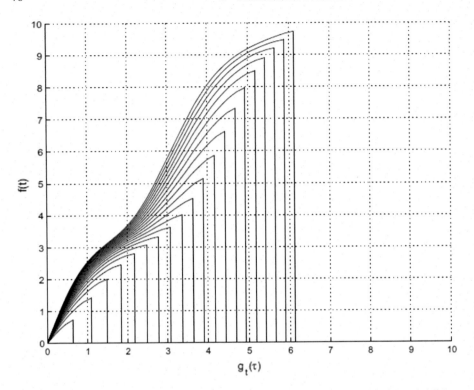

**Fig. 2.22** Snapshots of the changing "shadow" of changing "fence", with the time interval 0.5

# References

1. S.G. Samko, A.A. Kilbas, O.I. Marichev, Fractional integrals and derivatives: theory and applications [M] (Gordon and Breach, 1993)
2. I. Podlubny, *Fractional differential equations [M]* (Academic Press, San Diego, 1999)
3. A.A. Kilbas et al., *Theory and applications of fractional differential equations (North-Holland Mathematics Studies) [M]* (Elsevier, Amsterdam, 2006)
4. J.A. Connolly, The numerical solution of fractional and distributed order differential equations [D] (University of Liverpool, 2004)
5. A.A. Kilbas, H.M. Srivastava, J.J. Trujillo, Theory and applications of fractional differential equations [M] (North-Holland, 2006, p. 127)
6. G. Adomian, A review of the decomposition method in applied mathematics [J]. J. Math. Anal. Appl. **135**(2), 501–544 (1988)
7. K.G. Lin, Analysis and comparision of diferent definition about fractional integrals and derivatives [J]. J. Min Jiang Univ. **24**(5), 3–6 (2003). ((in Chinese))
8. C. Li, W. Deng, Remarks on fractional derivatives [J]. Appl. Math. Comput. **187**, 777–784 (2007)
9. F. Mainardi, G. Pagnini, The Wright functions as solutions of the time-fractional diffusion equation [J]. Appl. Math. Comput. **141**, 51–62 (2003)
10. W. Chen, H.G. Sun, Numerical algorithms of fractional derivative equations: status and problems [J]. Comput. Aided Eng. **19**(2), 1–2 (2010)
11. R. Balescu, V-Langevin equations, continuous time random walks and fractional diffusion [J]. Chaos Solitons Fractals **34**, 62–80 (2007)

12. F. Mainardi, Y. Luchko, G. Pagnini, The fundamental solution of the space-time fractional diffusion equation [J]. Fract. Calcul. Appl. Anal. **4**(2), 153–192 (2001)
13. G. Adomian, Solution of coupled nonlinear partial differential equations by decomposition [J]. Comput. Math. Appl. **31**(6), 117–120 (1996)
14. Q. Wang, Numerical solutions for fractional KdV-Burgers equation by Adomian decomposition method [J]. Appl. Math. Comput. **182**, 1048–1055 (2006)
15. S.A. El-Wakil, Adomian decomposition method for solving fractional nonlinear differential equations [J]. Appl. Math. Comput. **182**, 313–324 (2006)
16. S. Saha Ray, R.K. Bera, Analytical solution of the Bagley Torvik equation by Adomian decomposition method [J]. Appl. Math. Comput. **168**, 398–410 (2005)
17. N.T. Shawagfeh, Analytical approximate solutions for nonlinear fractional differential equations [J]. Appl. Math. Comput. **131**, 517–529 (2002)
18. J.H. He, Asymptotology by homotopy perturbation method [J]. Appl. Math. Comput. **156**, 591–596 (2004)
19. S.H. Hashemi, H.R. Mohammadi Daniali, D.D. Ganji, Numerical simulation of the generalized Huxley equation by He's homotopy perturbation method [J]. Appl. Math. Comput. **192**, 157–161 (2007)
20. M. Tavassoli Kajania, M. Ghasemib, E. Babolianb, Comparison between the homotopy perturbation method and the sine–cosine wavelet method for solving linear integro-differential equations [J]. Comput. Math. Appl. **54**, 1162–1168 (2007)
21. S.H. Hosein Nia et al., Maintaining the stability of nonlinear differential equations by the enhancement of HPM [J]. Phys. Lett. A **372**, 2855–2861 (2008)
22. H. Jafari, S. Momani, Solving fractional diffusion and wave equations by modified homotopy perturbation method [J]. Phys. Lett. A **370**, 388–396 (2007)
23. Q. Wang, Homotopy perturbation method for fractional KdV-Burgers equation [J]. Chaos Solitons Fractals **35**, 843–850 (2008)
24. Z. Odibat, S. Momani, Modified homotopy perturbation method: application to quadratic Riccati differential equation of fractional order [J]. Chaos Solitons Fractals **36**, 167–174 (2008)
25. O. Abdulaziz, I. Hashima, S. Momani, Solving systems of fractional differential equations by homotopy-perturbation method [J]. Phys. Lett. A **372**(4), 451–459 (2008)
26. S. Momani, Z. Odibat, Comparison between the homotopy perturbation method and the variational iteration method for linear fractional partial differential equations [J]. Comput. Math. Appl. **54**, 910–919 (2007)
27. Z.M. Odibat, Exact solitary solutions for variants of the KdV equations with fractional time derivatives [J]. Chaos Solitons Fractals **40**, 1264–1270 (2009)
28. Z.M. Odibat, Solitary solutions for the nonlinear dispersive $K(m, n)$ equations with fractional time derivatives [J]. Phys. Lett. A **370**, 295–301 (2007)
29. J.H. He, Homotopy perturbation technique [J]. Comput. Meth. Appl. Mech. Eng. **178**, 257–262 (1999)
30. S.J. Liao, Beyond perturbation: the basic concepts of the homotopy analysis method and its applications [J]. Adv. Mech. **38**(01), 1–34 (2008). ((in Chinese))
31. H. Xu, J. Cang, Analysis of a time fractional wave-like equation with the homotopy analysis method [J]. Phys. Lett. A **372**, 1250–1255 (2008)
32. H. Xu et al., Analysis of nonlinear fractional partial differential equations with the homotopy analysis method [J]. Commun. Nonlinear Sci. Numer. Simul. **14**(4), 1152–1156 (2009)
33. S. Momani, Z. Odibat, Analytical approach to linear fractional partial differential equations arising in fluid mechanics [J]. Phys. Lett. A **355**, 271–279 (2006)
34. A. Arikoglu, I. Ozkol, Solution of fractional differential equations by using differential transform method [J]. Chaos Solitons Fractals **34**, 1473–1481 (2007)
35. S. Momani, Z. Odibat, Numerical comparison of methods for solving linear differential equations of fractional order [J]. Chaos Solitons Fractals **31**, 1248–1255 (2007)
36. J. Biazar, H. Ghazvini, An analytical approximation to the solution of a wave equation by a variational iteration method [J]. Appl. Math. Lett. **21**, 780–785 (2008)

37. N.H. Sweilam, M.M. Khader, R.F. Al-Bar, Numerical studies for a multi-order fractional differential equation [J]. Phys. Lett. A **371**, 26–33 (2007)
38. S. Abbasbandy, An approximation solution of a nonlinear equation with Riemann–Liouville's fractional derivatives by He's variational iteration method [J]. J. Comput. Appl. Math. **207**, 53–58 (2007)
39. S. Bhalekar, V. Daftardar-Gejji, New iterative method: application to partial differential equations [J]. Appl. Math. Comput. **203**, 778–783 (2008)
40. S. Jespersen, R. Metzler, H.C. Fogedby, Lévy flights in external force fields: Langevin and fractional Fokker-Planck equations, and their solutions [J]. Phys. Rev. E **59**, 2736–2745 (1999)
41. W. Chen, Time-space fabric underlying anomalous diffusion [J]. Chaos Solitons Fractals **28**(4), 923–929 (2006)
42. K. Hoffmann, C. Essex, C. Schulzky, Fractional diffusion and entropy production [J]. J. Non-Equilib. Thermodyn. **23**, 166–175 (1998)
43. X. Li, Fractional calculus, fractal geometry, and stochastic processes [D], Ph.D thesis, University of Western Ontario, Canada (2003)
44. L. Nottale, Non-differentiable space-time and scale relativity [C, in ed. By D. Flament, Proceedings of International Colloquium Geometrie au XXe siecle (Paris, pp. 24–29, 2001.
45. R. Kanno, Representation of random walk in fractal space-time [J]. Phys. A **248**, 165–175 (1998)
46. W. Chen, S. Holm, Modified Szabo's wave equation models for lossy media obeying frequency power law [J]. J. Acoust. Soc. Am. **114**(5), 2570–2574 (2003)
47. T.L. Szabo, Time domain wave equations for lossy media obeying a frequency power law [J]. J. Acoust. Soc. Am. **96**(1), 491–500 (1994)
48. M.Y. Xu, W.C. Tan, Middle processes and critical phenomena-the theory, methods, progress, and applications of the fractional operator [J]. Sci. China Ser. G Phys. Mech. Astron. **36**(3), 225–238 (2006)
49. K.Q. Zhu, Some advances in non-Newtonian fluid mechancis [J]. Mech. Eng. **28**(4), 1–8 (2006). ((in Chinese))
50. D.K. Tong, R. Wang, Fractional flow analysis of non-Newtonian viscoelastic fluid [J]. Sci. China Ser. G Phys. Mech. Astron. **34**(1), 87–101 (2004)
51. R. Gorenflo et al., Discrete random walk models for space-time fractional diffusion [J]. Chem. Phys. **284**, 521–541 (2002)
52. R.L. Magin, Anomalous diffusion expressed through fractional order differential operators in the Bloch-Torrey equation [J]. J. Magn. Reson. **190**, 255–270 (2008)
53. L. Gaul, The influence of damping on waves and vibrations [J]. Mech. Syst. Signal Process. **13**(1), 1–30 (1999)
54. A. Tofighi, The intrinsic damping of the fractional oscillator [J]. Phys. A **329**, 29–34 (2003)
55. E. Cai, *Basis of viscoelasticity [M]* (Beihang University Press, Beijing, 1989). ((in Chinese))
56. L.I. Zhuo, B.Y. Xu, Equivalent viscous damping system for viscoelastic fractional derivative model [J]. J. Tsinghua Univ. (Sci. Technol.) **40**(11), 27–29 (2000)
57. S.G. Samko, Fractional integration and differentiation of variable order [J]. Anal. Math. **21**, 213–236 (1995)
58. Carlos F.M. Coimbra, Mechanics with variable-order differential operators [J]. Annalen der Physik (Leipzig) **12**(11–12), 692–703 (2003)
59. D. Ingman, J. Suzdalnitsky, Application of differential operator with servo-order function in model of viscoelastic deformation process [J]. J. Eng. Mech. 763–767 (2005)
60. C.F. Lorenzo, T.T. Hartley, Variable order and distributed order fractional operators [J]. Nonlinear Dyn. **29**, 57–98 (2002)
61. H.G. Sun, W. Chen, Y.Q. Chen, Variable-order fractional differential operators in anomalous diffusion modeling [J]. Phys. A **388**, 4586–4592 (2009)
62. H.G. Sun, Y.Q. Chen, W. Chen, Random-order fractional differential equation models [J]. Signal Process. **91**, 525–530 (2011)
63. N. Krepysheva, L.D. Pietro, N.C. Neel, Space-fractional advection-diffusion and reflective boundary condition [J]. Phys. Rev. E **73**, 021104 (2006)

64. S.V. Buldyrev, S. Havlin, A.Y. Kazakov, M.G.E. da Luz, E.P. Raposo, H.E. Stanley, G.M. Viswanathan, Average time spent by Lévy flights and walks on an interval with absorbing boundaries [J]. Rev. E **64**, 041108 (2001)

65. W. Chen, S. Holm, Fractional Laplacian time-space models for linear and nonlinear lossy media exhibiting arbitrary frequency dependency [J]. J. Acoust. Soc. Am. **115**(4), 1424–1430 (2004)

66. K. Diethelm, Fractional differential equations, theory and numerical treatment (2000)

67. A.D. Pierce, Acoustics, an introduction to its physical principles and applications [J]. Acoust. Soc. Am. (1989). Woodbury, NY

68. S. Jespersen, Anomalous diffusion. Progr. Rep. Inst. Phys. Astron. [J] (1999). University of Aarhus

69. M. Zahle, Fractional differentiation in the self-affine case. V. The local degree of differentiability [J]. Mathematische Nachrichten **185**, 279–306 (1997)

70. W. Feller, An introduction to probability theory and its applications [M] (2nd edn, Wiley, New York, 1971)

71. A. Hanyga, Multi-dimensional solutions of time-fractional diffusion-wave equations [J]. Proc. R. Soc. Lond. Ser. A Math. Phys. Eng. Sci. **458**, 429–450 (2002)

72. Q.Y. Guan, Z.M. Ma, Boundary problems for fractional Laplacians [J]. Stochast. Dyn. **5**, 385–424 (2005)

73. A. Zoia, A. Rosso, M. Kardar, Fractional Laplacian in bounded domains [J]. Phys. Rev. E **76**, 021116 (2007)

74. Y. Sire, E. Valdinoci, Fractional Laplacian phase transitions and boundary reactions: a geometric inequality and a symmetry result [J]. J. Funct. Anal. **256**, 1842–1864 (2009)

75. I. Podlubny, Geometric and physical interpretation of fractional integration and fractional differentiation [J]. Fract. Calcul. Appl. Anal. **5**(4), 367–386 (2002)

76. R.S. Rutman, On the paper by R.R. Nigmatullin "A fractional integral and its physical interpretation" [J]. Theoret. Math. Phys. **100**(3), 1154–1156 (1994)

77. M. Moshrefi-Torbati, J.K. Hammond, Physical and geometrical interpretation of fractional operators [J]. J. Franklin Inst. **335B**(6), 1077–1086 (1998)

78. F.J. Molz, G.J. Fix, S. Lu, A physical interpretation for the fractional derivative in Lévy diffusion [J]. Appl. Math. Lett. **15**, 907–911 (2002)

# Chapter 3
# Fractal and Fractional Calculus

It has been 2000 years since the third century BC, when Euclidean geometry was established by Euclid. This system has been considered as a definite, clear and logical geometrical system in people's minds. However, after the beginning of the twentieth century, with the continuous improvement of people's cognition and the development of mathematics, especially that a lot of new theories, new technologies and new areas of research are emerging, it is difficult to apply Euclidean geometry to describe the complex objects, such as fluctuating alpine landforms, the complex plant morphology and rough fracture surface. The limitation of Euclidean geometry prompted people to find a better tool or system for the description of the geometry.

In 1975, the American mathematician Mandelbrot asked such a question: "How long about the British coastline?" It has been found that with different scales to measure, the length of the coastline obtained is a very big difference. But actually, the length of the coastline is a determined value, and what is the reason for the different measured value? To solve this problem, Mandelbrot introduces the concept of the fractal.

Currently, the fractal theory has been widely applied in the fields of physics, materials science, geological sciences and life sciences, such as the simulation of turbulence, the simulation of the material surface, the analysis of the coastline contour and the computing of the area of the cerebral cortex. Fractal science has become an important tool in modern scientific research, and it is also an emerging branch of science.

Traditional scientific research and engineering design are based on Euclidean geometry, however, the geometry of the objects in nature are mostly irregular with fractal characteristics. Therefore, the fractal method is a powerful mathematical tool for the description of this irregular geometry. The fractal description also has the defect that it cannot satisfy the fine quantitative requirements in engineering and scientific research. This is one of the main reasons that fractal research is still at the theoretical level and has not been widely applied in practical engineering. Generally, to accurately describe and solve mechanics or physical problems, the calculus quantitative description must be established firstly. As we all know, the premise of

© Science Press 2022
W. Chen et al., *Fractional Derivative Modeling in Mechanics and Engineering*,
https://doi.org/10.1007/978-981-16-8802-7_3

classical mechanics and physical theory is continuous and integer dimension, and the corresponding quantitative description is the integer-order calculus. But for the discontinuous object and fractal media, the classic theory loses the foundation of the establishment. How to make a fine quantitative analysis and description for fractal media is a big problem of physical and mechanics studies recently.

As a new mathematical and mechanics method, fractional calculus received extensive attention in the past decade, and based on fractional calculus, some results in the quantitative description of the physical and mechanics behavior of fractal geometry have yielded. This chapter introduces the fractional calculus description of fractal media, and the relationship between fractal and fractional calculus. Due to the results of research in this area being relatively small, and limited to specific problems, not universal conclusion, therefore, this chapter mainly introduces the current research results.

## 3.1 Fractal Introduction and Application

The English name of "fractal" comes from the Latin adjective "fractus", coined by Mandelbrot in 1975. Fractal is used to describe a highly chaotic, complex and irregular set or organization.

Firstly, a similar dimension which is closely related to the fractal dimension will be introduced. Suppose graphic $A$, and a similar graphic $B$ is the $1/a$ part of graphic $A$. If $A$ equals $K$ times $B$, where $K = a^D$, on a measurable scale, a similar dimension can be expressed as [1, 2]

$$D_s = \frac{\lg K}{\lg a},$$ (3.1.1)

where $D_s$ can be both integer and fraction. If a similar dimension is a fraction, graphic **A** is a fractal graphic, and the fractal dimension is a similar dimension.

Geometric objects are known with the integer dimension. For example, point, line, surface and volume are, respectively, corresponding to zero-, one-, two- and three dimensions. However, to describe some natural objects, such as the meandering coastline, it is difficult and even impossible by integer-dimensional geometric system. Thus, the fractal whose dimension is a fraction is proposed. In view of integer dimension, the coastline and mountains and other natural features are complex and chaotic. Conversely, the fractal was able to describe these chaotic objects simply and clearly.

### 3.1.1 Simple Fractal Geometry

Before the introduction of fractal characteristics, several classical fractal geometries will be introduced first. All these fractal geometries strictly meet the basic properties of a fractal. In a long history, scientists generally agree that these fractal geometries are pathological phenomena and do not pay attention until the establishment of the fractal theory. Recently, in many fields of scientific research, these fractal models are frequently applied for simulation and auxiliary research. At the same time, through the understanding of these regular fractal graphics, the concept of fractal geometry can be deeply understood.

### 1  Triple Cantor Set

The generation of Triple Cantor Set: First, draw a segment with length L. Second, remove the 1/3 center part of the segment, and obtain two segments with length L/3. Then, remove the 1/3 center part of the two segments obtained in the previous step, and obtain four segments with length L/9. Repeat the operation to the segments obtained in the previous step, and the Triple Cantor Set will be plotted when the number of repetitions is infinite (Fig. 3.1).

The Triple Cantor Set is a set of infinite points, and its Euclidean length is approaching 0. Through similar dimension formula, its fractal dimension is

$$D_s = \frac{\lg 2}{\lg 3} = 0.6309.$$

Certainly, the operation for the segments can be different. The generalized case is applying the operation with random function $f(n, k, r)$, where $n$ denotes the $n$th division, $k$ denotes the number of divisions and $r$ denotes the division method. Then, the random Cantor Set will be obtained by applying the random function operation.

### 2  Koch Curve

The generation of the Koch Curve: First, draw a segment with length L. Second, replace the 1/3 center part of the segment by the broken line with angle 60°. Then repeat the operation to all the segments obtained in the previous step and the Koch

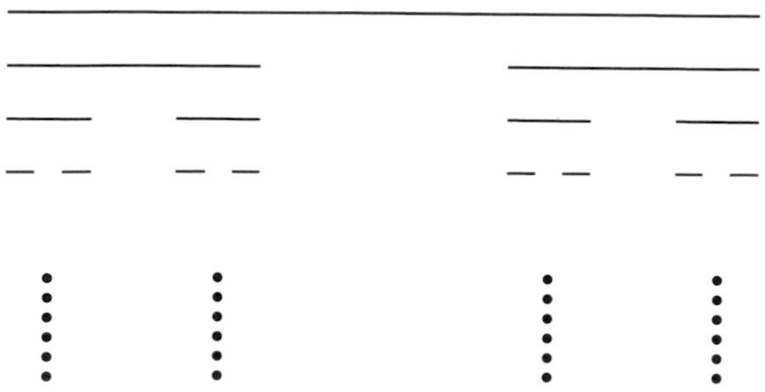

**Fig. 3.1** Triple Cantor Set

**Fig. 3.2** Koch curve

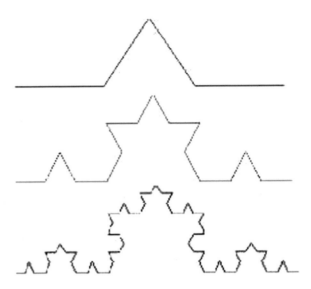

Curve will be plotted when the number of repetitions is infinite. If three Koch Curve are connected as in Fig. 3.3, the Koch Island which is similar to a snow flower is obtained.

The fractal dimension of the Koch Curve is obtained by the fractal dimension formula Eq. (3.1.1),

$$D_s = \frac{\lg 4}{\lg 3} = 1.2619.$$

The length of the Koch Island is infinite but the area is limited, which seems to run counter to the traditional geometric theory (Fig. 3.2).

Similar to the random Cantor Set, the random Koch Curve can be obtained by changing the direction of the broken line to a random direction, shown in Fig. 3.4. The Random Koch Curve can be employed to describe the random walk, such as the Brownian motion.

### 3   Sierpinski Carpet

The generation of the Sierpinski Carpet: First, draw a square with side length $L$. Second, divide it to nine equal small squares and remove the center one. Then, repeat the operation to all the squares obtained in the previous step and the Sierpinski Carpet will be plotted when the number of repetitions is infinite.

Observe the Sierpinski Carpet in Fig. 3.5; we find the area of it is approaching 0 and the length of the boundary in infinite. The fractal dimension of the Sierpinski Carpet is

$$D_s = \frac{\lg 8}{\lg 3} = 1.8928.$$

**Fig. 3.3** Koch island

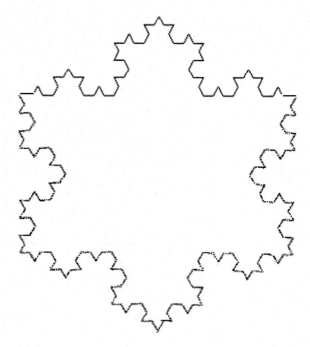

**Fig. 3.4** Random Koch Curve

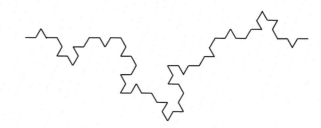

Similarly, the generation of the Sierpinski Gasket: First, draw an equilateral triangle with side length $L$. Second, divide it to four equal small equilateral triangles and remove the center one. Then, repeat the operation to all the equilateral triangles obtained in the previous step and the Sierpinski Gasket, shown in Fig. 3.6, will be plotted when the number of repetitions is infinite.

The property of the Sierpinski gasket is similar to the Sierpinski carpet, and its fractal dimension is

$$D_s = \frac{\lg 3}{\lg 2} = 1.5850.$$

The two above graphics are obtained by the transform of the plane graphic. If we do a similar transform of the Sierpinski Carpet to three-dimensional geometry, the Sierpinski Sponge (or Menger Sponge) whose fractal dimension is between 2

**Fig. 3.5**  Sierpinski Carpet

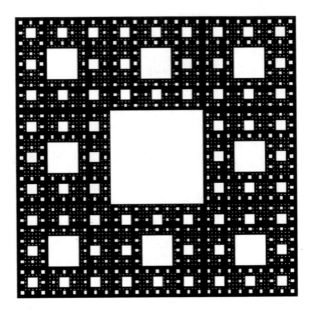

**Fig. 3.6**  Sierpinski Gasket

and 3 can be obtained. The generation of the Sierpinski Sponge: First, draw a cube with side length $L$. Second, divide it into twenty-seven equal small cubes and remove the center seven cubes. Then, repeat the operation for all the cubes obtained in the previous step. And the Sierpinski Sponge (Menger Sponge) shown in Fig. 3.7 will be got when the number of repetitions is infinite.

The surface of the Sierpinski Sponge is infinite but the volume is approaching 0. The fractal dimension of the Sierpinski Sponge is

**Fig. 3.7** Sierpinski Sponge (Menger Sponge) (*Note* from http://en.wikipedia.org/wiki/Menger_sponge)

$$D_s = \frac{\lg 20}{\lg 3} = 2.7768.$$

## 4   Peano Curve

The generation of the Peano Curve: First, draw a dotted square and divide it into 4 small squares. Second, draw three sections to connect the center points of the small squares and obtain the graphics $E_0$ shown in Fig. 3.8a. Third, repeat step 1 and step 2 in the small squares obtained in step 1, we obtain the graphics $E_1$ shown in Fig. 3.8b. Finally, the Peano Curve which can fill the whole dotted square is obtained; see Fig. 3.9.

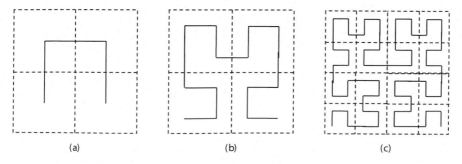

(a)                          (b)                          (c)

**Fig. 3.8** The plot of Peano Curve

**Fig. 3.9** Peano Curve (*Note* from http://qzc.zgz.cn/X-koch1.htm)

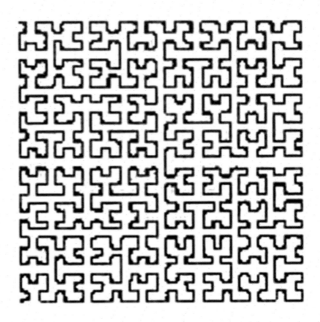

### 3.1.2 The Basic Characteristics of the Fractal

There are no strict instructions for the definition of the fractal and the description of the characteristics of the fractal. Mandelbrot proposed two fractal definitions in 1975 and 1986, respectively. Based on these two definitions, the characteristics of the fractal sets are described from a geometric point of view as follows.

(1)  The fractal dimension (Hausdorff dimension) of a fractal set is strictly greater than its topological dimension.

Mandelbrot proposed a definition of the fractal in 1975, which is described as follows: set the Hausdorff dimension of sets as $D$; if the Hausdorff dimension $D$ of the set $A$ is always greater than the topological dimension $D_T$ of the set $A$,

$$D > D_T. \tag{3.1.2}$$

Set $A$ is a fractal set, abbreviated as "fractal" [1].

For most sets, their topological dimension equals the Euclidean dimension. Therefore, the fractal dimension of fractal sets is greater than the Euclidean dimension. Take the Koch curve as an example; its topological dimension is one dimension, while the calculated fractal dimension is $1.2618 > 1$.

(2)  Fractal sets cannot be described by the traditional geometric because that these sets are neither the trajectory of the points which satisfy certain conditions nor the answer sets of some simple equations.

Some fractal structures seem like the collection of a bunch of scattered points or lines, which is a chaotic performance. Because of the irregularities of a fractal structure, it is difficult to use mathematical equations to describe it. And because the scale and dimension of a fractal structure go beyond the traditional definition of geometry, which is no longer an integral dimension, the traditional geometric cannot be applied to describe the fractal.

(3)　Fractal sets have the property of self-similar form, which includes approximate self-similar or statistical self-similar.

Another definition of a fractal is that "A fractal is a shape made of parts similar to the whole in some way" [1]. For a fractal structure, it is easy to find the part which is similar to the whole structure from a smaller scale of the structure. Such as the fractal geometries mentioned above, any part of the object can be considered as the whole structure multiplies a reduction factor. Although self-similarity is not a sufficient condition of a fractal, most of the fractal structure in nature has the property of statistical self-similarity. It is convenient to calculate the fractal dimension through the self-similar property of a fractal structure,0 and thus the study of this kind of fractal is much more.

(4)　Fractal sets have very fine structures on an arbitrarily small scale.

Take the triple Cantor set, for example: point set is obtained after infinite step, and the point is the reflection of the line in the infinitesimal scale. Therefore, the triple Cantor set has proportional details on the infinitesimal scale. Another example is the Koch curve that the curve obtained after the amplification for the infinitely small local is similar to the initial whole curve. Therefore, it is clear that the Koch curve has proportional details and a very fine structure on an arbitrarily small scale.

In addition to these typical fractal geometries, the fractal sets also have such properties. Take the Mandelbrot set, for example (shown in Fig. 3.8), and through the comparison of whole, local, microscopic view, it can be found that the Mandelbrot set is composed of a series of self-similar graphics, and graphics on the small scale have very fine structure.

(5)　Fractal sets can be defined by a very simple method, and produced by the algebraic method with changing parameters.

This property makes computer programming available and fractal application possible. The most representative sets of the fractal are the Mandelbrot set and Julia set. This property can be understood by observing their generation.

For a second-order mapping iteration on the complex plane, $f(z) = z^2 + C$, after a number of iterations for the point $z = z_0$ on the plane, the function value does not diverge. The set composed of point $z_0$ is the Julia set. For each specified $C$, there is a corresponding Julia set, denoted as $J(C)$, where $C$ is a complex number; or $J(a,b)$, where $a$ and $b$ are the real and imaginary parts, respectively.

The Mandelbrot set is the set of parameters $C$, with which the Julia set is connected. The Mandelbrot set was first obtained by Mandelbrot in 1980, which is considered as one of the most complex sets in mathematics. Because of the peculiarity of the

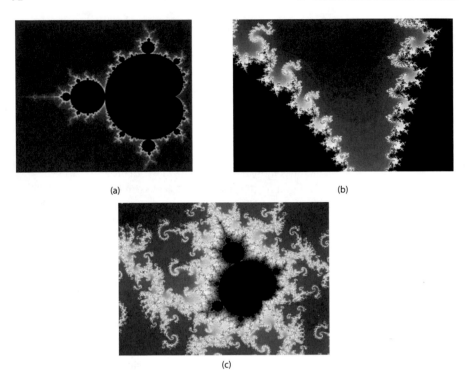

(a)                                                                                (b)

(c)

**Fig. 3.10** Mandelbrot Set. a. Mandelbrot Set in whole view, b. in local view, c. in microscopic view (*Note* From: http://www.wikilib.com/wiki?title=Image:Mandel_zoom_00_mandelbrot_set. jpg&variant=zh-cn)

Mandelbrot set, it attracts a large number of scientists and enthusiasts, and has become one of the most important symbols of chaos and fractal.

It's obvious that the fractal set is complex and chaotic, but virtually, it can be defined by simple formula and generated by iteration (Fig. 3.10).

### 3.1.3  The Measurement of Fractal Dimension

The dimension is the most important concept to describe fractal geometry, which is used to indicate the proportion of the space. Before the introduction of fractal dimension, the Hausdorff measure and dimension are introduced first [3].

Given $U$ is a nonempty set in $\mathbb{R}^n$, the diameter of $U$ is defined as the supremum of the distances between two arbitrary points in the set, i.e.

$$|U| = \sup\{|x - y||(\forall)x, y \in U\}. \tag{3.1.3}$$

A denumerable (or finite) set of $U_i$ whose diameter does not exceed $\delta$, namely $\{U_i: |U_i| \leq \delta\}$, is called the $\delta$-cover of the set $F \subset \mathbb{R}^n$, if the union of $U_i$ covers $F$, i.e. $F \subset \overset{\infty}{\underset{i=1}{\cup}} U_i$.

For a non-negative number $s$ and $\delta > 0$, define

$$\mathcal{H}_\delta^s(F) = \inf_{\{U_i\} \in \Psi} \left\{ \sum_{i=1}^\infty |U_i|^s \right\}, \tag{3.1.4}$$

where $\Psi$ is the set of all the possible $\delta$-covers of $F$.

One can see that when $\delta$ gets smaller, the number of covers in $\Psi$ decreases. As a result, the infimum $\mathcal{H}_\delta^s(F)$ increases and tends to a limit $\delta \to 0$, denoted by

$$\mathcal{H}^s(F) = \lim_{\delta \leftarrow 0} \mathcal{H}_\delta^s(F). \tag{3.1.5}$$

For arbitrary $F$, this limit always exists, being two independent values, 0 and $\infty$. We call $\mathcal{H}^s(F)$ the $s$-dimensional Hausdorff measure of $F$.

Equation (3.1.4) indicates that for any $F$ and $\delta < 1$, $\mathcal{H}_\delta^s(F)$ is a non-increasing function with respect to $s$. From (3.1.5), it follows that $\mathcal{H}^s(F)$ is also non-increasing.

If $t > s$ and $\{U_i\} \in \Psi$, then it holds that

$$\sum_i |U_i|^t \leq \sum_i |U_i|^{t-s} |U_i|^s \leq \delta^{t-s} \sum_i |U_i|^s. \tag{3.1.6}$$

Take infimums of each term in (3.1.6) to obtain $\mathcal{H}_\delta^t(F) \leq \delta^{t-s}\mathcal{H}_\delta^s(F)$. Let $\delta \to 0$ to see that for $t > s$, if $\mathcal{H}^s(F) < \infty$, then it must hold $\mathcal{H}^t(F) = 0$. Hence, there exists a critical point $s$ such that $\mathcal{H}^s(F)$ exhibits a sharp decrease from $\infty$ to 0 when $s$ increases, as shown in Fig. 3.11. Theoretical $s$ is called the Hausdorff dimension of $F$, denoted by $\dim_H F$.

This definition can be rigorously written as

**Fig. 3.11** Plot of $\mathcal{H}^s(F)$ versus $s$: Hausdorff dimension is the point in the $s$-axis where the sharp changeoccurs

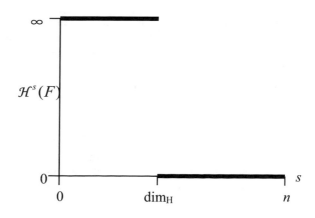

$$\dim_H F = \inf\{s \geq 0 : \mathcal{H}^s(F) = 0\} = \sup\{s : \mathcal{H}^s(F) = \infty\}, \qquad (3.1.7)$$

provided the supremum of an empty set is taken as 0. The relation of the Hausdorff measure to dimension is as follows:

$$\mathcal{H}^s(F) = \begin{cases} \infty & 0 \leq s < \dim_H F, \\ 0 & s > \dim_H F. \end{cases} \qquad (3.1.8)$$

At the point of $s = \dim_H F$, $\mathcal{H}^s(F)$ can be an arbitrary non-negative number.

The fractal dimension is an important parameter in the analysis of the characteristics of a fractal structure. Because there is neither an accurate method nor a universal method for the fractal dimension measurement, the theoretical, experimental and other appropriate methods are applied to the dimension measurements of different fractal structures. Three categories of the commonly used methods for the fractal dimension measurements are introduced as follows, which are experimental measurement method, observational comparison method and statistical calculation method.

## 1    Experimental measurement method

In this method, the auxiliary curves are drawn on the graphics in different scales, respectively. The fractal dimension is obtained by fitting the data points. The common methods are the Grade Regulation Method, Box Counting Method and Area–Perimeter Method.

### (i)    Grade Regulation Method

As the name suggests, in this method, the circles of the same radius are drawn along the measured curve one by one. Several groups of measurement results can be recorded for different circle radiuses. Then, through the scaling law analysis, the fractal dimension can be calculated by the results recorded. Of course, in the actual measurement, for different graphics, the different basic elements with a characteristic length can be applied for the measurement, such as a line segment, circle, square, sphere and cube.

Example: the fractal dimension measurement of the curve in Fig. 3.12. The numbers of the measured circles are obtained corresponding to different radiuses. If we measure the curve by the circle with radius $r$ and the number of the circles is $N$, the relationship of the circle radius $r$, the measurement number $N$ and the fractal dimension $D$ of the curve is stated as

$$N = kr^{-D}, \qquad (3.1.9)$$

where $k$ is the measurement scale constant. By marking the data recorded on the circle of different radiuses in the double logarithmic coordinate system, and applying linear regression to the data, the negative value of the slope of the fitting line $-\alpha$ is the fractal dimension $D$ of the curve.

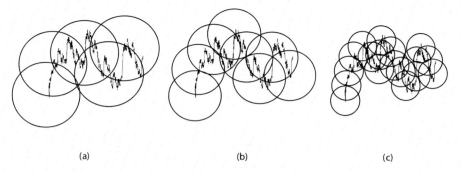

| (a) | (b) | (c) |

**Fig. 3.12** The measurement of the fractal dimension by Grade Regulation Method

If the measurement element changes to the grid with the side length $\delta$, and the number of the grid curved by the curve is $N$, the method for the fractal dimension measurement is through the relation between the number $N$ and the side length is called the Box Counting Method. In this method, the relationship of side length $\delta$, the number of the grid and the fractal dimension $D$ is stated as

$$N(\delta) = k\delta^{-D}. \tag{3.1.10}$$

Because the principles and steps of the Box Counting Method are similar to that of the Grade Regulation Method, we will not discuss the detailed examples.

(ii)   **Area–Perimeter Method**

The area–perimeter method is mainly used for the fractal dimension measurement of individual or groups of irregular dispersed graphics. The method is based on a certain relationship between the lengths of the graphical boundary and their area, which is determined by the fractal dimension of the graphics.

Based on the perspective of the analysis, this method can be divided into the method based on area and the method based on perimeter. As shown in Fig. 3.13, the method based on area is to measure the extent of the area occupied by each small graphic in the corresponding rectangular region; the method based on perimeter is to measure the filling degree of the graphic boundary in the whole plane.

Applying the relation between the perimeter and area [4]

$$P = kS^{\frac{D}{2}}, \tag{3.1.11}$$

the fractal dimension of the boundary can be measured, where $P$ denotes the perimeter of the fractal island, $S$ denotes the area of the fractal island and $k$ is the constant of the measurement scale.

Using the relation between the length and area [4]:

$$S = L^{D}. \tag{3.1.12}$$

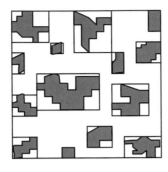   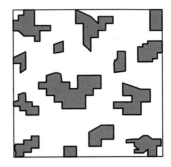

(a)   the method based on area                    (b)   the method based on perimeter

**Fig. 3.13** The Area–Perimeter Method for the measurement of fractal dimension of the irregular dispersed islands

The fractal dimension of the area can be measured, where $L$ is the maximum length scale for all the dispersed islands.

## 2  Observational Method

According to the self-similar characteristic of fractal curves, the fractal dimension can be measured by observing the similarity degree of the curve. For the graphics with approximate self-similar and statistical self-similar characteristics, the dimension can be obtained by measuring the approximate similarity degree and statistics similarity degree. Take the Koch curve as an example; if we make $1/n$ portion of the original curve $m$ times larger, and the curve obtained equals the original curve, the fractal dimension of this curve is stated by the formula

$$D = \frac{\lg n}{\lg m}. \tag{3.1.13}$$

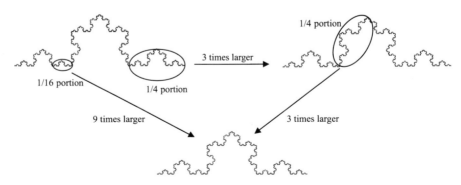

**Fig. 3.14** The measurement of fractal dimension by Observational Method

In Fig. 3.14, a similar curve like the original one is obtained by making the 1/4 portion of the original curve 3 times larger. In the same way, the curve obtained by making 9 times larger to the 1/16 portion of the original curve is also similar to the original curve. Thus, the fractal dimension of the Koch curve is expressed as $D = \frac{\lg 4}{\lg 3} = \frac{\lg 16}{\lg 9} = 1.2619$. This method is suitable for other fractal graphics. For a class of statistical self-similarity or approximate self-similar graphics, the enlarged graphic is statistically or approximately similar to the original one, but the fractal dimension can also be obtained by this method.

## 3 Statistical Calculation Method

This method is based on the correlation function, which expresses the fluctuations of the physical quantities between the two points in the space scale or two moments in the time scale in the system. The correlation function of the fluctuation for the physical quantity $A(r)$ between the two different spatial positions $x$ and $x + r$ is defined as

$$C(r) = < A(x)A(x + r) >, \tag{3.1.14}$$

where symbol $< >$ denotes the mean value of the system.

The exponential, Gaussian formulas which have characteristic scale cannot be applied to simulate the fractal distribution. Here, the power function is employed to represent the correlation function

$$C(r) \propto r^{-a}. \tag{3.1.15}$$

Therefore, the relation between the power exponent $a$ and fractal dimension $D$ is

$$a = d - D, \tag{3.1.16}$$

where $d$ is the Euclidean space dimension [5].

The structure function method is one of the correlation functions [6]. The surface contour in the structure function method is considered as a space series $z(x)$. The structure function of the sampled data can be simulated by the space series with fractal characteristics

$$S(\tau) = [z(x + \tau) - z(x)]^2 = k\tau^{4-2D}. \tag{3.1.17}$$

In addition to the methods introduced above, the spectrum method, the distribution function method, the probability density method and other new methods generated with the development of computers are able to solve these problems. Due to irregular characteristics, different methods of measurement are elected for different forms of fractal graphics. Each method has its own advantages, but its disadvantages are also obvious. Therefore, before the use of the fractal measure method, a full understanding of the properties of the graphic should be made first.

### *3.1.4  The Application of Fractal*

Because the fractal is advanced in the description and simulation of complex geometric structure objects, it is widely applied in many disciplines and research areas, particularly the complexity and chaos phenomena, such as the non-smooth surface material, chaotic turbulence, crack extension and the irregular diffusion in heterogeneous media.

**1    The application of fractal in the modeling of the material surface**

The roughness of the surface of the material is a very important issue in the field of tribology research. No matter how fine the surface polished is, it is unable to get a perfectly smooth surface. The fluctuation of the material surface is called roughness. The study showed that the profile height of the rough surface is a non-stationary random process. It has great significance to find a roughness parameter, which is independent of the measurement scale, in tribology and materials science research. In recent years, the fractal theory which can describe the scaling law phenomena, and is independent of the measurement scale, has been widely applied in the field of tribology.

In order to express the surface roughness simply which is on the two-dimensional space, the fractal dimension $D_s$ of the rough surface is stated as $D_s = D_p + 1$, where $D_p$ represents the fractal dimension of the profile curve [7]. Therefore, the measurement of surface roughness and the description of fractal characteristics can be converted to the measurement of the fractal dimension of the profile.

Because the fractal dimension of the surface for different structures may be the same, the surface roughness cannot be described by fractal dimension only. Shirong Ge [7] applied the fractal dimension and the scale parameter $\tau$ to describe the surface. A certain power-law relation the measure $M(\tau)$ and the scale $\tau$ is stated as

$$M(\tau) = C\tau^{2-D}, \tag{3.1.18}$$

where $D$ is the fractal dimension of the surface profile, and $C$ denotes the scale parameter.

Rough surface obeys the scaling law in the mean square root measure of multi-scale measurement in a certain scale range. Within the given scope of the study, the fractal dimension of the grinding surfaces and turning surfaces are mostly reduced as the value of the surface roughness R increases. The relationship between the fractal dimension and roughness is expressed as a negative exponential function [7].

$$D = 1.515R_a^{-0.088}. \tag{3.1.19}$$

In addition to describing the roughness of the surface by fractal dimension and measurement scale parameter, a class of fractal functions can also be used to describe the rough contour surface. The functions applied frequently and maturely are the Weierstrass curve or Weierstrass–Mandelbrot curve. The roughness of the surface and the rough profile can be extremely finely described by the Weierstrass–Mandelbrot curve which is defined as

$$Z(x) = G^{D-1} \sum_{n=n_1}^{\infty} \frac{\cos(2\pi \gamma^n x + \phi_n)}{\gamma^{(2-D)n}}, \quad 1 < D < 2, \ \gamma > 1, \quad (3.1.20)$$

where $Z(x)$ denotes the random surface profile height, $x$ denotes the position coordinates of the profile, $G$ is the characteristic scale coefficient, $D$ is the fractal dimension, $\gamma^n$ is the spatial frequency of the profile, $\gamma$ is the constant larger than 1 and $\phi$ denotes the phase. Wang et al. [8] apply the Weierstrass–Mandelbrot curve to simulate the profile curve. The profiles obtained in different resolution are similar, which show that the shape of the profile simulated by the Weierstrass–Mandelbrot curve is independent of the measure scale. And they also obtained the result that the roughness degree of the surface simulated by the Weierstrass–Mandelbrot curve will be larger with the increase of the fractal dimension.

## 2   The application of fractal in the simulation of crack propagation in rock

Generally, cracks usually occur when the load on the material exceeds the allowable range. The direction of crack propagation and its change with time are the problems concerned in engineering. In recent studies, researchers and scholars noted that the traditional methods are difficult to describe and simulate the crack propagation path because it has usually very irregular curves. Meanwhile, the crack propagation path was found that it characterized a certain geometric features, such as fine structure, detailed proportion in the small-scale, self-similarity of the crack branch. Therefore, it is considered to analyze and simulate the development of crack by the fractal approach.

For the fractal curve in Fig. 3.15, which is similar to the crack shape, Mandelbrot [9] gave the evaluation formula for the length as

$$L(\varepsilon) = L_0^D \varepsilon^{(1-D)}, \quad (3.1.21)$$

where $D$ is the dimension of fractal curve, $L_0^D$ denotes the straight length of the fractal domain and $\varepsilon$ denotes the measure scale.

**Fig. 3.15** The model of fractal crack propagation, from [10]

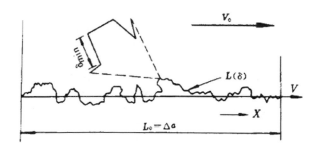

Although the cracks of different materials are in different forms, all of have similar shape to the fractal curve (Fig. 3.15). With respect to other materials, the development of the crack of the brittle material is relatively faster and much dangerous. Here, we will introduce the fractal applications in fracture simulation by the example of rock cracks. Because there are not only lattice structures and layered structures but also small impurity particles and voids, the trunk of the cracks in the rocks develop along the structure fracture, and the turning or bifurcation is generated in the influence of local impurity particles and voids. Thus, when the cracks develop in accordance with the L-type or Z-type, the local part of cracks has smaller L-type or Z-type characteristics. The experimental results show that the roughness and irregularity of the fracture material surface are mainly caused by an intergranular fracture and transgranular fracture. The irregular path of crack propagation simulated by fractal theory is

$$l = L_0^D \delta^{1-D}, \tag{3.1.22}$$

where $L_0$ denotes the crack length in macroscopic, and $\delta$ denotes the measure scale. If the measure scale is infinitesimal, the crack length obtained will be infinity large which is in contradiction with nature that the crack exists in a finite length. Hence, the measure scale should be confined to a limited range of values $[\delta_{min}, \delta_{max}]$. Because the roughness of the cracks is mainly caused by an intergranular fracture and transgranular fracture, $\delta$ can be considered as the grain size of the material $d$. The Eq. (3.1.22) changes to

$$l/L_0 = (d/L_0)^{1-D}. \tag{3.1.23}$$

After the numerator and denominator of the left part of Eq. (3.1.23), divided by the time $t$, respectively, we have the ratio relation between the fractal crack growth speed and the crack propagation speed in macroscopic,

$$V/V_0 = (d/\Delta\alpha)^{1-D}. \tag{3.1.24}$$

Here, $L_0$ is replaced by $\Delta\alpha$, which represents the increase step along the direction of macroscopic crack propagation. $d/\Delta\alpha$ is considered as the parameter of the structural performance of the material, $V$ denotes the crack actual propagation speed and $V_0$ denotes the macroscopic crack propagation speed [10].

The ratio of actual crack propagation speed to macroscopic crack propagation speed is shown in Fig. 3.16. We can find that the actual propagation speed is greater than the macroscopic propagation speed, and the ratio becomes larger with the increase of fractal dimension. For different materials, because the structural performance parameters $d/\Delta\alpha$, which depend on the structural properties of the material, are different, the ratio $V/V_0$ is also different. This is a very important problem in the monitoring of the actual engineering project, because the crack propagation speed is actually faster than that observed in macroscale.

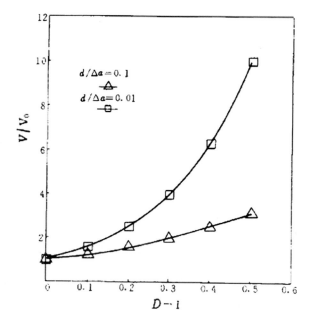

**Fig. 3.16** The change of $V/V_0$ with the fractal dimension $D$, from [10]

Because the fracture surface is not smooth, the actual area of the fracture surface is much larger than the section area. In fact, the fracture surface is a fractal surface, the relation between its actual area and section area can be described by the relationship of the actual crack length and straight crack length as follows:

$$A = [L(\varepsilon)/L_0(\varepsilon)]A', \qquad (3.1.25)$$

where $L(\varepsilon)$ denotes the actual crack length, $L_0(\varepsilon)$ denotes the straight crack length and $A'$ denotes the section area. Griffith and Irwin supposed that the crack propagates straightly, and proposed the crack critical extension force law which stated as

$$G'_{crit} = 2r_s, \qquad (3.1.26)$$

where $r_s$ denotes the surface energy of fracture area for unit macro-measure. The supposition in this formula is not consistent with the actual situation that the crack propagates along the irregular bending curve. The fracture surface is roughness and both the curve and surface are self-similar, therefore, based on Eq. (3.1.25), the crack critical extension force law can be modified to

$$G_{crit} = 2[L(\varepsilon)/L_0(\varepsilon)]r_s, \qquad (3.1.27)$$

where the proportion relationship between the actual fracture area and section area is added. Because the actual track of crack is a fractal curve, by Eqs. (3.1.21) and

(3.1.27) the expression of the critical extension force is stated as [11]

$$G_{crit} = 2r_s\varepsilon^{1-D}.\qquad\qquad(3.1.28)$$

### 3  Fractal Growth and its Application

The fractal growth researched now is the process of the growth of the fractal structure from nucleation, such as discharge phenomena of dielectric medium, diffusion, tumor growth and crack propagation. Therefore, fractal growth has a wide application in chemistry, physics, biology, medicine and materials science.

In 1981, Witten and Sander [12] proposed the diffusion-limited aggregation model (DLA) by studying the diffusion-limited aggregation process. Two years later, they further studied the relationship between the DLA model and the diffusion equation and let the diffusion-limited aggregation process satisfy the Fick diffusion equation. In 1984, Niemeyer [13] studied the discharge phenomenon of SF6 between two parallel glass plates, simulated the twig-like discharge graphic by computing and proposed the dielectric breakdown model.

The basic idea of the DLA model is a simple nucleus increasing process. Put a fixed particle which is called seed in the center of a two-dimensional grid plane with a boundary. Then, send a free particle, which can move along with any direction of the grid, in the arbitrary position of the plane. If the free particle moves to the fixed particle position, it will adhere to the fixed particle and be part of the new nucleus. On the other hand, if the free particle moves to the boundary, it will disappear. Then, send the next free particle, and follow the above determinant conditions. Finally, when the nucleus is big enough, the aggregation model with fractal property is obtained [1].

## 3.2  The Relationship Between Fractional Calculus and Fractal

As is known, calculus is the basic tool in mathematics, mechanics and physics. And this classical integer-order calculus is established on the Euclidean spacetime geometry. In recent years, fractional calculus is successfully applied in the description of complex mechanics and physical behaviors, such as the acoustics in soft material and the anomalous diffusion in complex media. We want to find if there is somewhat a relationship between the fractal which is developed by Mandelbrot and fractional calculus. This is a fascinating research topic that can fractal phenomena be simulated and described by fractional calculus? Although a long work has been done on this topic, the results are immature and not systemic. In the following, we will introduce the main work and discuss the relationship between fractal geometry and fractional calculus.

### 3.2.1 The Fractional Derivative of a Class of Fractal Function

The most classical fractal function is the Weierstrass function or Weierstrass–Mandelbrot function, which is the basic research object for many scholars [14, 15]. The characteristic of these functions is continuous but non-differentiable and non-integrable in the definition domain. Therefore, the traditional integer-order calculus methods cannot be applied to study and analyze the mathematic characteristic of these fractal functions. The definition of the Weierstrass function is

$$W(t) = \sum_{j \geq 1} \lambda^{-\mu j} \sin(\lambda^j t) \quad (0 < \mu < 1, \lambda > 1). \tag{3.2.1}$$

The fractional calculus of sine and cosine functions is stated as [16]

$$D^{-\alpha} \sin at = \frac{1}{\Gamma(\alpha)} \int_0^t x^{\alpha-1} \sin a(t-x) dx := S_t(\alpha, a),$$

$$D^{-\alpha} \cos at = \frac{1}{\Gamma(\alpha)} \int_0^t x^{\alpha-1} \cos a(t-x) dx := C_t(\alpha, a), \tag{3.2.2}$$

where the relation between the two formulas in Eq. (3.2.2) is

$$S_t(\alpha, a) = aC_t(\alpha + 1, a), \tag{3.2.3}$$

$$DS_t(\alpha, a) := S_t(\alpha - 1, a),$$
$$DC_t(\alpha, a) := C_t(\alpha - 1, a). \tag{3.2.4}$$

The fractional-order integral and differential of the Weierstrass function obtained by (3.2.2), (3.2.3) and (3.2.4) are stated in (3.2.5) and (3.2.6), respectively [16], where $0 < \mu < 1, \lambda > 1, 0 < \alpha < 1$ and $0 < \beta < 1$.

$$D^{-\alpha}(W(t)) = D^{-\alpha}\left(\sum_{j\geq1} \lambda^{-\mu j} \sin(\lambda^j t)\right) = \sum_{j\geq1} \lambda^{-\mu j} \frac{1}{\Gamma(\alpha)} \int_0^t x^{\alpha-1} \sin a(\lambda^j t) dx$$

$$= \sum_{j\geq1} \lambda^{-\mu j} S_t(\alpha, \lambda^j), \tag{3.2.5}$$

$$D^{\beta}(W(t)) = D\left(D^{\beta-1}(W(t))\right) = D\left(\sum_{j\geq1} \lambda^{-\mu j} S_t(1-\beta, \lambda^j)\right) = \sum_{j\geq1} \lambda^{(1-\beta)j} C_t(1-\beta, \lambda^j). \tag{3.2.6}$$

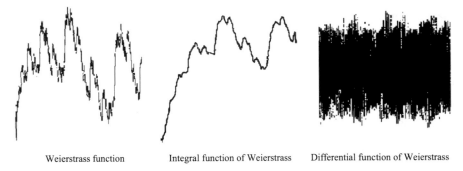

Weierstrass function          Integral function of Weierstrass    Differential function of Weierstrass

**Fig. 3.17**  $\mu = 0.5, \lambda = 2, \alpha = 0.2, \beta = 0.3$ from [16]

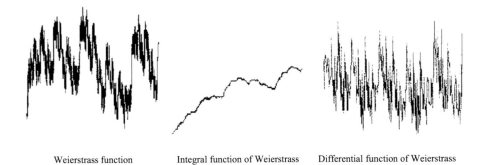

Weierstrass function          Integral function of Weierstrass    Differential function of Weierstrass

**Fig. 3.18**  $\mu = 0.3, \lambda = 2, \alpha = 0.6$ and $\beta = 0.6$ from [17]

The Weierstrass function, its integral function and its differential function are shown in Figs. 3.17 and 3.18, respectively, corresponding to the parameters $\mu = 0.5, \lambda = 2, \alpha = 0.2, \beta = 0.3$ and $\mu = 0.3, \lambda = 2, \alpha = 0.6$ and $\beta = 0.6$. From the comparison of these two figures and the comparison of the initial function and its integral and differential functions, we can find that the fluctuation of the differential function of the Weierstrass function is much larger than the initial function, and the non-smoothness becomes greater as the fractional-order $\mu$ increases. On the other hand, the integral function of the Weierstrass function is more gentle than the original function, but there are also varying degrees of fluctuation locally.

### 3.2.2  The Relationship Between the Dimension of Fractal Function and the Order of Fractional Calculus

In recent years, many researches indicated that there is a certain relationship between the dimension of fractal function and the order of fractional calculus. Yongshun Liang analyzed the fractional calculus of the Besicovitch function and obtained the relation between the relationship between the dimension and the fractional order [26].

The Besicovitch function is defined as

$$B(t) = \sum_{n \geq 1} \lambda_n^{-\alpha} \sin(\lambda_n t), \quad 0 < \alpha < 1, \quad \lambda_n \to +\infty. \tag{3.2.7}$$

The $v$-order integral $g(t)$ and u-order differential $m(t)$ are calculated by

$$g(t) := D^{-v} B(t) = \sum_{n \geq 1} \lambda_n^{-\alpha} S_t(v, \lambda_n), \tag{3.2.8}$$

$$m(t) := D^u B(t) = \sum_{n \geq 1} \lambda_n^{1-\alpha} C_t(1 - u, \lambda_n). \tag{3.2.9}$$

In [18], let $\lambda_n = n_n$ and $\alpha = 0.5$; we have a Besicovitch function,

$$B(t) = \sum_{n \geq 1} n^{-\frac{1}{2}n} \sin(n^n t), \tag{3.2.10}$$

Whose dimension is 1.5. As shown in Fig. 3.19, the first one is the Besicovitch function curve, the second one is the 1/6-order integral of $B(t)$ function and the third one is the 1/3-order differential of $B(t)$ function.

In order to obtain the relationship between the order of fractional calculus and the curve dimension, let $u$ and $v$ equal 0.1, 0.2, 0.3 and 0.4, respectively. The Hausdorff Dimension and Connectivity Dimension are shown in Table 3.1.

Using the data in Table 3.1, the relationship between the dimension (Hausdorff and Connectivity) and $v$ or $u$ is shown in Fig. 3.20. We found that the relationship between the Hausdorff dimension of fractional calculus of $B(t)$ and the dimension of $B(t)$ is linear,

$$\dim_H \Gamma(g, l) = \dim_H \Gamma(B, l) - v,$$

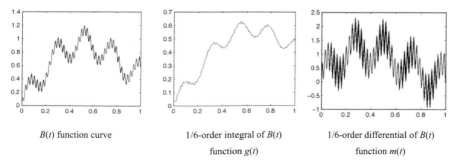

|  |  |  |
|---|---|---|
| $B(t)$ function curve | 1/6-order integral of $B(t)$ function $g(t)$ | 1/6-order differential of $B(t)$ function $m(t)$ |

**Fig. 3.19** Liang and Su from [18]

**Table 3.1** Liang and Su [18]

| $v$ | $\dim_H \Gamma(g,l)$ | $\dim_C \Gamma()$ | $u$ | $\dim_H \Gamma(m,l)$ | $\dim_C \Gamma(m,l)$ |
|---|---|---|---|---|---|
| 0 | 1.5 | $g,l$1.5850 | 0 | 1.5 | 1.5850 |
| 0.1 | 1.4 | 1.4854 | 0.1 | 1.6 | 1.6781 |
| 0.2 | 1.3 | 1.3785 | 0.2 | 1.7 | 1.7655 |
| 0.3 | 1.2 | 1.2630 | 0.3 | 1.8 | 1.8480 |
| 0.4 | 1.1 | 1.1375 | 0.4 | 1.9 | 1.9260 |

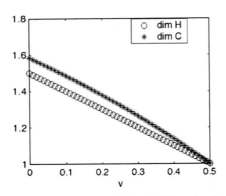

Relationship between the Hausdoff dimension &
Connectivity dimension for integral function $g(t)$ and
integral order $v$

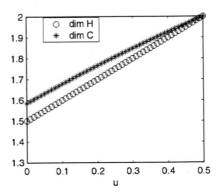

Relationship between the Hausdoff dimension &
Connectivity dimension for differential function
$m(t)$ and differential order $u$

**Fig. 3.20** Liang and Su from [18]

$$\dim_H \Gamma(m,l) = \dim_H \Gamma(B,l) + u. \tag{3.2.11}$$

Similarly, the linear relationships were also found for the Connectivity dimension. But the absolute value of the relation curve slope does not equal 1. The relation curve slope for integral function $g(t)$ and $v$ equals -1.11875, and the slope for differential function $m(t)$ and $v$ equals 0.8525.

For other fractal functions, the same characteristic with the Besicovitch function is obtained. Andrea Rocco and Bruce J. West [19] analyzed the Weierstrass function, and obtained results which are similar to Eq. (3.2.11)

$$\mathbf{Dim}\left[D^{(-\beta)}W\right] = Dim[W] - \beta,$$
$$\mathbf{Dim}\left[D^{(\beta)}W\right] = Dim[W] + \beta. \tag{3.2.12}$$

The above equations show that the fractional calculus of the Weierstrass function linearly relates to its fractal dimension. Kui Yao et al. [17] also researched the fractional calculus of the Weierstrass function and obtained the relationships between the dimension of the calculus function and the fractional order. From the relationships shown in Fig. 3.21, we can find that the relation curves are similarly linear.

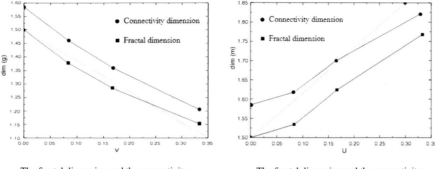

The fractal dimension and the connectivity dimension for the integral of Weierstras function

The fractal dimension and the connectivity dimension for the differential of Weierstras function

**Fig. 3.21** The fractal dimension and the connectivity dimension for the integral of Weierstras function

## 3.2.3 The Application of Fractional Calculus in the Description of Constitutive Relation for Fractal Media

In this section, the application of fractional derivatives in the constitutive modeling for fractal media will be introduced. The fractal media in nature, such as the porous medium, are considered to be an extremely complex and irregular geometric meso-scopic (or microscopic) structure. For fractal media, because of the discontinuity and complexity properties of the geometry structure, the continuum theory cannot be applied to describe its physical and mechanics behavior. The new method, which is called the fractional-order derivative, should be used to describe the constitutive relation of fractal media. In this section, we mainly introduce the research of Carpinteri and Cornetti [20].

Detailed research has been done by scientists for the fractional calculus of the two types of fractal functions, the Weierstrass function, Cantor staircase, which is also known as the devil staircase. The Weierstrass function is continuous but a non-integer-order differential everywhere. The fractal dimension of it is 2-s, where $0 < s < 1$. For the Weierstrass function, the continuous fractional derivative of the order less than s exists, and the relationship between the fractal dimension and the fractional derivative order is definitely. For the Cantor staircase function which is similar to the Cantor Fractal introduced in Sect. 3.1.1 (1), it is proved that the fractional derivative with order less than the fractal dimension exists.

Here, a new fractional-order calculus, Local Fractional Derivative (LFD) is used, which is defined as

$$D^q f(y) = \lim_{x \to y} \frac{d^q (f(x) - f(y))}{(d(x - y))^q}, \quad 0 < q \le 1, \tag{3.2.13}$$

Proposed in [21], the local fractional differentiable of the Weierstrass function depends on a critical order $\alpha$, where $0 < \alpha < 1$. When the derivative order is less than $\alpha$, the local fractional derivative is 0. On the other hand, the local fractional derivative does not exist when the order is greater than $\alpha$. Therefore, the local fractional derivative exists and does not equal 0 only when the fractional order equals $\alpha$.

For the differential equation $df/dx = 1_{[0,x]}$, the solution is a line between 0 and $x$. Kolwankar made an extension to this simple example, which deduced the measure method for fractal by the inverse form of the local fractional derivative. The discrete inverse form of the local fractional derivative in domain $[a, b]$ is

$$
{}_aD_b^{-\alpha} f(y) \equiv \lim_{N \to \infty} \sum_{i=0}^{N-1} f(x_i^*) \frac{d^{-\alpha} 1_{dx_i}(x)}{(d(x_{i+1} - x_i))^{-\alpha}},  \tag{3.2.14}
$$

where $\left[x_i, x_{i+1}\right]$ is sub-domain, $i = 0, \cdots, N-1$, $x_0 = a$, $x_N = b$, $(i = 0, \cdots, N-1)$, $x_0 = a$, $x_N = b$, $x_i^*$ is a point in the sub-domian and $1_{dx_i}(x)$ is the unit function defined in the sub-domian.

For the simple local fractal derivative equation $D^\alpha f(x) = g(x)$, it has no solution when $g(x)$ equals constant. When $g(x)$ is fractal and the fractal dimension equals fractional derivative order, the solution exists. For example, suppose $1_C(x)$ is the Cantor set in the domain $[0,1]$; when the fractional order equals its fractal dimension $\ln 2 / \ln 3$, the solution of the equation $D^\alpha f(x) = g(x)$ is $f(x) = {}_0D_x^{-\alpha} 1_C(x)$. Let $x_0 = 0$, $x_N = 1$ in Eq. (3.2.14) and elect $x_i^*$ as the maximum in sub-domain, we obtain

$$
f(y) \equiv {}_0D_x^{-\alpha} 1_C(x) = \lim_{N \to \infty} \sum_{i=0}^{N-1} F_C^i \frac{(x_{i+1} - x_i)^\alpha}{\Gamma(1 + \alpha)} = \frac{S(x)}{\Gamma(1 + \alpha)}  \tag{3.2.15}
$$

where $F_C^i$ is the flag function whose value equals 1 when the points of the Cantor set are in sub-domian $\left[x_i, x_{i+1}\right]$ and equals 0 otherwise, and $S(x)$ is the Cantor staircase function.

The measure method for the fractional derivative of the Cantor set is by $\mathcal{F}^\alpha(C) = {}_0D_1^{-\alpha} 1_C(x)$. The value of $\mathcal{F}^\alpha(C)$ is infinite when $\alpha < d$ and 0 when $a > d$. $\mathcal{F}^\alpha(C) = 1/\Gamma(1 + \alpha)$ only when $a = d$. The fractal dimension obtained by this method is consistent with the Hausdorff dimension.

Suppose a bar with the shape shown in Fig. 3.22, and a tension $F$ along the $x$-axis; based on the equilibrium of force and the distribution of the Cantor staircase, we obtain

$$
u = \varepsilon b = \varepsilon_1 b_1 = \varepsilon_2 b_2 = \cdots = \varepsilon_i b_i = \varepsilon^* b^*,  \tag{3.2.16}
$$

where $b_i$ denotes the length of the $i$th section and $\varepsilon_i$ denotes the strain. The limiting case existed like $\varepsilon_i \to \infty$, $b_i \to 0$. Because the displacement of the bar is expressed

**Fig. 3.22** The relation between strain and stress for Cantor staircase bar, from [20]

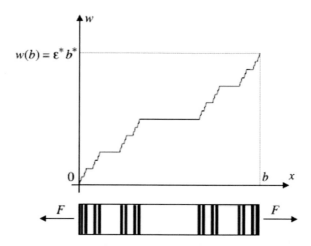

as $u = \varepsilon^* b^*$, to obtain the expression of strain $\varepsilon^*$ of fractal material, we will first analyze the characteristic of the section size. The section size obeys the Cantor staircase distribution and its fractal dimension $[L]^\alpha$ is $b^\alpha / \Gamma(1 + \alpha)$. Therefore, the dimension of fractal strain $\varepsilon^*$ is $[L]^{1-\alpha}$. Because of the local fractal characteristic of strain, the relationship between the fractal strain and the distance is the Cantor staircase shown in Fig. 3.22. For the strain field, only in the singular point, the value is not zero. This singular point can be described by the local fractional derivative with the order equals the fractal dimension. Therefore, we consider the strain of fractal media $\varepsilon^*$ is the local fractional derivative of displacement

$$\varepsilon^*(x) = D^\alpha u(x). \tag{3.2.17}$$

### 3.2.4 The Kinetic Equations of Fractal Media

As part of the classical mechanics, a class of basic equations has been established to describe the dynamical problems of the continuum media, such as continuity equation, momentum conservation equation and energy conservation equation. But these equations are powerless to solve the problem of the fractal medium, because the fractal medium is not continuous and the dimension of the medium is the fractional number. To describe the dynamical problems of such a class of medium, the basic equation should be reconsidered to satisfy the geometric properties of fractal media. In this section, we will introduce the work of Tarasov [22]. The classical kinetic equation was expanded by a fractional-order integral equation to describe the fractal properties.

The properties of the fractal media, such as the mass, obey the power law

$$M(R) = kR^D \quad (D < 3), \tag{3.2.18}$$

where $M$ denotes the mass of fractal media, $R$ denotes the box size or sphere radius and $D$ is the fractal dimension of media. Equation (3.2.18) means that the mass of the box with side length $R$ or the sphere with radius $R$ satisfies the power law, which can be deduced by fractional calculus.

The mass of the ball region $W$ of fractal medium obeys the power-law relation

$$M_D(W) = M_0 \left( \frac{R}{R_p} \right)^D, \tag{3.2.19}$$

where $D < 3$ and $R$ is the radius of the ball region. In the general case, scaling law relation is obtained as

$$dM_D(\lambda M) = \lambda^D dM_D(W), \tag{3.2.20}$$

where $\lambda M = \{ \lambda x, x \in W \}$.

The mass of an object with integer dimension in three-dimensional space can be obtained by the integral of density,

$$M_3(W) = \int_W \rho(\mathbf{r}) \, d^3\mathbf{r}. \tag{3.2.21}$$

For the fractal media in three dimension, the mass can be obtained by the fractional extension of Eq. (3.2.21). The fractional integral in Euclidean space in the Riesz form is defined as [23]

$$(I^\alpha f)(x) = \frac{1}{\gamma_n(\alpha)} \int_{R^n} \frac{f(t)dt}{|x - t|^{n-\alpha}}, \quad (R(\alpha) > 0; \alpha - n \neq 0, 2, 4, \cdots), \tag{3.2.22}$$

where

$$\gamma_n(\alpha) := \begin{cases} 2^\alpha \pi^{n/2} \Gamma(\alpha/2)(\Gamma((n-\alpha)/2))^{-1} & \alpha - n \neq 0, 2, 4, \cdots \\ (-1)^{(n-\alpha)/2} 2^{\alpha-1} \pi^{n/2} \Gamma(1 + (\alpha - n)/2)\Gamma(\alpha/2) & \alpha - n = 0, 2, 4, \cdots \end{cases}.$$

The fractional extension of Eq. (3.2.21) is

$$(I^D \rho)(\mathbf{r}_0) = \int_W \rho(\mathbf{r}) \, dV_D, \tag{3.2.23}$$

where $dV_D = c_3(D, r, r_0)d^3\mathbf{r}$, $c_3(D, r, r_0) = \frac{2^{3-D}\Gamma(3/2)}{\Gamma(D/2)|\mathbf{r}-\mathbf{r}_0|^{3-D}}$, $|\mathbf{r} - \mathbf{r}_0| = \sqrt{\sum_{k=1}^{3}(x_k - x_{k0})^2}$ and $\mathbf{r}_0 \in W$ is the initial point of fractional integral. When the dimension $D$ equals 3, $c_3(D, r, r_0) = 1$ and Eq. (3.2.23) is equivalent to Eq. (3.2.21), which means the mass formula of integer dimension media $M_3(W) = (I^3\rho)(\mathbf{r}_0)$ is the special case of fractal dimension formula. Therefore, the fractional extension of this equation can be defined in the form

$$M_D(W) = (I^D\rho)(\mathbf{r}_0) = \frac{2^{3-D}\Gamma(3/2)}{\Gamma(D/2)}\int_W \rho(\mathbf{r})|\mathbf{r} - \mathbf{r}_0|^{D-3}d^3\mathbf{r}. \qquad (3.2.24)$$

If we consider the homogeneous fractal medium, $\rho(\mathbf{r}) = \rho_0 = const$, we obtain

$$M_D(W) = \rho_0\frac{2^{3-D}\Gamma(3/2)}{\Gamma(D/2)}\int_W |\mathbf{R}|^{D-3}d^3\mathbf{R}, \qquad (3.2.25)$$

where $\mathbf{R} = \mathbf{r} - \mathbf{r}_0$. Applying the spherical coordinates, we have

$$M_D(W) = \rho_0\pi\frac{2^{5-D}\Gamma(3/2)}{\Gamma(D/2)}\int_W R^{D-1}dR = \rho_0\pi\frac{2^{5-D}\Gamma(3/2)}{D\Gamma(D/2)}R^D, \qquad (3.2.26)$$

where $R = |\mathbf{R}|$. The result obtained is equivalent to Eq. (3.2.18). Therefore, the fractal media can be described by the fractional calculus with the order $D$.

In the following, the kinetic equations of fractal media will be introduced. The fractal media, such as porous material, which are not continuity media can be considered as fractional continuity media. The equation of balance of mass density and the equation of continuity will be deduced in the following for the fractal media. Considering the fractal media in region $W$, the boundary is $\partial W$, the dimension of media is $D$ and the dimension of boundary is $d$. Then, the equation of balance of mass density can be described as

$$\frac{dM_D(W)}{dt} = 0. \qquad (3.2.27)$$

Through Eqs. (3.2.23) and (3.2.27), we have

$$\frac{d}{dt}\int_W \rho(\mathbf{R}, t)dV_D = 0, \qquad (3.2.28)$$

where $dV_D = \frac{2^{3-D}\Gamma(3/2)}{\Gamma(D/2)}|\mathbf{R}|^{D-3}dV_3$, $dV_3 = d^3\mathbf{R}$.

The definition of the time derivative of the volume integral is

$$\frac{d}{dt} \int_W A dV_D = \int_W \frac{\partial A}{\partial t} dV_D + \int_{\partial W} A u_n dS_d \qquad (3.2.29)$$

where $u_n = (\mathbf{u}, \mathbf{n}) = u_k n_k$, $\mathbf{u}$ is the vector of speed $u_k \mathbf{e}_k$ and $\mathbf{n}$ is the vertical vector $n_k \mathbf{e}_k$. Applying the fractional extension of the Gauss formula $\int_{\partial W} A u_n \, dS_2 = \int_W \text{div}(A\mathbf{u}) dV_3$,

$$\int_{\partial W} A u_n \, dS_d = \int_W c_3^{-1}(D, R) \text{div}(c_2(d, R) A\mathbf{u}) dV_D, \qquad (3.2.30)$$

and the time derivative of volume integral is obtained as

$$\frac{d}{dt} \int_W A dV_D = \int_W \left( \frac{\partial A}{\partial t} + c_3^{-1}(D, R) \text{div}(c_2(d, R) A\mathbf{u}) \right) dV_D. \qquad (3.2.31)$$

Define

$$\left( \frac{d}{dt} \right)_D = \frac{\partial}{\partial t} + c(D, d, R) u_k \frac{\partial}{\partial x_k}, \qquad (3.2.32)$$

where parameter $c(D, d, R) = c_3^{-1}(D, R) c_2(d, R) = \frac{2^{D-d-1}\Gamma(D/2)}{\Gamma(3/2)\Gamma(d/2)} |\mathbf{R}|^{d+1-D}$.
Define the generalization of the divergence formula as

$$\text{div}_D(\mathbf{u}) = c_3^{-1}(D, R) \frac{\partial}{\partial \mathbf{R}} (c_2(d, R)\mathbf{u}) = \frac{2^{D-d-1}\Gamma(D/2)}{\Gamma(3/2)\Gamma(d/2)} |\mathbf{R}|^{3-D} \text{div}(|\mathbf{R}|^{d-2}\mathbf{u}). \qquad (3.2.33)$$

The equilibrium form of the time derivative of volume integral is

$$\frac{d}{dt} \int_W A dV_D = \int_W \left( \left( \frac{\partial}{\partial t} \right)_D A + A \, \text{div}_D(\mathbf{u}) \right) dV_D. \qquad (3.2.34)$$

By substituting $A = \rho(\mathbf{R}, t)$ into Eqs. (3.2.34) and (3.2.28), the equation of balance of mass density is deduced as

$$\frac{d}{dt} \int_W \rho dV_D = \int_W \left( \left( \frac{d}{dt} \right)_D \rho + \rho \, \text{div}_D(\mathbf{u}) \right) dV_D = 0. \qquad (3.2.35)$$

Because the region W is arbitrary, the fractional continuity equation is obtained as

$$\left(\frac{d}{dt}\right)_D \rho + \rho \, \mathrm{div}_D(\mathbf{u}) = 0. \tag{3.2.36}$$

Likewise, the fractional equation of balance of momentum density in fractal media is stated as

$$\rho\left(\frac{d}{dt}\right)_D u_k + u_k\left(\left(\frac{d}{dt}\right)_D \rho + \rho \, \mathrm{div}_D(\mathbf{u})\right) - \rho f_k - \nabla_l^D p_{kl} = 0, \tag{3.2.37}$$

where $f(R, t)$ is the function with time and space, which means the force condition of point $R$ in time $t$, and $p(R, t)$ represents the density function of the surface force. By the continuity equation, we have

$$\rho\left(\frac{d}{dt}\right)_D u_k = \rho f_k + \nabla_l^D p_{kl}. \tag{3.2.38}$$

The fractional equation of balance of energy density is obtained as

$$\rho\left(\frac{d}{dt}\right)_D e = c(D, d, R) p_{kl}\frac{\partial u_k}{\partial x_l} + \nabla_k^D q_k, \tag{3.2.39}$$

where $e(\mathbf{R}, t)$ denotes the internal energy function, and $q$ denotes the density of heat flow.

Additionally, based on the above works, the fractional generalized Navier–Stokes equation is obtained as

$$\rho\left(\frac{\partial u_k}{\partial t} + c(D, d, R)u_l\frac{\partial u_k}{\partial x_l}\right) = \rho f_k - \nabla_k^D p + \mu\nabla_l^D\frac{\partial u_k}{\partial x_l} + \mu\nabla_l^D\frac{\partial u_l}{\partial x_k}$$

$$+ \left(\xi - \frac{2}{3}\mu\right)\nabla_k^D\frac{\partial u_l}{\partial x_l}. \tag{3.2.40}$$

The uncertain relationship between fractional calculus and fractal prompts many scientists to study it. And a considerable part of scholars believe that there is an inevitable relationship between the two. Therefore, whether there is a relationship and what is the relationship between fractional calculus and fractal are the major problems in present scientific research.

# References

1. L.K. Dong, *Fractal theory and application [M]* (Liaoning Science and Technology Publishing House, Shenyang, 1990)
2. J.Z. Zhang, *Fractal [M]* (Tsinghua University Press, Beijing, 1995)

3. K. Falconer (W.Q. Zeng Translate), *Fractal Geometry: Mathematical Foundations and Applications* [M] (Posts & Telecom Press, Beijing, 2007)
4. S.R. Ge, H. Zhu, *The Fractal of Tribology [M]* (China Machine Press, Beijing, 2005)
5. T.X. Han, The application of fractal theory to wood fracture behavior [D]. Master's Thesis (Northeast Forestry University, Haerbin, 2005.8)
6. W. Wang, The application of fractal theory in surface roughness non-contact measurement [D], . Master's Thesis, vol. 18 (Jilin University, Jilin, 2006)
7. S.R. Ge, K. Tonder, The fractal behavior and fractal characterization of rough surfaces [J]. Tribology **17**(1), 73–80 (1997)
8. W.J. Wang, Z.T. Wu, L.X. Chen, L. Cao, Fractal simulations and analysis of 2-D surface roughness [J]. Mech. Sci. Technol. Aerosp. Eng. *16*(6), 1059–1062 (1997)
9. B.B. Mandelbrot, *The Fractal Geometry of Nature [M]* (W. H. Freman, New York, 1982)
10. H.P. Xie, Fractal effects of dynamic crack propagation [J]. Mech. Sci. Technol. Aerosp. Eng. **27**(1), 18–27 (1995)
11. H.P. Xie, F. Gao, H.W. Zhou, J.P. Zuo, Fractal effects of dynamic crack propagation [J]. J. Disaster Prev. Mitig. Engineeing **23**(4), 1–9 (2003)
12. T.A. Witten, L.M. Sander, Diffusion-limited aggregation [J]. Phys. Rev. Lett. **47**, 1400–1411 (1981)
13. L. Niemeger L. Pietronero, H. J. Wiesmann, Fractal dimension of dielectric breakdown [J]. Phys. Rev. Lett. **52**, 1033–1036 (1984)
14. S.R. Ge, S.F. Suo, The computation methods for the fractal dimension of surface profiles [J]. Tribology **17**(4), 354–362 (1997)
15. E. Doege, B. Laackman, B. Kischnick, Fractal geometry used for the characterization of sheet surfaces [J]. CIRP Ann. Manuf. Technol. **44**(1), 197–200 (1995)
16. K. Yao, W.Y. Su, On the fractional calculus functions of a type of weierstrass function [J]. Chin. Ann. Math. **25A**(6), 711–716 (2004)
17. K. Yao, W.Y. Su, S.P. Zhou, On the fractional calculus functions of a fractal function [J]. Appl. Math.-A J. Chin. Univ. **17**(4), 377–381 (2002)
18. Y.S. Liang, W.Y. Su, The relationship between the fractal dimensions of a type of fractal functions and the order of their fractional calculus [J]. Chaos Solitons Fractals **34**, 682–692 (2007)
19. A. Rocco, B.J. West, Fractional calculus and the evolution of fractal phenomena [J]. Phys. A **265**, 535–546 (1999)
20. A. Carpinteri, P. Cornetti, A fractional calculus approach to the description of stress and strain localization in fractal media [J]. Chaos Solitons Fractals **13**, 85–94 (2002)
21. K.M. Kolwankar, A.D. Gangal, Fractional differentiability of nowhere differentiable functions and dimensions [J]. Chaos **6**, 505–523 (1996)
22. V.E. Tarasov, Fractional hydrodynamic equations for fractal media [J]. Ann. Phys. **318**, 286–307 (2005)
23. S.G. Samko, A.A. Kilbas, O.I. Marichev, *Fractional Integrals and Derivatives Theory and Applications [M]* (Gordon and Breach Science Publishers, New York, 1993)
24. L.K. Dong, *Fractal dynamics [M]* (Liaoning Science and Technology Publishing House, Shenyang, 1994)
25. S.X. Qu, J.H. Zhang, *Fractal Theory and Application of Complex Systems [M]* (Shanxi People's Publishing House, Xian, 1996)
26. Y.S. Liang, Application of fractional calculus in fractal analysis [D]. PhD Dissertation (Nanjing University, Nanjing, 2007.6)

# Chapter 4
# Fractional Diffusion Model, Anomalous Statistics and Random Process

The phenomenon of spontaneous migration of particles (e.g. atoms, molecules or molecules) is called "diffusion". Diffusion can happen in the same material or different kinds of solid, liquid and gas. The physical mechanism of the diffusion phenomenon is the existence of density or temperature gradient [1].

Diffusion phenomena are widespread in nature and industrial projects. It is an extremely important physical and mechanics process in material migration and transport. For example, the diffusion process of harmful gases, water vapor and solid in the atmosphere; or the diffusion and transport of various salts, organics and heavy metals in soil and other types of diffusion phenomena in semiconductors, metallurgy, oil refining, mining, etc. Complex diffusion is also a key topic in environmental fluid mechanics. For example, seawater erosion, nuclear waste disposal, unfinished ore pollution and concrete corrosion. Many problems on heat conduction and particle transport are related to the diffusion process actually. The above-mentioned physical processes are mainly describing the transfer of quality, thermal or electron from one spatial position to another, so as to achieve the uniform distribution of quality, concentration, temperature or electron.

In 1855, Fick proposed famous Fick's first and second laws to describe the diffusion process, based on the Fourier heat conduction equation. Hereby classical diffusion is usually called as Fickian diffusion. Einstein rederived the diffusion equation from the Brownian motion of molecular in 1905 and obtained the famous Einstein relation formula on the macroscale diffusion coefficient and micromolecular parameters.

Anomalous diffusion is named on the basis of the classical Fickian diffusion ("normal" diffusion). Richardson [2] opened the mathematical and mechanics modeling research on anomalous diffusion phenomena in turbulence and proposed the famous Richardson diffusion equation model in the 1920s to 1930s. Since the 1960s, the non-crystalline electron anomalous transport phenomena and their physical modeling received widespread attention. Thereafter, anomalous diffusion has become a hotspot in the areas of seepage in porous media, non-crystalline semiconductor material, geophysics, polymer materials, laser cooling, biomechanics, medical engineering

© Science Press 2022
W. Chen et al., *Fractional Derivative Modeling in Mechanics and Engineering*,
https://doi.org/10.1007/978-981-16-8802-7_4

and other fields. An important feature of anomalous diffusion is that the flux at a spatial position is not only related to the physical quantity in the neighborhood (for example, concentration or temperature), but also related to spatial global correlation and its history, and even its initial values. Hence, it exhibits strong memory behavior and long-range correlation and is also called as the history and path dependence (history- and path-dependency) process. It is a non-Markov process from the microscopic statistics viewpoint.

There are mainly seven methods on anomalous diffusion modeling [3]: (1) fractional Brownian motion method; (2) generalized heat statistical mechanics; (3) continuous-time random walk model; (4) the Langevin equation; (5) the generalized Langevin equation; (6) fractional derivative anomalous diffusion equation; (7) integer-order derivative generalized diffusion equation. Continuous-time random walk (CTRW) model is one of the most powerful statistical models for the analysis of anomalous diffusion. Different types of anomalous diffusion models can be obtained by assuming the probability density function of particle jump length and waiting time.

The sixth and seventh methods are phenomenological partial differential equation models. The drawbacks of the integer-order derivative nonlinear generalized diffusion equation model are it is computationally expensive and difficult to analyze and select the model parameters. The representative model of integer-order nonlinear equations is the time multi-relaxation model [4]. Although this model has some successful applications, it is difficult to get the relaxation parameters from experiments and cannot accurately describe anomalous diffusion processes. In recent years, the fractional derivative anomalous diffusion equation model has been considered as a promising approach. Hence, this chapter will systematically introduce fractional derivative anomalous diffusion equation models and further explore the intrinsic link between stochastic processes and statistical distributions in anomalous diffusion.

## 4.1 The Fractional Derivative Anomalous Diffusion Equation

Anomalous diffusion is the diffusion behavior that cannot be accurately described by Fick's laws, which usually manifests the properties of slow or rapid diffusion over time, long-range space correlation. Fractional differential equations can exactly characterize physical and mechanics processes. In recent years, fractional diffusion equation has been used to describe anomalous diffusion processes in the fractal media or complex multi-phase medium. For example, anomalous diffusion processes in porous media (oil production, geotechnical engineering seepage, etc.), coagulation state physics, medicine drug release in the polymer matrix [5–8], nuclear magnetic resonance, porous media, polymers, solid surface diffusion, colloid transport, quantum optics, molecular spectroscopy, economic and financial [7], etc.

### 4.1.1 Statistical Description of Anomalous Diffusion Problems

The classical model expressed by the integer-order calculus equations can not explain the so-called "complex phenomenon" experimental observations and empirical formulas in the phenomena of turbulence, plasma diffusion under high temperature, changes in the financial markets, polymer dynamics, thermal conductivity, diffusion and electron transport in soft matter, in which adjusting the parameters of classic models or applying the nonlinear equation alone are of no avail. However, the anomalous diffusion equation constructed by the fractional derivative mentioned previously is in good agreement with a lot of these experimental observations. In addition, the Gaussian distribution and Markov process cannot describe the above phenomenon in a large number of spatial distribution and time-series observation data; the Lévy statistics and fractional Brownian motion, however, can be well fitted these data, and internal close contact with the anomalous diffusion equation. The substantive problem is that such complex "anomalous" physical behavior is non-local in space and has the property of strong memory in time (long-range correlation); the local limit definition of the integer-order calculus cannot reflect the non-local time–space process. In recent years, research in this area has gradually become a new research focus in the international community.

We can use the Feller equation or classic integer-order derivative diffusion equation to describe the normal diffusion

$$\frac{\partial C(x,t)}{\partial t} = D\frac{\partial^2 C(x,t)}{\partial x^2}. \tag{4.1.1}$$

The statistical interpretation of the analytical solution is that the particles obey normal distribution density function in its unlimited space domain and the no-memory Markov process in time. The random motion of the normal diffusion can be described by the theory of Brownian motion, and the mean square displacement is a linear function of time $t$:

$$< r^2(t) > \propto t, \tag{4.1.2}$$

where $r(t)$ denotes the displacement, $t$ the time interval and $< >$ the mean value. A very effective statistical description of the diffusion phenomena is the continuous (or discrete)-time random walk model. If the waiting time obeys a Poisson distribution and the jump step is under the assumption of a Gaussian distribution, the standard diffusion equation can also be deduced by the CTRW model.

For anomalous diffusion, the mean square displacement is a nonlinear function of the migration time

$$< r^2(t) > \propto t^\eta, \tag{4.1.3}$$

where $\eta < 1$ denotes the sub-diffusion process and $\eta > 1$ means the super-diffusion process. $\eta = 1$ indicates the normal diffusion process following Fick's second law. From a mathematical point of view, anomalous diffusion involves non-locality in time and space. There are diverse methods to model this movement, i.e. fractional Brownian motion method, generalized diffusion equation, continuous-time random walk model, generalized diffusion equation, nonlinear diffusion equation method, etc. Fractional calculus is given by the form of the convolution integral transformation, which reflects the concept of "non-local nature", and the actual calculation illustrates the fractional diffusion equation can better reflect the nature of (4.1.3). In the continuous-time random walk model, the different assumptions on the waiting time and jump step can help us obtain the corresponding time or space or time–space fractional diffusion equation.

When $\eta$ is not equal to 1, the laws of fractional Brownian motion will be different. And then, we can speculate that the laws of diffusion of particles in space will also be very different. The diffusion coefficient can be expressed as

$$D \propto t^{\eta-1}, \tag{4.1.4}$$

where $\eta = 1$ represents the normal diffusion process; $\eta < 1$ the sub-diffusion process, which means that the diffusion process is slow and occurs in disordered media, and $\eta > 1$ means the super-diffusion process, such as the particles in turbulence, in which their each jump step can be very long and diffuse very fast.

### 4.1.2   Fractional Anomalous Diffusion Equation

Here are some often used fractional anomalous diffusion equations, and the definition of fractional derivative can be selected according to the specific conditions:

Space fractional derivative diffusion Eq. ($1 < \alpha \leq 2$):

$$\frac{\partial C(x,t)}{\partial t} = D \frac{\partial^\alpha C(x,t)}{\partial x^\alpha}, \tag{4.1.5}$$

Time-fractional derivative diffusion Eq. ($0 < \gamma \leq 1$):

$$\frac{\partial^\gamma C(x,t)}{\partial t^\gamma} = D \frac{\partial^2 C(x,t)}{\partial x^2}, \tag{4.1.6}$$

Time and space fractional derivative diffusion mixed Eq. ($0 < \gamma \leq 1, 1 < \alpha \leq 2$):

$$\frac{\partial^\gamma C(x,t)}{\partial t^\gamma} = D \frac{\partial^\alpha C(x,t)}{\partial x^\alpha}. \tag{4.1.7}$$

One-dimensional fractional derivative diffusion equation with the external inter-action term (source term or absorption items) and variable coefficients ($1 < \alpha \leq 2$) can be written as

$$\frac{\partial u(x,t)}{\partial t} = d(x)\frac{\partial^\alpha u(x,t)}{\partial x^\alpha} + q(x,t). \tag{4.1.8}$$

The appearance of variable coefficients in Eq. (4.1.8) is attributed to the spatial difference of diffusion rate in different spatial locations. Thus, using variable coefficients can more accurately describe the actual diffusion process.

Fractional derivative diffusion wave equation [8]:

$$\begin{cases} D_t^\alpha u(x,t) = a^2 D_x^\beta u(x,t), \ t > 0, 0 < x < l, 0 < \alpha \leq 2, 0 < \beta \leq 2, \\ u(0,t) = 0, u(l,t) = \theta(t), \ t \geq 0, \\ u(x,0) = \Phi(x), \qquad\quad 0 \leq x \leq l, (if\ 0 < \alpha \leq 1), \\ u_t(x,0) = 0, \qquad\qquad 0 \leq x \leq l, (if\ 1 < \alpha \leq 2). \end{cases} \tag{4.1.9}$$

The fractional diffusion wave equation is an extension of the classical diffusion equation (or wave equation), which is used to describe the phenomenon of anomalous diffusion in porous media with fractal structure. When $0 < \alpha \leq 1$, the equation is called a fractional diffusion equation, and when $1 < \alpha \leq 2$, the equation is called fractional-order wave equation.

### 4.1.3 Fractional Fick's Law [7]

The classic mathematical expression of Fick's law is as follows:

$$J = -D\frac{\partial C}{\partial x}, \tag{4.1.10}$$

where $J$ is the transmission rate of a physical quantity of some substance in the unit area, i.e. the so-called flux; $D$ the diffusion coefficient and $C$ the concentration of the diffusion substance.

Fick's law has a specific name in different physical and mechanics processes, i.e. when $C$ is the temperature variable, Fick's law is known as the Fourier law; Darcy's law when $C$ is the pressure; and when $C$ is the concentration of particles, Fick's law. Fick's diffusion law is a phenomenological relation from the experimental summary, rather than the basic first principles and its corollaries. This differs from the continuous equation derived from the law of conservation of mass. The law of conservation of mass is the fundamental law of nature. Fick's diffusion law illustrates that particle flux $J$ at a point in the space is proportional to the concentration gradient of the near-local domain.

In fact, for the anomalous diffusion equation,

$$_0^C D_t^\gamma C(x,t) = D \times {}_x D_\theta^\alpha C(x,t), \; 0 < \gamma \le 1, \; 1 < \alpha \le 2, \quad (4.1.11)$$

where $D$ is the diffusion coefficient, the operator $_0^C D_t^\gamma$ denotes the $\gamma$th derivative with respect to time $t$ under the Caputo definition. There is the following relationship between the Caputo fractional derivative and the commonly used Riemann–Liouville fractional derivative:

$$_0^C D_t^\gamma f(t) = {}_0^{RL} D_t^\gamma f(t) - f(0) \frac{t^{-\gamma}}{\Gamma(1-\gamma)}. \quad (4.1.12)$$

And $_x D_\theta^\alpha C(x,t)$ represents the spatial Riesz–Feller fractional derivative:

$$_x D_\theta^\alpha f(x) = \left[ C_+ \frac{\partial^\alpha}{\partial x^\alpha} + C_- \frac{\partial^\alpha}{\partial(-x)^\alpha} \right] C(x,t), \quad (4.1.13)$$

where

$$C_\pm = \frac{\sin[(\alpha \, m \, \theta)\pi/2]}{\sin \alpha \pi}, \quad (4.1.14)$$

and

$$\frac{d^\alpha}{dx^\alpha} f(x) = \frac{1}{\Gamma(n-\alpha)} \frac{d^n}{dx^n} \int_{-\infty}^{x} \frac{1}{(x-x')^{\alpha-n+1}} f(x') dx', \quad (4.1.15)$$

$$\frac{d^\alpha}{d(-x)^\alpha} f(x) = \frac{1}{\Gamma(n-\alpha)} \frac{d^n}{dx^n} \int_{x}^{+\infty} \frac{1}{(x-x')^{\alpha-n+1}} f(x') dx' \quad (4.1.16)$$

are called as the positive and negative axle $\alpha$-order Riemann–Liouville fractional derivative (or the Weyl fractional derivative), in which $n$ is the smallest integer greater than $\alpha$, $\alpha < n < \alpha + 1$.

According to the property of the Caputo fractional derivative, we can obtain the similar form of Eq. (4.1.11), that is,

$$\frac{\partial C(x,t)}{\partial t} = D \times {}_0^{RL} D_t^{1-\gamma} ({}_x D_\theta^\alpha C(x,t)). \quad (4.1.17)$$

Then, the flux $J$ can be redefined as

$$J = D \times {}_0^{RL} D_t^{1-\gamma} ({}_x D_\theta^{\alpha-1} C(x,t)), \quad (4.1.18)$$

which can be seen as a promotion of Fick's diffusion law. The flux is the fractional derivative of concentration to time and space; therefore, it is called fractional Fick's diffusion law. Adding the convection term to Eqs. (4.1.17), (4.1.18) can be rewritten as

$$J = (-\upsilon + D \times {}_0^{RL}D_t^{1-\gamma}{}_x D_\theta^{\alpha-1})C(x, t). \qquad (4.1.19)$$

In order to deeply understand, more details about the generalized fractional Fick law of diffusion (4.1.18) will be discussed below.

When $\alpha = 2$, $\gamma = 1$, as defined by the Riemann–Liouville fractional derivative and the Caputo fractional derivative, Eq. (4.1.18) represents traditional Fick's law of diffusion (4.1.4). Therefore, Fick's law of diffusion can also be seen as a special case of fractional Fick's diffusion law.

When $\alpha = 2$, $0 < \gamma < 1$, according to the Caputo fractional derivative definition, Eq. (4.1.18) can be expressed as

$$J(x, t) = \frac{D}{\Gamma(\gamma)} \frac{\partial}{\partial x}[\frac{\partial}{\partial t} \int_0^t (t - \tau)^{\gamma-1}C(x, \tau)d\tau]. \qquad (4.1.20)$$

As can be seen from the above equation, the flux $J$ is the derivative of convolution of concentration diffusion kernel function, that is, which is related to the course of the history of the concentration changes. The diffusion process shows the property of memory, and the motion of the particles is of non-Markov nature [9, 10].

When $\gamma = 1$, according to the definition of Riemann–Liouville fractional derivative [11], we have

$$J(x, t) = D[\frac{C_+}{\Gamma(n - \alpha)} \frac{d^n}{dx^n} \int_{-\infty}^x \frac{C(x', t)}{(x - x')^{\alpha-n}} dx'$$

$$+ \frac{C_+}{\Gamma(n - \alpha)} \frac{d^n}{dx^n} \int_x^\infty \frac{C(x', t)}{(x - x')^{\alpha-n}} dx'], \qquad (4.1.21)$$

where $n$ is the smallest integer greater than $\alpha$. As shown in Eq. (4.1.21), the spatial non-local relationship between flux $J$ and concentration is the $n$-order derivative of the concentration with respect to diffusion and function convolution.

### 4.1.4   Fractional Derivative Advection–Diffusion Equation

1.   **The classical convection–diffusion equation**

$$\frac{\partial C(x,t)}{\partial t} = -\upsilon \frac{\partial C(x,t)}{\partial x} + D \frac{\partial^2 C(x,t)}{\partial x^2}. \tag{4.1.22}$$

The basic solution of the above equation can be obtained by performing the Fourier transform to Eq. (4.1.22), then

$$\frac{d\widehat{C}}{dt} = -\upsilon(ik)\widehat{C} + D(ik)^2\widehat{C}. \tag{4.1.23}$$

Equation (4.1.23) can be further simplified to

$$\widehat{C} = \exp(-\upsilon(ik)t + D(ik)^2 t). \tag{4.1.24}$$

Applying the inverse Fourier transform, we have

$$C(x,t) = \frac{1}{\sqrt{2\pi\sigma^2 t}} \exp(-\frac{(x-\upsilon t)^2}{2\sigma^2}). \tag{4.1.25}$$

As is evident from the above equation, Eq. (4.1.25) is the Galilei offset Gaussian-type concentration distribution function of the classic convection–diffusion equation.

2   **The fractional convection–diffusion equation**

Giona and Benson had ever discussed the anomalous diffusion equation with time and space fractional derivatives. When considering the time and spatial correlation at the same time, then the time and space fractional convection–diffusion equations are obtained.

The traditional diffusion processes can be described by the following second-order convection–diffusion equation:

$$\frac{\partial C(x,t)}{\partial t} = \nabla(-\upsilon C(x,t) + D\nabla C(x,t)), \tag{4.1.26}$$

where $C(x,t)$ is the particle concentration at $(x,t)$, $\upsilon$ the convection velocity and $D$ the diffusion coefficient.

Fick's law is essentially localized diffusion, i.e. the flux $J$ of a certain point is proportional to the concentration gradient within a small range of the space location, without considering the effects of other local particle migration and also the influence of the history. However, in complex systems, the particle motion in the different moments and different points in space influence each other. Thus, when studying the particle motion of a spatial location at a certain moment, it is necessary to take into account the particle motion at other times and other points in space. Here, considering the correlation in time and space, respectively, but without the coupling effect of time and space, we can obtain the fractional derivative anomalous diffusion equations. For the convenience of explanation, the one-dimensional case is firstly considered, and the results can be extended to higher dimension situations. Thus, the local flux expressions are modified as the relationship between the particle flux and concentration:

$$\int_0^t d\tau \int_0^x J(x', t) \, dx' = \int_0^t d\tau \int_0^x k(x, x'; t, \tau) \, C(x', \tau) \, dx', \tag{4.1.27}$$

where $k(x, x'; t, \tau)$ is the diffusion kernel function. Because we do not consider the effect of time and space on the coupling, the diffusion kernel function can be written as $k(x, x'; t, \tau) = k_x(x, x') k_t(t, \tau)$. Without loss of generality, it is considered that the diffusion is statistically uniform in space and stationary random in time. Therefore, the spatial diffusion kernel function $k_x(x, x')$ will only be the function of $(x - x')$, and the time diffusion kernel function $k_t(t, \tau)$ will only be the function with respect to $(t - \tau)$. Assuming the form of a negative power-law function,

$$k_x(x, x') = \frac{D}{\Gamma(-\alpha)} \frac{1}{(x - x')^{\alpha - 1}}, \tag{4.1.28}$$

$$k_t(t, \tau) = \frac{1}{\Gamma(\gamma)} \frac{1}{(t - \tau)^{1-\gamma}}, \tag{4.1.29}$$

where $D, \alpha, \gamma$ are all the constants, $\Gamma(-\alpha), \Gamma(\gamma)$ are Gamma functions. Substituting Eqs. (4.1.28) and (4.1.29) into Eq. (4.1.27) and performing derivation with respect to $x$ and $t$ in both sides of Eq. (4.1.28), then we have

$$J = D \frac{1}{\Gamma(\gamma)} \frac{\partial}{\partial t} \int_0^t \frac{1}{(t - \tau)^{1-\gamma}} d\tau \frac{1}{\Gamma(-\alpha)} \frac{\partial}{\partial x} \int_0^x \frac{1}{(x - x')^{\alpha - 1}} C(x', \tau) \, dx'. \tag{4.1.30}$$

According to the definition of the Riemann–Liouville fractional derivative, we have

$$\frac{\partial C(x, t)}{\partial t} = {}_0^{RL} D_t^{1-\gamma} (D \frac{\partial^{\alpha - 1}}{\partial x^{\alpha - 1}} C(x, t)). \tag{4.1.31}$$

Substituting Eq. (4.1.31) into Eq. (4.1.10), we obtain the fractional diffusion equation

$$\frac{\partial C(x, t)}{\partial t} = {}_0^{RL} D_t^{1-\gamma} (D \frac{\partial^\alpha}{\partial x^\alpha} C(x, t)), \tag{4.1.32}$$

According to the property of the fractional derivative, we have

$$
{}^{C}_{0}D^{\gamma}_{t}C(x,t) = D\frac{\partial^{\alpha}}{\partial x^{\alpha}}C(x,t). \tag{4.1.33}
$$

Thus, we obtain the anomalous diffusion equation without considering the convection term. In this equation, the diffusion term and the derivative with respect to time are replaced by the fractional derivatives [12, 13].

### 3  Space fractional derivative advection–diffusion equation

$$
{}^{C}_{0}D^{\gamma}_{t}C(x,t) = -\frac{\partial}{\partial x}[vC(x,t)] + D_{x}D^{\alpha}_{\theta}C(x,t), \tag{4.1.34}
$$

where the parameters $0 < \gamma \leq 1$, $0 < \alpha \leq 2$, $|\theta| \leq \alpha$. When $\gamma = 1, \alpha = 2$, the above equation reduces to the traditional second-order convection–diffusion equation, and when $\lambda = 1, 0 < \alpha < 2$, the equation describing the Lévy transition space fractional advection–diffusion equation:

$$
\frac{\partial C(x,t)}{\partial t} = -\frac{\partial}{\partial x}[vC(x,t)] + D_{x}D^{\alpha}_{\theta}C(x,t). \tag{4.1.35}
$$

The solution of this equation is the stable distribution (or the Lévy distribution) density function. And when $\alpha = 2, 0 < \gamma < 1$, Eq. (4.1.34) is the time-fractional advection–diffusion equation:

$$
{}^{C}_{0}D^{\gamma}_{t}C(x,t) = -\frac{\partial}{\partial x}[vC(x,t)] + D\frac{\partial^{2}}{\partial x^{2}}C(x,t). \tag{4.1.36}
$$

Considering the spatial and time correlations of the diffusion process and using the non-local method, we can obtain the fractional derivative convection–diffusion equation. And then, we can describe the anomalous diffusion with it. According to the fractional convection–diffusion equation, traditional Fick's diffusion law has been extended as generalized fractional Fick's law.

Performing the Fourier transform to Eq. (4.1.35): $\frac{d\widehat{C}}{dt} = -v(ik)\widehat{C} + D|k|^{\alpha}\widehat{C}$ and solving the above equation, we can get $\widehat{C} = \exp(-v(ik)t + D|k|^{\alpha}t)$. The above expression and the definition of the Lévy distribution show that the solution of the original equation is a stable Lévy distribution density function, with a long tail dissipation feature.

### 4  Space fractional advection–diffusion equation containing the external interaction term

$$
\frac{\partial c(x,t)}{\partial t} = -v(x)\frac{\partial c(x,t)}{\partial x} + d(x)\frac{\partial^{\alpha}c(x,t)}{\partial x^{\alpha}} + f(x,t), \tag{4.1.37}
$$

where $\alpha$ is the space fractional derivative and the external term $f(x, t)$ denotes the effect of diffusion source or sink. In some problems, because fluid particle diffusion and convection are related to the particle location, the equation contains diffusion coefficient and convection coefficient with position changes [14, 15].

In the numerical solution study of the above equation, we can apply the following space fractional derivative definition:

$$\frac{\partial^{\alpha} c(x, t)}{\partial x^{\alpha}} = \frac{1}{\Gamma(-\alpha)} \lim_{M \to \infty} \frac{1}{h^{\alpha}} \sum_{k=0}^{M} \frac{\Gamma(k - \alpha)}{\Gamma(k + 1)} c(x - kh, t), \qquad (4.1.38)$$

where $M$ is the numbers of spatial nodes, $h$ the space step and $\Gamma()$ the Gamma function. Thus, we can get the numerical calculation recurrence formula of Eq. (4.1.37) (explicit difference method in space and time):

$$c_i^{n+1} = (1 - \frac{\Delta t}{h} v_i + \frac{\Delta t}{h^{\alpha}} d_i) c_i^n + (\frac{v_i}{h} - \frac{\alpha}{h^{\alpha}} d_i) \Delta t c_{i-1}^n + \frac{d_i \Delta t}{h^{\alpha}} \sum_{k=2}^{i} g_k c_{i-k}^n + f_i^n \Delta t.$$
$$(4.1.39)$$

## 5 Porous medium contaminant anomalous diffusion and penetration equation

$$\frac{\partial p}{\partial t} = -\frac{\partial}{\partial x}(Ap) + \overline{D}\frac{1 + \beta}{2} \frac{\partial^{\alpha}}{\partial x^{\alpha}}(B^{\alpha}p) + \overline{D}\frac{1 - \beta}{2} \frac{\partial^{\alpha}}{\partial(-x)^{\alpha}}(B^{\alpha}p), \quad (4.1.40)$$

where $1 < \alpha < 2$, $A$ is the flux term, $B$ the diffusion parameter, $\alpha$ the standard Lévy order and $\beta$ the offset. The above model takes into account the global impact of spatial diffusion phenomena. Therefore, it can more accurately describe the actual diffusion and seepage.

In super anomalous diffusion case $\beta = 1$, the simplified equation

$$\frac{\partial p}{\partial t} = -\frac{\partial}{\partial x}(Ap) + \overline{D}\frac{\partial^{\alpha}}{\partial x^{\alpha}}(B^{\alpha}p). \qquad (4.1.41)$$

The Leibniz formula of fractional derivative used in the numerical calculation of the above equation can be expressed as

$$\frac{\partial^{\alpha-1}}{\partial(-x)^{\alpha-1}}[D(x)\frac{\partial p}{\partial x}] = \sum_{n=0}^{\infty} \binom{\alpha - 1}{n} \frac{\partial^n D(x)}{\partial(-x)^n} \frac{\partial^{\alpha-n}}{\partial(-x)^{\alpha-n}},$$
$$\binom{\alpha}{n} = \frac{\Gamma(1 + \alpha)}{\Gamma(1 + \alpha - n)n!}. \qquad (4.1.42)$$

## 4.2  Statistical Model of the Acceleration Distribution of Turbulence Particle

In theoretical physics, developing a multi-scale model describing fully developed turbulence is an open issue, which involves various spatial and temporal features covering many scales [16, 17]. The difficulty is cross-scale coupling simulation. The principle of fluid particle trajectories is the key issue in turbulence diffusion study when existing gradient outside pressure. It is theoretically easy to measure fluid particle trajectories by seeding tracer particles into a turbulent flow and employ an imaging system to follow their motions. However, it is a very challenging task in performing. One reason is that fully developed turbulence is difficult to achieve, another is the accuracy of camera equipment is low, which cannot meet the measurement standards.

In 2001, La Porta et al. [18] had done an up-now most accurate measurement of particle transverse accelerations in fully developed turbulence using the advanced technology. Experiment parameter: $\tau_\eta = (\upsilon/\varepsilon)^{1/2}$ is the Kolmogorov time, where $\upsilon$ is the kinematic viscosity, $\varepsilon$ is the turbulent energy dissipation, $\eta$ represents the Kolmogorov distance and $\tau_n$ equals 0.93 ms. This practice successfully shows three-dimensional, time-resolved trajectory of a tracer practical undergoing accelerations in violent turbulent water flow. Some significant work has been done to find the appropriate statistical model fitting the experimental data. La Porta et al. [18] first provided a stretched exponential distribution $p(x) = C \exp(-x^2/((1+|xa/c|)^b c^2))$. In 2004, Mordant et al. [19] refined the parameters $a = 0.513 \pm 0.003$, $b = 1.600 \pm 0.003$, $c = 0.563 \pm 0.02$, $C = 0.733$ and $R_\lambda = 690$. When $|x| \rightarrow \infty$, $p(x) \propto \exp[-|x|^{0.4}]$.

But the above statistical model requires three obscure parameters, which have vague physical interpretation, and is simply a phenomenological description [20]. To overcome these drawbacks, Beck et al. [21–24] employed the Tsallis distribution to refit the experimental data via the Tsallis entropy in 2001. Later on, Arimitsu et al. [20, 25] proposed a multifractal model via the concept of non-extensive statistical mechanics, which has high accuracy, but with complex mathematical expression. Most importantly, the physical mechanism behind the experimental data is still unclear.

This section will give a comprehensive description of the existing statistical models for fully developed turbulence particle accelerations and introduce a fractional diffusion model proposed by the authors through the turbulence multi-scale property and then make a comparative study of the mentioned models.

## 4.2.1  Existing Models

### 1  Lévy stable distribution model

The Lévy stable distributions have been playing an increasing role in diverse scientific and engineering fields, such as fractional Brownian motion, anomalous diffusion, stock market and the scale structure of protein molecular, thanks to its great success in fitting a broad range of empirical data exhibiting a heavy tail. The Lévy stable distributions have been successfully used to describe chaos-induced turbulent diffusion [17, 26]. To reveal the physical mechanism of the Lévy stable distributions applied in turbulence, Chen [27] developed a Reynolds equation via the innovative fractional derivative approach and gave the reasons that the Lévy stable distributions can depict turbulence via the solution of the equation.

The Kolmogorov $-5/3$ scaling $E(k) = C\varepsilon^{\frac{2}{3}}k^{-\frac{5}{3}}$ is the most significant theoretical finding in the turbulence field in the twentieth century. $E(k)$ is the energy spectrum in terms of wave number $k$, $\varepsilon$ denotes the kinetic energy dissipation and $C$ denotes the Kolmogorov constant rate per unit mass. To some extent, the scaling law has been validated by numerous experimental and numerical data of sufficiently high Reynolds number turbulence. However, a clear departure from the $-5/3$ scaling exponent is also often observed in various turbulence experiments at finite Reynolds numbers, i.e. the so-called intermittency. In fact, intermittency manifests a non-Gaussian velocity distribution. This argument is controversial since many believe that the Kolmogorov theory does not assume the velocity increment Gaussianity.

Several researchers discover that some physical quantities in turbulence obey the Lévy stable law. Yet, Gaussian distribution is the special case of the Lévy stable distributions with stability index 2. It is well known that the classical Fokker–Planck equation has a corresponding relationship with the Navier–Stokes equation via statistical physics. Consequently, the author attempted to develop a fractional derivative model to depict turbulence. The following is the Navier–Stokes equation for an incompressible fluid:

$$\frac{\partial \mathbf{u}}{\partial t} + \mathbf{u} \cdot \nabla \mathbf{u} = -\frac{1}{\rho}\nabla p + \frac{1}{Re}\Delta \mathbf{u}, \qquad (4.2.1\text{a})$$

$$\nabla \cdot \mathbf{u} = 0, \qquad (4.2.1\text{b})$$

where $Re$ is the Reynolds number, $\mathbf{u}$ is the velocity vector and $p$ represents pressure. The following Reynolds, velocity and pressure can be decomposed as a sum of mean flow components $\bar{u}$, $\bar{p}$ and small-scale fluctuating components $\tilde{u}$, $\tilde{p}$. The mean value of fluctuating quantities is considered to be zero. Substituting the decomposition of velocity and pressure into Eq. (4.2.1), we have the following Reynolds equations:

$$\frac{\partial \bar{u}_i}{\partial t} + \bar{u}_j \cdot \frac{\partial \bar{u}_i}{\partial x_j} = -\frac{1}{\rho}\nabla \bar{p} + \upsilon \Delta \bar{u}_i - \frac{\partial}{\partial x_j}\langle \tilde{u}_i \tilde{u}_j \rangle, \qquad (4.2.2\text{a})$$

$$\nabla \cdot \overline{u}_i = 0. \tag{4.2.2b}$$

The nonlinear fluctuation term $\partial \langle u_i u_j \rangle / \partial x_j$ gives rise to the controversial closure problem in the Reynolds equations. For the fully developed homogeneous isotropic turbulence, the fluctuating velocity components are considered to exhibit a variety of universal features, namely statistically homogeneous isotropy and self-similar eddy structures, corresponding to the Richardson and Kolmogorov picture of cascade transport of kinetic energy in the inertial range of scales. Intermittency is interpreted as the joint action of the mean zero random velocity field and molecular diffusion on the large scale and long times. By analogizing with the statistical Fokker–Planck with the Laplacian operator and the Kolmogorov $-5/3$ scaling, Wen Chen [27] conjectured a representation of these universal characteristics of the Reynolds nonlinear fluctuation interactions by

$$\frac{\partial}{\partial x_j} \langle \tilde{u}_i \tilde{u}_j \rangle = \frac{1}{\widetilde{Re}} (-\Delta)^{1/3} \overline{u}_i, \tag{4.2.3}$$

where $\widetilde{Re}$ is the Reynolds number related to the 1/3-order fractional derivative. It is noted that Eq. (4.2.3) is different from the traditional eddy effective diffusivity of empirical turbulence models in that it underlies Gaussian velocity increments and agrees with Kolmogorov's key hypothesis that the small-scale structures of turbulence flows, away from boundaries, are independent of the large-scale configuration. Then we present the fractional derivative Reynolds equation

$$\frac{\partial \overline{u}_i}{\partial t} + \overline{u}_j \cdot \frac{\partial \overline{u}_i}{\partial x_j} = -\frac{1}{\rho} \nabla \overline{p} + \upsilon \Delta \overline{u}_i - \frac{1}{\widetilde{Re}} (-\Delta)^{1/3} \overline{u}_i. \tag{4.2.4}$$

Here the fractional Laplacian $(-\Delta)^{1/3}$ serves as a stochastic driver underlying statistical self-similarity in the inertial range and guarantees the positive definiteness of energy dissipation. The molecular diffusivity is a property of fluids [11], while the inertial diffusivity is a characteristic of flows where the fractional Laplacian reflects the long-range correlation in chaotic turbulence motions, apparently resembling an inherent property of non-Newtonian fluids.

Let $T$, $L$, $V_\infty$ and $P$ represent the characteristic time, length, velocity and pressure of the fluid flow. Then, we have the dimensionless expression of the fractional Reynolds equation:

$$St \frac{\partial \overline{\mathbf{u}}^*}{\partial t} + \overline{\mathbf{u}}^* \cdot \nabla \overline{\mathbf{u}}^* = -Et \nabla \overline{p}^* + \frac{1}{Re} \Delta \overline{\mathbf{u}}^* - \frac{\hat{\gamma}}{Re^{1/3}} (-\Delta)^{1/3} \overline{\mathbf{u}}^*, \tag{4.2.5}$$

where $\overline{\mathbf{u}}^*$ and $\overline{p}^*$ are dimensionless velocity and pressure, $St$ is a constant, $Et$ denotes the Euler number and $Re$ represents the Reynolds number. Equation (4.2.5) shows that the coefficient of the inertial chaos diffusion is three orders of magnitude greater than that of molecular diffusion. For instance, the inertial diffusion constant has a denominator only around 100 in a Reynolds number $10^6$ flows.

The Reynolds Eq. (4.2.4) is deterministic, but its solution has many attributes of random processes, thanks to both the Laplacian and fractional Laplacian viscous terms, and satisfies the same scale invariance of the standard Navier–Stokes equation,

$$x' = \lambda x, \ t' = \lambda^{2/3} t, \ u' = \lambda^{1/3} u,$$
$$(p/\rho)' = \lambda^{2/3}(p/\rho), \ \upsilon' = \lambda^{4/3}\upsilon, \ \gamma' = \gamma. \tag{4.2.6}$$

The characteristic function of the Lévy stable distributions regarding the fractional Laplacian operator is stated as

$$\hat{l}_\alpha(c, k) = e^{-c|k|^\alpha}, \ (0 < \alpha \leq 2), \tag{4.2.7}$$

where $\alpha$ is the stability index of the Lévy stable distributions. We refer the interested reader to Sect. 4.3 in this book.

## 2  Tsallis distribution model

The Tsallis distribution has recently been used to describe the probability density function (PDF) of Lagrangian turbulence particle accelerations. Its physical interpretation of turbulence is not satisfactory and related to the well-known Tsallis entropy [24, 27]. The Tsallis distribution has the following form in the case $q > 1$:

$$P_q = \frac{1}{\delta} \left[ \frac{q-1}{\pi(3-q)} \right]^{1/2} \frac{\Gamma(\frac{1}{q-1})}{\Gamma(\frac{3-q}{2(q-1)})} \frac{1}{[1 + \frac{q-1}{3-q}\frac{x^2}{\delta^2}]^{1/(q-1)}}. \tag{4.2.8}$$

## 3  Stretched exponential distribution model

The model proposed by La Porta et al. is stated below, and the parameters are given before.

$$p(x) = C \exp(-x^2/((1 + |xa/c|)^b c^2)). \tag{4.2.9}$$

This model is the first statistical model for the acceleration distribution of turbulence particle.

### 4.2.2  Power-Stretched Gaussian Distribution Model

Normal diffusion process in molecular scale and anomalous diffusion process in vortex scale are the two main factors considered in the practical simulation. This model is obtained based on the two scales.

(1)    Normal diffusion process in molecular scale

In the molecular scale of turbulence, the collision and friction between particles can be expressed by stress variable which determines normal diffusion in molecular scale. Considering the normal diffusion circumstance, the PDF of velocities on the molecular scale is described by the following formula:

$$\partial u / \partial t = D_1 \nabla^2 u, \tag{4.2.10}$$

where $D_1$ is the diffusion coefficient. In the normal diffusion process, the velocities of random walkers obey Gaussian distribution [28]. Using theoretical deduction or the Monte Carlo simulation, we get the particle accelerations distribution in the molecular scale normal diffusion process.

$$p_1(a) = C_1 / \sqrt{2\pi\sigma_1^2} \exp(-a^2 / 2\sigma_1^2). \tag{4.2.11}$$

(2)    Anomalous diffusion process in vortex scale

Anomalous diffusion exists in several physical and engineering fields, and many new models have been proposed to describe the process [4]. In vortex scale, turbulence diffusion exhibits statistical distribution with a heavy tail, which is an evident feature of anomalous diffusion. The coupling motion of large particles forms the vortex with fractal structures in different spatial and time scales. Therefore, anomalous diffusion in the vortex scale is the main factor leading to the random character of turbulence [25, 29].

Fractional derivative and fractal derivative are two important modeling tools for anomalous diffusion. The fractal derivative is suitable to characterize anomalous diffusion in the media with fractal structure. However, the fractional derivative is inappropriate to analyze anomalous diffusion in turbulence by numerical simulation. As turbulence exhibits fractal structure, the fractal derivative model has the ability to picture turbulence anomalous diffusion.

$$\frac{\partial u}{\partial t} = D \frac{\partial}{\partial x^{\beta/2}} \left( \frac{\partial u}{\partial x^{\beta/2}} \right), \ (0 < \beta \le 2). \tag{4.2.12}$$

The solution of the above Eq. (4.2.12) is of stretched Gaussian distribution function. And thus the expression of velocity increments can be written as

$$p_2(a) = C_2 / \sqrt{2\pi\sigma_2^\beta} \exp(-a^\beta / 2\sigma_2^\beta), \tag{4.2.13}$$

where $\beta$ is the order of fractal derivative, $\sigma_2$ variance and $C_2$ constant.

Here, we modify the fractional Laplace–Reynolds equation, then get the following equation.

$$\frac{\partial \overline{u}_i}{\partial t} + \overline{u}_j \cdot \frac{\partial \overline{u}_i}{\partial x_j} = -\frac{1}{\rho} \nabla \overline{p} + \upsilon \Delta \overline{u}_i + \kappa_0 \Delta_F^{1/3} \overline{u}_i, \qquad (4.2.14)$$

where $\kappa_0 \Delta_F^{1/3} \overline{u}$ is the Laplace operator with fractal derivative.

(3)   Power-stretched Gaussian distribution model

Molecular scale normal diffusion and vortex scale anomalous diffusion are regarded as the two essential mechanisms behind complex turbulence behaviors. The cross-scale effect of the two scales determines the distribution shape of particle accelerations. In terms of acceleration experiment data reported [27, 28], the new PDF, named after Power-Stretched Gaussian distribution (PSGD), is given as

$$p(a) = \frac{1}{\sqrt{2\pi}\sigma^\beta} \frac{1}{(|a| + a_0)^2} \exp(\frac{-(|a| + a_0)^\beta}{2\sigma^\beta}). \qquad (4.2.15)$$

The values of the parameters in this model are labeled in Fig. 4.1.

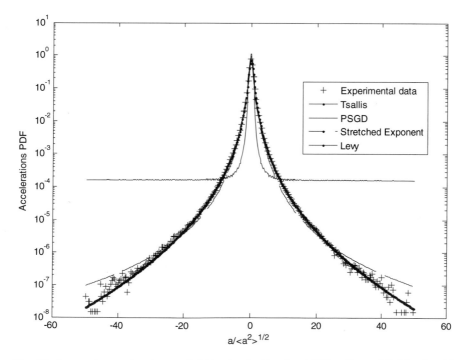

**Fig. 4.1** Lagrangian accelerations probability density function (PDF) at $R_\lambda = 690$. Lévy fit with parameters: $\alpha = 1.9$, $a = 0.1$; Tsallis fit with parameters: $q = 1.5$, $\delta = 0.5270$; Power-stretched Gaussian fit with parameters: $\beta = 0.5401$, $\sigma = 0.2211$, Coordinate offset: $x_0 = 0.3467$ and stretched exponential fit with parameters: $a = 0.513$, $b = 1.600$, $c = 0.563$

### 4.2.3　Comparisons and Discussions

Among the anomalous statistical distributions, this section chooses the four most classical statistical models to depict turbulence particle accelerations (Table 4.1).

The Lévy stable distributions have been successfully used to describe chaos-induced turbulent diffusion, but the Lévy stable distributions are not suitable to describe turbulence particle accelerations for their too slow decay rate which is far away from experimental data.

Beck employed the Tsallis distribution model via $\chi^2$ distribution. It reveals the relationship between statistical theory of turbulence and non-extensive entropy; however, the physical interpretation is not satisfactory. From Fig. 4.1, we find the Tsallis distribution does not fit the experimental data well. It is stressed that the fourth-order moment of the Tsallis fitting differs dramatically from the experiment one.

It is noted that the stretched exponential distribution model, the first statistical model used to fit the distribution of turbulence particle accelerations [30], makes excellent results in the smallest error rate. The downside is that this model requires three free parameters with vague physics interpretation. The model is merely a phenomenological description.

The power-stretched Gaussian model is built on the cross-scale coupling of molecular scale normal diffusion and vortex scale anomalous diffusion. From Table 4.1, Figs. 4.1 and 4.2, we find that it fits the data very well. In addition, the model only requires two free parameters with clear physical meaning.

However, its connection to the Navier–Stokes equations is still tenuous [31–34]. Therefore, further investigation of the present approach and model is still under way [27, 34, 35].

## 4.3　Lévy Stable Distributions

The Lévy stable distributions have received great interest across disciplines such as fractional Brownian motion, anomalous diffusion, stock market and heart rate dynamics [36–39]. In 1963, Mandelbrot suggested that it is a powerful tool to model financial time series.

**Table 4.1** Comparison of the four models

| Models | Parameters | Physical interpretation | Mean square error |
|---|---|---|---|
| Lévy stable distributions | 2 | Accelerated diffusion | 1.3972E + 03 |
| Tsallis distribution | 2 | Tsallis entropy | 19.9716 |
| Stretched exponential distribution | 3 | – | 3.3373 |
| PSGD | 2 | Multi-scale | 3.5695 |

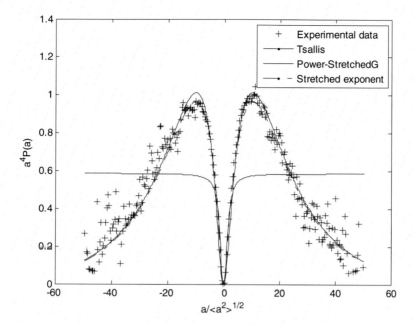

**Fig. 4.2** Comparison of the experimental accelerations PDF at the fourth-order moment, the parameters are the same as those in Fig. 4.1

### 4.3.1  General Expression of Lévy Stable Distributions

The Lévy stable distributions contain a rich class of non-Gaussian probability distributions, and its second moment or variance is infinite except for its extreme case, Gaussian distribution. Since the closed-form densities of most Lévy stable distributions are not available, the characteristic function (i.e. the Fourier transform of PDF) is often used to specify these distributions.

$$
\begin{aligned}
p_{\alpha,\beta}(k; \mu, \sigma) &= F\{p_{\alpha,\beta}(x; \mu, \sigma)\} \\
&= \exp\left[i\mu k - \sigma^\alpha |k|^\alpha \left(1 - i\beta \frac{k}{|k|}\omega(k, \alpha)\right)\right],
\end{aligned} \tag{4.3.1}
$$

where

$$
\omega(k, \alpha) = \begin{cases} \tan\frac{\pi\alpha}{2} & \alpha \neq 1, 0 < \alpha < 2, \\ -\frac{2}{\pi}\ln|k| & \alpha = 1, \end{cases}
$$

the four parameters: stability index $\alpha$, skewness parameter $\beta$, scale parameter $\gamma$ and location parameter $\delta$ with varying ranges: $0 < \alpha \leq 2$, $-1 \leq \beta \leq 1$, $\gamma > 0, \delta \in v$. Parameters $\alpha$ and $\beta$ determine the taper off rate at the tail and symmetric character of PDF, respectively. The last two parameters $\gamma$ and $\delta$ are usual scale and location parameters, respectively. By the following transform, two parameters $\alpha$ and $\beta$ can characterize the Lévy stable distributions.

$$p_{\alpha,\beta}(k; \mu, \sigma) = \frac{1}{\sigma} p_{\alpha,\beta}(\frac{x - \mu}{\sigma}; 0, 1).$$

The asymptotic behavior is described with $\beta = 0$

$$p_{\alpha,0}(x) \approx \frac{\sin(\frac{\alpha\pi}{2})\Gamma(\alpha + 1)}{\pi} \frac{1}{|x|^{\alpha+1}}, \quad x \to \pm\infty. \tag{4.3.2}$$

The Fourier transform of the above PDF of the Lévy stable distributions is stated as

$$\tilde{p}_{\alpha,0}(k) = \exp(-|k|^{\alpha}). \tag{4.3.3}$$

The Fox H-function gives the analytical form of the Lévy stable distributions with the whole range of $\alpha$, but the expression has complicated mathematics. We refer the interested reader to the work given by Sanjana [40]. Figure 4.3 shows the PDF curves of the Lévy stable distributions with four special cases.

Cauchy distribution: $p_{1,0}(x) = \frac{1}{\pi(1+x^2)}$;

Gaussian distribution: $p_{2,0}(x) = \frac{1}{\sqrt{4\pi}} \exp\left(-\frac{x^2}{4}\right)$.

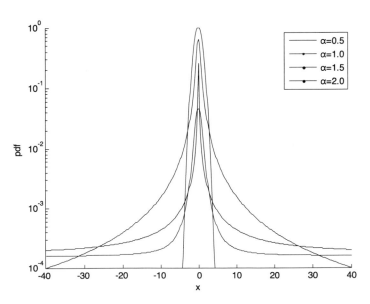

**Fig. 4.3** Plots of the PDF of the Lévy stable distributions with different $\alpha$

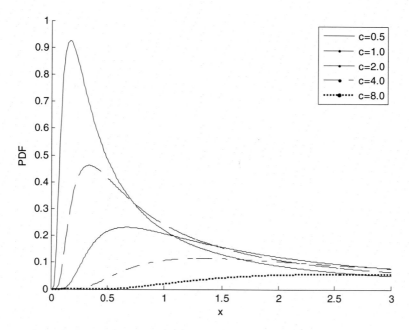

**Fig. 4.4** Plots of the PDF of the Lévy stable distributions with different parameter $c$

## 4.3.2 Lévy Stable Distributions with X ≥ 0 [41]

The Lévy stable distributions are skewed when $\beta = 1$, and the plots of $p_{0.5,1}(x; 0, c)$ are in the following.

The PDF over the domain x ≥ 0 is $\sqrt{\frac{c}{2\pi}} \frac{e^{-c/2x}}{x^{3/2}}$; (see Fig. 4.4).

The cumulative density function (CDF) over the domain x ≥ 0 is $erfc(\sqrt{c/2x})$; (see Fig. 4.5).

The median value is $c/2(erf^{-1}(1/2))^2$.

The entropy is $\frac{1+3\gamma+\ln(16\pi c^2)}{2}$.

The properties of the Lévy stable distributions:

1. The variance is infinite except for its extreme case, Gaussian distribution.
2. A sum of two independent random variables of the Lévy stable distributions with the same index $\alpha$ is again the Lévy stable distributions with the same index $\alpha$.

## 4.3.3 Stable Distributions

**Definition 1.** A random variable $X$ is stable or generalized stable if and only if two independent variables $X_1$, $X_2$ of $X$ satisfy the following equation:

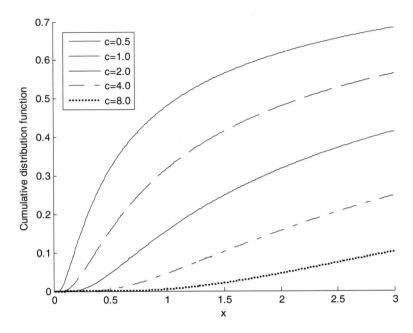

**Fig. 4.5** Plots of the CDF of the Lévy stable distributions with different c

$$aX_1 + bX_2 \overset{d}{=} cX + d, \tag{4.3.4}$$

where $a$, $b$ and $c$ are constants larger than 0 and $d$ is a real number [42].

If $d$ equals zero in Eq. (4.3.4), the random variable $X$ is strict stable or stable in the narrow sense. If $X \overset{d}{=} -X$, it is symmetric stable.

**Definition 2.** A non-degenerate random variable $X$ is stable if and only if when $n > 1$, there exist a constants $c_n > 0$ and $d_n \in R$ that satisfy the following equation:

$$X_1 + \ldots X_n \overset{d}{=} c_n X + d_n, \tag{4.3.5}$$

where $X_1, \ldots X_n$ are independent and identically distributed (i.d.d) random variables sharing the same distribution with $X$. The random variable $X$ is strict stable if and only if $d_n = 0$ with an arbitrary value of $n$.

**Definition 3.** A random variable $X$ is stable if and only if $X \overset{d}{=} aZ + b, 0 < \alpha \leq 2$, $-1 \leq \beta \leq 1, a > 0, b \in R$ and the characteristic function of the random variable $Z$ has the following expression:

$$E \exp(iuZ) = \begin{cases} \exp(-|u|^{\alpha}[1 - i\beta \tan \dfrac{\pi \alpha}{2}(signu)]) & \alpha \neq 1 \\[3mm] \exp(-|u|[1 + i\beta \dfrac{2}{\pi}(signu) \ln|u|]) & \alpha = 1 \end{cases}. \qquad (4.3.6)$$

When $\beta = 0, b = 0$, the stable distribution is symmetric and the characteristic function of $aZ$ has the following formula:

$$\phi(u) = e^{-a^{\alpha}|u|^{\alpha}}. \qquad (4.3.7)$$

## 4.4  Stretched Gaussian Distribution

In some cases, statistical phenomenon and distribution of the experimental data are difficult to be characterized by Gaussian distribution. It encourages researchers to attempt to find a more general distribution. Extending the Gaussian distribution to Stretched Gaussian distribution is a very natural way.

Stretched Gaussian distribution has played an increasing role in characterizing anomalous diffusion and turbulence, especially the anomalous diffusion in porous media with fractal structure. Using the fractal derivative in accordance with the characteristics of fractal media to solve the anomalous diffusion equation, the form of the solution is similar to Stretched Gaussian distribution. The PDF of Stretched Gaussian distribution is

$$f_{\beta}(x) = \frac{\beta}{2^{1+1/\beta}\Gamma(1/\beta)\sigma} e^{-\frac{1}{2}\left|\frac{x-a}{\sigma}\right|^{\beta}}, \ (-\infty < x < +\infty, \beta > 0).$$

When $\beta = 2$, $f_{\beta}(x)$ becomes Gaussian distribution (Fig. 4.6).

## 4.5  Tsallis Distribution

Based on quantum mechanics, Tsallis extended the entropy in microscopic scale to overcome the limitations of Newtonian mechanics and derived the Tsallis distribution.

### 4.5.1  Tsallis Entropy

Since the last century, the Boltzmann entropy has been applied to analyze the macro and microphysical phenomenon. The two pillars of the thermodynamic are

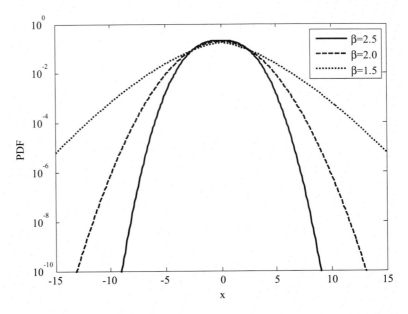

**Fig. 4.6** Plots of the PDF of the Lévy stable distributions and stretched Gaussian distribution with different values of $\beta$ ($\sigma = 2$)

energy and entropy. As it were, the Boltzmann entropy is one pillar of modern thermodynamic. However, the expression of the Boltzmann entropy has certain limitations.

The Boltzmann–Gibbs statistical mechanics is based on the following formula:

$$S_{BG} = -k \sum_{i=1}^{W} p_i \ln p_i, \qquad (4.5.1)$$

where $k$ is a positive constant and $W$ is the total number of microscopic possibilities of the system. The Boltzmann entropy is an indispensable tool to explore the physical mechanism of systems, such as short-range dependent systems and Markov system, which almost meet ergodic conditions. If the microscopic dynamics of the system is complicated, the researchers naturally hope that the phase space has a relatively simple structure, such as multifractal or hierarchical geometry structures. Researchers have attempted to use a similar approach with the standard statistical mechanics. However, the Boltzmann–Gibbs statistical mechanics and the classical thermal dynamics expose serious limitations in long-range interaction, long-range microscopic memory (non-Markov stochastic process), temporal and spatial correlation conservative or dissipative systems including two-dimensional electron plasma turbulence and Lévy anomalous diffusion. Nolan [43] listed several systems which do not meet the ergodic assumption, but can be handled via a new definition of entropy.

$$S_q = k\frac{1 - \sum_{i=1}^{w} p_i^q}{q-1}, \quad q \in R, \ S_1 = S_{BG}. \tag{4.5.2}$$

The theory is also known as the non-extensive statistical mechanics method [44]. Then, the generalized $q$ entropy is given by

$$S_q[p(x, v)] = \frac{1 - \int p^q(x, v)dxdv}{q-1}, \tag{4.5.3}$$

which satisfies the constrain equations:

$$\int p(x, v)\, dxdv = 1, \quad \int p^q(x, v)E(x, v)dxdv = U,$$

where $p(x, v)$, $E(x, v)$ are the density and energy of the particles, respectively. $U$ represents potential. The discrete form of Eq. (4.5.3) is Eq. (4.5.2). When $q \to 1$, $S_q$ is equivalent to the Boltzmann entropy.

Tsallis entropy has received great interest in non-extensive statistic, such as restructuring of the unfolded protein, cosmic rays stream, turbulence, electron–positron annihilation and finance [45].

### 4.5.2 Tsallis Distribution

Using the normal maximization principle of entropy, we can obtain the exponential form of the Boltzmann distribution, while the Tsallis statistic gives the form of the power-law distribution of Pareto. The power-law form has frequently been used in academic research and engineering, thanks to its success in fitting the distribution exhibiting a heavy tail.

The PDF of Tsallis distribution:

$$p_q(x) = \frac{1}{\delta}[\frac{q-1}{\pi(3-q)}]^{1/2}\frac{\Gamma(\frac{1}{q-1})}{\Gamma(\frac{3-q}{2(q-1)})}\frac{1}{[1 + \frac{q-1}{3-q}\frac{x^2}{\delta^2}]^{1/(q-1)}}, \quad (q > 1), \tag{4.5.4}$$

$$p_q(x) = \frac{1}{\delta}[\frac{1}{2\pi}]^{1/2}e^{-(x/\delta)^2/2}, \quad (q = 1), \tag{4.5.5}$$

$$p_q(x) = \frac{1}{\delta}[\frac{1-q}{\pi(3-q)}]^{1/2}\frac{\Gamma(\frac{5-3q}{2(1-q)})}{\Gamma(\frac{2-q}{1-q})}\frac{1}{[1 - \frac{1-q}{3-q}\frac{x^2}{\delta^2}]^{1/(q-1)}}, \quad (q < 1). \tag{4.5.6}$$

The Tsallis distribution is widely employed to simulate turbulence and solve the fractional calculus with a specific form. The noises of the Tsallis distribution and the Lévy stable distributions are different, the generating mechanism of the noise as

**Fig. 4.7** Plots of the PDF of
the Tsallis distribution with
$q > 1, \delta = 0.72$

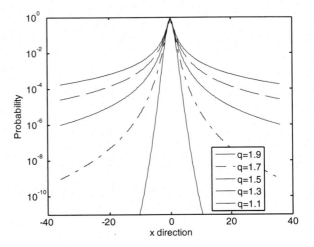

well. The former relies on the diffusion coefficient related to the distribution density
function, while the infinite second moment of the distribution density function leads
to the heavy tail (Fig. 4.7).

## 4.6  Ito Formula

### 4.6.1  Ito Integral

In this section, we define the stochastic integral of the Brownian motion
$\int_a^b X(t)\mathrm{d}B(t)$, where $\{B(t), t \geq 0\}$ is Brownian motion and $X(\mathrm{t})$ is a stochastic
process. It is well acknowledged that the Brownian motion trajectory is special,
which is continuous everywhere and non-differentiable $\forall t \geq 0$. $\int_a^b X(t)\mathrm{d}B(t)$ cannot
be calculated by transforming into $\int_a^b X(t)B'(t)\,\mathrm{d}t$; hence, a new definition for
stochastic integral is required. Generally speaking, the Darboux summation of the
Riemann–Stieltjes (R-S) does not converge to a certain limit with probability 1;
however, the mean square limit of the Darboux summation exists under appropriate
conditions, which was used by Ito to define the stochastic integral of Brownian
motion.

The main idea of Ito integral: if the stochastic process $X(\mathrm{t})$ only depends on
$\{B_u : u \leq t\}$, then the series $J_n(t) = \sum_{k=0}^{l_n-1} X(t_k)(B(t_{k+1}) - B(t_k))$ can prove its
convergence in probability, where $l_n$ is derived from an interval division of $[a,b]$, i.e.
$a = t_0^n < \ldots < t_{l_n}^n = b$. The limit of $J_n(t)$ in probability is the stochastic integral of
Brownian motion. That is, for $\forall \varepsilon > 0$,

$$\lim_{n \to \infty} P(|J_n(t) - \eta(t)| > \varepsilon) = 0 \qquad (4.6.1)$$

holds. Hence, $\int\limits_0^t X(\tau)\mathrm{d}B(\tau)$ is named Ito integral, $\int\limits_0^t X\mathrm{d}B$ for short.

**Remark:** $J_n(t)$ in the definition of Ito integral is different from the ordinary integral $J_n(t) = \sum\limits_{k=0}^{l_n-1} X(t_k')(B(t_{k+1}) - B(t_k))$, $t_k' \in [t_k, t_{k+1}]$. Because for $\forall t' \in [t_k, t_{k+1}]$, the mean square limit of $J_n(t)$ does not exist, let $t_k' = t_k$ in Ito integral, i.e. the left terminal of $[t_k, t_{k+1}]$.)

Ito integral satisfies linear behavior and integration interval additivity. The main properties of Ito integral are as follows:

(1)  $\int_a^b (af(t) + bg(t))\mathrm{d}B(t) = a \int_a^b f(t)\,\mathrm{d}B(t) + b \int_a^b g(t)\mathrm{d}B(t)$;
(2)  $E[\int_0^t X(\tau)\,\mathrm{d}B(\tau)] = 0$;
(3)  $E[(\int_0^t X(\tau)\,\mathrm{d}B(\tau))^2] = \int_0^t E(X^2(\tau))\mathrm{d}t$;
(4)  $\int_0^t B(\tau)\mathrm{d}B(\tau) = \frac{1}{2}B^2(t) - \frac{1}{2}t$. (That is the difference between Ito integral and ordinary integral [46].)

### 4.6.2  Ito Formula [47]

First, we define an Ito stochastic process and then introduce the Ito formula.

**Definition** If stochastic $X = \{X(t), t > 0\}$ satisfies the following Ito integral: $\forall 0 \le a < t < T$, it has

$$X(t) - X(t_0) = \int\limits_{t_0}^t b(s, X(s))\mathrm{d}s + \int\limits_{t_0}^t \sigma(s, X(s))\mathrm{d}B(s), \qquad (4.6.2)$$

or convert it into a differential form

$$\mathrm{d}X(t) = b(t, X(t))\mathrm{d}t + \sigma(t, X(t))\mathrm{d}B(t), \qquad (4.6.3)$$

where $\int\limits_{t_0}^t b(s, X(s))\mathrm{d}s$ is the ordinary mean square integral and $\int\limits_{t_0}^t \sigma(s, X(s))\mathrm{d}B(s)$ is the Ito integral. $b(t, x), \sigma(t, x)$ are binary continuous functions and $\forall x \in R, |b(t)|^{1/2}, \sigma(t)$ meets measurable conditions [47]. $X = \{X(t), t \ge 0\}$ is named Ito stochastic process.

**Theorem** If $X = \{X(t), t \ge 0\}$ is an Ito stochastic process, $y = f(x, t)$ is a binary continuous function that has continuous partial derivatives $\frac{\partial f}{\partial t}, \frac{\partial f}{\partial x}, \frac{\partial^2 f}{\partial x^2}$. Let $Y(t) \overset{\Delta}{=} f(t, X(t))$, then we have a stochastic process $Y = \{Y(t), t \ge 0\}$ satisfying the following for $\forall 0 \le t_0 < t$

$$Y(t) - Y(t_0) = \int_{t_0}^{t} [\frac{\partial f}{\partial t} + b\frac{\partial f}{\partial x} + \frac{\sigma^2}{2}\frac{\partial^2 f}{\partial x^2}](s, X(s))ds + \int_{t_0}^{t} \sigma\frac{\partial f}{\partial x}(s, X(s))dB(s),$$

$$(4.6.4)$$

or written as a differential form

$$dY(t) = (\frac{\partial f}{\partial t} + b\frac{\partial f}{\partial x} + \frac{\sigma^2}{2}\frac{\partial^2 f}{\partial x^2})(t, X(t))dt + \sigma\frac{\partial f}{\partial x}(t, X(t))dB(t). \quad (4.6.5)$$

Two Eqs. (4.6.4) and (4.6.5) are known as the Ito formula.

**Remark** The core of understanding and proof of Ito formula is: $(dB(t))^2 = dt$, i.e. $dB(t) = (dt)^{1/2}$ which will be replaced by $E\left|\sum_{k}(\Delta B(t_k))^2 - t\right|^2 \to 0$ with a rigorous proof.

**Example** If $\sigma, b$ are constants, a Gaussian process $\eta_t = e^{-bt}[\eta_0 + \sigma\int_0^t e^{bs}dB_s]$, then we have

$$d\eta_t = -b\eta_t dt + \sigma dB_t. \quad (4.6.6)$$

Proof: let $\varepsilon_t = \eta_0 + \sigma\int_0^t e^{bs}dB_s$, then $\eta_t = e^{-bt}\varepsilon_t$. Using the Ito formula, we get

$$d\eta_t = e^{-bt}d\varepsilon_t - be^{-bt}\varepsilon_t dt = \sigma B_t - b\eta_t dt.$$

**Remark** $\eta_t$ satisfies the Langevin stochastic differential equation which is first applied in theoretical physics.

## 4.7 Random Walk Model

Particle diffusion and correlated process are widely applied in engineering science, such as oil production engineering, hydrology and geophysics, chemical engineering and environmental engineering. The research on the particle diffusion process has academic significance and application value, and it has become an interdisciplinary hotspot.

The random walk model is the most direct and effective approach to study particle diffusion process, and the ideology is derived from the numerical simulation of particle Brownian motion. The jump length and waiting time of a particle have certain restrictions in the numerical simulation of the Brownian motion. General assumptions are the statistical distribution of jump length obeys Gaussian distribution and the waiting time is a uniform distribution. A specific anomalous diffusion process can be simulated by setting the statistical distributions of jump length and waiting time in the random walk model.

Montroll, Scher and Weiss first introduced the random walk model into statistical mechanics. The idea of this approach is from the microscopic point of view, which extends the normal diffusion transition probability to anomalous diffusion transition probability. The random walk model derived from statistical mechanics is divided into discrete random walk model and continuous random walk model [48–55].

### 4.7.1  Discrete Random Walk Model

In this section, we will introduce the model based on the space fractional anomalous diffusion process and employ the Riesz–Feller pattern to define the space fractional derivative in the following equation.

$$\begin{cases} \frac{\partial u(x,t)}{\partial t} = \mathrm{d}D_\theta^\alpha u(x,t) \\ u(x,0) = u_0 \end{cases}. \tag{4.7.1}$$

Definition of a random variable:

$$S_n = hY_1 + hY_2 + hY_3 + \ldots + hY_n, \ (n \in N), \tag{4.7.2}$$

where $h$ represents the basic space step, $Y_i$ represents the random number of jump length and $S_n$ represents the location after the $n$th jump. In random walk simulation of numerous particles, the density of a point is stated as

$$y_j(t_{n+1}) = \sum_{k=-\infty}^{+\infty} p_k y_{j-k}(t_n), \ (j \in Z, n \in N). \tag{4.7.3}$$

Here, we use the generating function to determine the transition probability $p_k$. There are many methods for constructing a generating function, and this section just introduces one of them.

$$\tilde{p}(z) = \sum_{j=-\infty}^{+\infty} p_j z^j, \ \tilde{y}(z) = \sum_{j=-\infty}^{+\infty} y_j(t_n)z^j, \ |z| = 1, \tag{4.7.4}$$

where $p_k$ satisfies $\sum_{k=-\infty}^{+\infty} p_k = 1, \ k = 0, \pm 1, \pm 2 \ldots$.

Here, we only discuss the approach to construct generating function under $\theta = 0$. $\alpha$ is divided into two cases in accordance with its range.

(a) $0 < \alpha < 1$; (b) $1 < \alpha \leq 2$.

$$\tilde{p}(z) = \begin{cases} 1 - \frac{\mu}{2\cos(\alpha\pi/2)}[(1-z)^\alpha + (1-z^{-1})^\alpha], & (0 < \alpha < 1) \\ 1 - \frac{\mu}{2\cos(\alpha\pi/2)}[z^{-1}(1-z)^\alpha + z(1-z^{-1})^\alpha], & (1 < \alpha \leq 2) \end{cases}. \qquad (4.7.5)$$

In case (a), $0 < \mu \leq \cos(\alpha\pi/2)$ holds and we have

$$\begin{cases} p_0 = 1 - \frac{\mu}{\cos(\alpha\pi/2)}, \\ p_{\pm k} = (-1)^{k+1} \frac{\mu}{\cos(\alpha\pi/2)} \binom{\alpha}{k}, & k = 1, 2, 3 \ldots\ldots \end{cases} \qquad (4.7.6)$$

In case (b), $0 < \mu \leq |\cos(\alpha\pi/2)|/\alpha$ holds and we have

$$\begin{cases} p_0 = 1 - \frac{\mu\alpha}{|\cos(\alpha\pi/2)|}, \quad p_{\pm 1} = \frac{\mu}{2|\cos(\alpha\pi/2)|}[\binom{\alpha}{2} + 1], \\ p_{\pm k} = (-1)^{k+1} \frac{\mu}{2|\cos(\alpha\pi/2)|} \binom{\alpha}{k+1}, \quad k = 2, 3, 4 \ldots\ldots \end{cases} \qquad (4.7.7)$$

When $\alpha = 2$, the generating function is

$$\tilde{p}(z) = 1 + \mu[z - 2 + z^{-1}], \quad (\mu = \frac{d\tau}{h^2}). \qquad (4.7.8)$$

The differential discrete form of the normal diffusion equation can be derived from Eq. (4.7.8).

The advantages of this method are as follows:

(1)  It is a numerical difference scheme for solving equations;
(2)  It can simulate the random walk process of a certain particle;
(3)  It can simulate the position distribution of certain numerous particles after a period of time.

Here, we give the following example to illustrate the process of this approach.

**Example 1**

$$\begin{cases} \frac{\partial u(x,t)}{\partial t} = D_0^{1.5} u(x, t), \quad t > 0; 0 \leq x \leq 1 \\ u(x, 0) = \sin x \end{cases}. \qquad (4.7.9)$$

Solution idea: the deviation proportion of Eq. (4.7.9) is $\theta = 0$. The order of the spatial fractional derivative is 1.5. Firstly, set the space step $h = 0.01$, time step

**Fig. 4.8** Plot of the anomalous diffusion ($\alpha = 1.5$), the time step is 0.00001, space step is 0.01, time span is 500 and space span is 100

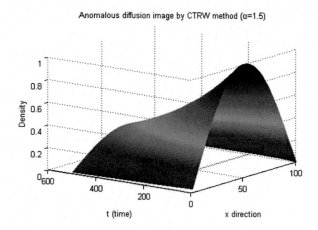

$\tau = 0.00001$, then $\mu = \tau/h^{1.5} = 0.00001/(0.01)^{1.5} = 0.01$; $|\cos(\alpha\pi/2)|/\alpha = |\cos(3\pi/4)|/1.5 = \sqrt{2}/3; 0 < \mu \le |\cos(\alpha\pi/2)|/\alpha$.

The transition probability can be obtained by substituting the above parameters into the formula (4.7.7), then we obtain the iterative format of space fractional derivative by substituting the transition probability into Eq. (4.7.3).

Figure 4.8 depicts the anomalous diffusion using the random walk model.

### 4.7.2 Continuous Random Walk Model (CTRW)

The primary coverage of the continuous random walk model is to construct the PDF of waiting time $\phi(t)$ and PDF of jump length $\omega(x)$. If the joint probability density function of waiting time and jump length is decomposable, then the CTRW model can be represented by $\phi(t)$ and $\omega(x)$. Let $p(x, t)$ be the probability of the random particles found at the time $t$ and on the position $x$, we can get the following expression:

$$p(x, t) = \delta(x)\Psi(t) + \int_{-\infty}^{+\infty} d\xi \int_{0}^{t} \varphi(\xi, \tau)p(x - \xi, t - \tau)d\tau, \qquad (4.7.10)$$

where $\Psi(t) = 1 - \int_{-\infty}^{+\infty} d\xi \int_{0}^{t} \varphi(\xi, \tau) \, d\tau$.

The initial condition $x(0) = 0$ is implicated in $\delta(x)\Psi(t)$. $\varphi(\xi, \tau)$ represents the joint probability density function of waiting time and jump length. Generally speaking, $\phi(t)$ and $\omega(x)$ are dependent, and the stochastic process will be easy to handle if $\phi(t)$ and $\omega(x)$ are independent.

The Fourier–Laplace transform of $p(x, t)$ satisfies the following equation given by Montroll and Weiss.

$$P(k, s) = \frac{1 - \phi(s)}{s} \frac{1}{1 - \phi(s)\omega(k)}. \tag{4.7.11}$$

In simulating the anomalous diffusion process, the $\phi(t)$ usually obeys the Lévy stable distributions and $\omega(x)$ obeys the Mittag–Leffler distribution. A famous algorithm to generate the Lévy random variables is proposed by Chamber, Mallows and Stuck, which is stated as

$$\xi_\alpha = \gamma_x \left( \frac{-\ln u \cos \theta}{\cos((1 - \alpha)\theta)} \right)^{1 - 1/\alpha} \frac{\sin(\alpha\theta)}{\cos \theta}, \tag{4.7.12}$$

where $\theta = \pi(\upsilon - 1/2)$ and $u$ and $\upsilon$ are two independent random variables uniformly distributed on the interval $(0,1)$. The stability index $\alpha$ equals the order of space fractional derivative in the fractional anomalous diffusion equation. $\gamma_x$ is the spatial scale parameter.

An algorithm to generate the Mittag–Leffler random variables is proposed by Kozubowski and Rachev, which is stated as

$$\tau_\beta = -\gamma_t \ln u \left( \frac{\sin(\beta\pi)}{\tan(\beta\pi\upsilon)} - \cos(\beta\pi) \right)^{1/\beta}, \tag{4.7.13}$$

where $\gamma_t$ represents the spatial scale parameters which have the relation with spatial scale parameter $\gamma_x = \gamma_t^{\beta/\alpha}$; $\beta$ the parameter of the Mittag–Leffler distribution, which is equivalent to the order of time-fractional derivative and $u$ and $\upsilon$ are two independent random variables uniformly distributed on the interval $(0,1)$.

Figure 4.9 presents the plots of particle trajectories under different parameters based on the Eqs. (4.7.12) and (4.7.13). In general cases, the larger $\alpha$ is, the bigger the jump length will be, and the smaller $\beta$ is, the shorter waiting time will be.

Figure 4.10 gives the histograms of the Lévy random numbers, which exhibit the character of the Lévy stable distributions.

Figure 4.11 shows the histograms of the Mittag–Leffler random numbers, which exhibit the character of power-law decay.

### 4.7.3  The Relationship Between Random Walk Model and Fractional Diffusion Equation

The relationship between random walk model and fractional diffusion equation will be explored in this section.

(1)  The Fourier–Laplace transform of the following fractional anomalous diffusion equation is stated as

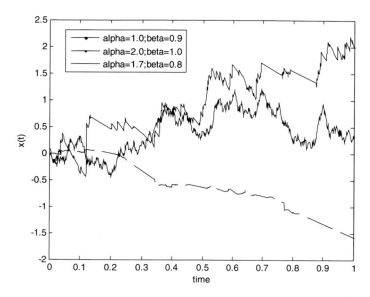

**Fig. 4.9**  Plots of the particle trajectories, the scale parameters $\gamma_t = 0.001$, $\gamma_x = \gamma_t^{\beta/\alpha}$

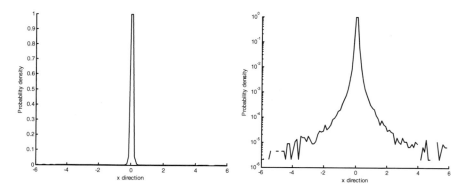

**Fig. 4.10**  The histograms of the Lévy random numbers

$$\begin{cases} \frac{\partial^\beta u}{\partial t^\beta} = \frac{\partial^\alpha u}{\partial |x|^\alpha}, & -\infty < x < +\infty, \ t > 0, \\ u(x, 0) = \delta(x). \end{cases} \tag{4.7.14}$$

$$s^\beta \tilde{\bar{u}}(k, s) - s^{\beta-1} = -|k|^\alpha \tilde{\bar{u}}(k, s), \ 0 < \alpha \le 2, \ 0 < \beta \le 1. \tag{4.7.15}$$

Then simplifies Eq. (4.7.15)

$$\tilde{\bar{u}}(k, s) = \frac{s^{\beta-1}}{s^\beta + |k|^\alpha}, \ s > 0, \ k \in R. \tag{4.7.16}$$

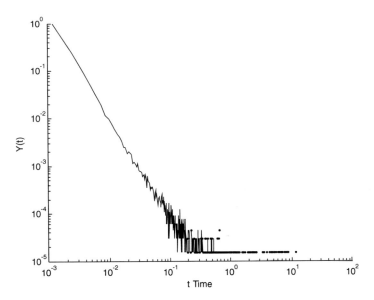

**Fig. 4.11** The histogram of the Mittag–Leffler random numbers; the sample size is 100000

(2)    The approximation of the random walk model
       Let PDF of jump length function satisfies

$$\omega(x) > 0, \ \omega(x) = \omega(-x) \text{ for } x \in R, \ \int_{-\infty}^{+\infty} \omega(x)\mathrm{d}x = 1, \qquad (4.7.17)$$

and satisfies one of the following conditions

(a)    $\sigma^2 := \int_{-\infty}^{+\infty} x^2 \omega(x)dx < \infty$, if $\alpha = 2$,

(b)    $\omega_x \propto b|x|^{-(\alpha+1)}$, for $|x| \to \infty$, $\alpha \in (0, 2)$, $b > 0$.

Then we get
$\mu = \frac{\sigma^2}{2}$ (a); $\mu = \frac{b\pi}{\Gamma(\alpha+1)\sin(\alpha\pi/2)}$ (b).
The Fourier transform of $\omega(x)$ is

$$\bar\omega(k) = 1 - \mu|k|^\alpha + o(|k|^\alpha) \text{ for } k \to 0. \qquad (4.7.18)$$

Similarly, let the PDF of the waiting time function satisfy

$$\phi \geq 0 \text{ for } t \geq 0, \ \int_{0}^{\infty} \phi(t)\mathrm{d}t = 1,$$

and also satisfy one of the following conditions

(a)   $\rho := \int_0^\infty t\phi(t)dt < \infty$, if $\beta = 1$,

(b)   $\phi_t \propto ct^{-(\beta+1)}$, for $t \to \infty$, $\beta \in (0, 1)$, $c > 0$.

Then we get
$\lambda = \rho$ (a); $\lambda = \frac{c\Gamma(1-\beta)}{\beta}$ (b).
The Laplace transform of $\phi(t)$ is s

$$\widetilde{\phi}(s) = 1 - \lambda s^\beta + o(s^\beta) \text{ for } 0 < s \to 0. \tag{4.7.19}$$

$h, \tau$ are the space and time steps, respectively, then Eqs. (4.7.18) and (4.7.19) are transformed into the following equations:

$$\overline{w}(kh) = 1 - \mu|kh|^\alpha + o(|h|^\alpha) \text{ for } h \to 0, \tag{4.7.20}$$

$$\widetilde{\phi}(s\tau) = 1 - \lambda(s\tau)^\beta + o(\tau^\beta) \text{ for } \tau \to 0. \tag{4.7.21}$$

Considering the scale transform of the diffusion process

$$\lambda \tau^\beta = \mu h^\alpha. \tag{4.7.22}$$

The Fourier–Laplace transform of $p(x, t)$ satisfies the following formula.

$$P(k, s) \to \frac{s^{\beta-1}}{s^\beta + |k|^\alpha}. \tag{4.7.23}$$

(3)   Analysis and discussion

Based on the random walk model mentioned above, we get

$$P(k, s) \to \widehat{\widetilde{u}}(k, s), \quad h, \tau \to 0. \tag{4.7.24}$$

Equation (4.7.24) shows that the fractional diffusion equation is approximately equivalent to the random walk model.

If $\phi(t)$ obeys Gaussian distributions and $\omega(x)$ obeys the Mittag–Leffler distribution, then the following hold

$$\overline{w}(k) = 1 - \mu k^2 + o(k^2) \text{ for } k \to 0,$$
$$P(k, s) \to \frac{s^{\beta-1}}{s^\beta + k^2}. \tag{4.7.25}$$

The corresponding master equation of fractional diffusion is

$$\frac{\partial^\beta u}{\partial t^\beta} = \frac{\partial^2 u}{\partial^2 x}. \tag{4.7.26}$$

Then, the random walk model of the space fractional derivative equation can be obtained.

$$P(k, s) \rightarrow \frac{1}{s + |k|^\alpha} \rightarrow \frac{\partial u}{\partial t} = \frac{\partial^\alpha u}{\partial |x|^\alpha}. \tag{4.7.27}$$

Similarly, by altering the approximate forms of $\bar{\omega}(k)$ and $\tilde{\phi}(s)$, we can also obtain different forms of the fractional advection–diffusion equation and the Fokker–Plank–Kolmogorov equation [56].

## 4.8   Discussion

### 4.8.1   The Applications of Statistical Distribution

The statistical distributions of many natural phenomena obey Gaussian distribution, such as measurement error, height and weight of a given species, particles Brownian motion in fluid, noise in a large number of electronic devices and Fickian diffusion of pollutants in homogeneous isotropic media.

The anomalous statistical distributions mentioned in this chapter have been widely applied in science and engineering. The Lévy stable distributions can depict anomalous diffusion and fluctuations in the stock market. Stretched Gaussian distribution can characterize the anomalous diffusion in porous media with fractal structure. The Tsallis distribution has received great interest in non-extensive statistics, such as restructuring of the unfolded protein, cosmic rays stream and energy dissipation in turbulence.

The anomalous statistical distributions equipped with explicit physical interpretation have become the academic frontier of multidisciplinary, such as mathematics, physics and mechanics.

### 4.8.2   The Relationship Between Statistical Distribution and Differential Equation

For a specific physical or mechanics problem, macroscopic differential equations and microscopic statistical distributions are the two general modeling approaches. If a problem can be handled by the two methods, a corresponding relationship may exist between them. Here, we give an example of anomalous diffusion.

In analyzing the process of Brownian motion or normal diffusion, integer-order diffusion equation is employed

$$
\begin{cases}
\frac{\partial u(x,y,z,t)}{\partial t} = d\Delta u(x,y,z,t) \\
u(x,y,z,0) = \delta(x,y,z)
\end{cases}
\tag{4.8.1}
$$

where $u$ represents velocity, density or temperature; $d$ is the diffusion coefficient, $\delta(x,y,z)$ is the Dirac delta function and $\Delta$ represents the Laplace operator.

The solution of Eq. (4.8.1) has the same form as the Gaussian distribution; thus, Gaussian distribution is usually used to depict the physical or mechanics process liking normal diffusion process.

A representative differential equation to model anomalous diffusion is fractional anomalous diffusion.

$$
\begin{cases}
\frac{\partial^{\beta} u(x,y,z,t)}{\partial t^{\beta}} = d\Delta^{\alpha} u(x,y,z,t) \\
u(x,y,z,0) = \delta(x,y,z)
\end{cases}
\tag{4.8.2}
$$

where $\Delta^{\alpha}$ represents the fractional Laplace operator, density or temperature. $\beta$ is the order of time-fractional derivative.

The explicit solution of Eq. (4.8.2) exists in the one-dimensional case (See Chap. 2 of this book), which can predict the solution of Eq. (4.8.2) satisfying the Lévy stable distributions. The corresponding statistical method to characterize anomalous diffusion is the random walk model which has a close relationship with the Lévy stable distributions. Therefore, many researchers point out that under specific conditions, the time random walk model is equivalent to the fractional diffusion equation.

Meanwhile, stretched Gaussian distribution plays an increasing role in characterizing anomalous diffusion, especially the anomalous diffusion in porous media with fractal structure. Using the fractal derivative in accordance with the characteristics of fractal media to solve the anomalous diffusion equation, the form of the solution is similar to stretched Gaussian distribution. In the one-dimensional case, the diffusion equation with fractal derivative is stated as follows:

$$
\begin{cases}
\frac{\partial u(x,t)}{\partial t^{\eta}} = D \frac{\partial}{\partial x^{\gamma/2}} \left( \frac{\partial u(x,t)}{\partial x^{\gamma/2}} \right) \\
u(x,0) = \delta(x)
\end{cases}
\tag{4.8.3}
$$

where $\eta$ is the order of time fractal derivative. $\gamma$ is the order of space fractal derivative. The solution of Eq. (4.8.3) has a similar form with Stretched Gaussian distribution.

The Tsallis distribution, which is derived by analyzing the entropy of the energy dissipation process, has succeeded in fitting the turbulence and certain anomalous diffusion. It reveals the relationship between turbulence and non-extensive entropy; however, the equivalent derivative equation with the Tsallis statistic is still not reported. The topic should be concerned in future research.

**Table 4.2** Various statistical distributions have relationships with the stochastic process, diffusion process, entropy, derivative equation, geometry and algebra

| Types | 1 | 2 | 3 | 4 | 5 | 6 |
|---|---|---|---|---|---|---|
| PDF | WT[1]: Poisson JL[2]: Gaussian | WT: Poisson JL: Lévy stable | WT: Poisson JL: Stretched Gaussian | WT: Poisson JL: Tsallis | WT: Power law JL: Gaussian | Coupled WT with JL |
| Diffusion Process | Brownian motion, normal diffusion | Lévy flight, Sub-diffusion | Heavy tail, super diffusion | Heavy tail, super diffusion | Fractional Brownian motion, Sub-diffusion | Lévy walk, superdiffusion |
| Entropy | Boltzmann–Gibbsentropy | To be determined[3] | / | Tsallis entropy | To be determined | / |
| Differential Equation | Fick diffusion equation | Fractional Laplacian equation | Fractal derivative space diffusion equation | Nonlinear integer diffusion Eq.[4] | Time-fractional diffusion equation, time fractal derivative, multi-relaxation–diffusion equation | / |
| Geometry | Euclidean | Space fractal | Space fractal | Euclidean | Time fractal | Space and time fractal |
| Fourier transform | Yes | Yes | / | Fractional Fourier transform | To be determined | Yes |
| Taylor expansion | Yes | Yes | / | / | To be determined | Yes |

Table 4.2 gives the relationship between statistical distribution and differential equation in connection with diffusion progress. A lot of work remains to be further studied.

Remarks:

1. WT(waiting time) represents the waiting time in the random walk model.
2. JL(Jump length) represents the jump length.
3. "To be determined" means the conclusion is not determined according to the present study.
4. It has been reported in the literature, but arouses rare research.

### 4.8.3  Lévy Stable Distributions and [0, 2] Power-Law Dependence of Acoustic Absorption on Frequency

This section will analyze the internal relationship between the Lévy stable distributions and [0, 2] power-law dependence of acoustic absorption on frequency, to reveal the intrinsic link of statistical mechanics, deterministic differential equation model and physical phenomenon. The contents of this section are mainly taken from [57].

The effect of the dissipative attenuation of acoustic wave propagation over a finite range of frequency is typically characterized by a measured power-law function of frequency $\alpha_0 \omega^\eta$ [7, 8], where $\eta \in [0,2]$, $\omega$ denotes angular frequency and $\alpha_0$ and $\eta$ are non-negative media-dependent constants. It is apparent that the higher the frequency, the faster the dissipation. For the ideal solid and liquid materials, $\eta = 0 \sim 2$, i.e. energy dissipation does not depend on the frequency or is not proportional to its square, such as the $\eta$ of the ultrasonic propagation in water dissipation is 2. However, in most complex soft condensed matter, such as human organism, polymer, liquid crystal, soil, colloid, particulate matter, porous rock, foam, fabric and petroleum, the dissipation parameter $0 < \eta < 2$. The $\eta$ of layered porous rocks and sediments underwater approximates 1. The $\eta$ of a variety of human tissues varies from 1.1 to 1.7, fat 1.7, carcinoma tissue 1.3 and connective tissue 1.5. Thus, the cancer tissue and normal tissue can be separated by $\eta$.

Section 5.5.6 introduces the fractional Laplacian dissipative acoustic wave Eq. (5.5.63) which can depict the arbitrary frequency-dependent dissipative processes. Under certain conditions, the fractional derivative acoustic wave equation can be simplified as an anomalous diffusion equation which correlates with the Lévy stable distributions.

The Cauchy problem of the one-dimensional anomalous diffusion equation is expressed as

$$\frac{\partial p}{\partial t} + \kappa \left( -\frac{\partial^2}{\partial x^2} \right)^{\eta/2} p = 0, \qquad (4.8.4)$$

$$p(x,0) = \delta(x), \quad -\infty p \; x \; p\infty, \qquad (4.8.5)$$

where $\kappa$ represents the diffusion coefficient, and $\delta(x)$ is the Dirac delta function. The solution of the above Eqs. (4.8.4) and (4.8.5) is [58]

$$p(x,t) = \frac{1}{t^{1/\eta}} w_y \left( \frac{x}{t^{1/\eta}} \right), \qquad (4.8.6)$$

where

$$w_\eta(\xi) = \frac{1}{2\pi} \int\limits_{-\infty}^{\infty} e^{-iq\xi} W_\eta(k) dk, \, \xi = x/t^\eta. \qquad (4.8.7)$$

$W_\eta$ is the characteristic function of $w_\eta$

$$W(k) = e^{-\kappa k^\eta}. \qquad (4.8.8)$$

Equation (4.8.8) is also the Fourier transform of the probability density function of the $\eta$-stable Lévy distributions. The anomalous diffusion equation is thus considered underlying the Lévy stable distributions [38, 58]. In the limiting case $\eta = 2$ for the standard diffusion equation, the solution is the explicit Gaussian probability density function

$$p(x,t) = \frac{1}{\sqrt{4\pi\kappa t}} e^{-x^2/4\kappa t}. \qquad (4.8.9)$$

Saichev and Zaslavsky [58] pointed out that in order to satisfy the positive probability density function, the Lévy stable index y must obey

$$0 \prec \eta \leq 2. \qquad (4.8.10)$$

Namely, the $\eta$-stable distribution requires the power y to be positive but not greater than 2 [59]. In particular, $\eta = 1$ corresponds to the Cauchy distribution [58]. In terms of this statistical theory, the media having $\eta > 2$ power-law attenuation are not statistically stable in nature. In other words, the corresponding probability density function is no longer positively defined. It is noted that the Lévy process does not include $\eta = 0$. This means that the media obeying absolutely frequency-independent attenuation is simply an ideal approximation. For acoustic wave propagations, all media exhibit more or less degree of absorption dependence on frequency.

As shown in power-law formula $\alpha_0 \omega^\eta$, exponent $\eta$ obtained by experimental data fitting has always been observed within the finite scope in between 0 and 2 for all media. The above analysis shows that the Lévy stable distribution theory provides a mathematical interpretation of empirical [0, 2] power dependence of the absorption coefficient y on the frequency. The stability index of the Lévy stable distributions is

(0,2] proved mathematically which presents a theoretical explanation that the power-law dependence of acoustic absorption on the frequency of a variety of substances always ranges from 0 to 2, from the viewpoint of mesoscopic statistics stability.

The dissipative index $\eta \in [0,2]$ has widely been observed not only in acoustics but also in many other physical behaviors such as vibrational damping, dielectrics, thermoviscosity and fluid thermoviscous dissipation.

# References

1. http://zhidao.baidu.com/question/9108763
2. L.F. Richardson. Atmospheric diffusion shown on a distance-neighbour graph. Proceedings of the Royal Society of London. Series A, Containing Papers of a Mathematical and Physical Character **110**, 709–737 (1926)
3. J. Crank, *Free and moving boundary problems* (Clarendon Press, Oxford, 1987)
4. A.I. Nachman, J. Smith, R.C. Waag, An equation for acoustic propagation in inhomogeneous media with relaxation losses. J. Acoust. Soc. America **88**(3), 1584–1595 (1990)
5. J.D. Bao, Fractional Brownian motion and anomalous diffusion. Prog. Phys. **5**(4), 259–367 (2005)
6. Y. Zhang, D.A. Benson, M.M. Meerschaert, H.P. Scheffler, On using random walks to solve the space-fractional advection-dispersion equations. J. Stat. Phys. **123**(1), 89–100 (2006)
7. F.X. Chang, J. Chen, W. Huang, Anomalous diffusion and fractional advection-diffusion equation. Acta Physica Sinica **54**(3), 1113–1117 (2005)
8. S.Q. Zhang, Fractional diffusion wave equation on finite interval. J. Northw. Normal Univ. (Natural Science Edition) **41**(2), 10–13 (2005)
9. X.Z. Lu, F.W. Liu, Time fractional diffusion-reaction equation [J]. Numerical Math. A J. Chinese Univ. **27**(3), 267–273 (2005)
10. J.S. Duan, M.Y. Xu, Solution of semiboundless mixed problem of fractional diffusion equation. Appl. Math. A J. Chin. Univ. **18**(3), 259–266 (2003)
11. I. Podlubny, *Fractional differential equations* (Academic Press, San Diego, 1999)
12. C. Tadjeran, M.M. Meerschaert, H.P. Scheffler, A second-order accurate numerical approximation for the fractional diffusion equation. J. Comput. Phys. **213**, 205–213 (2006)
13. M. Meerschaert. Heavy Tails: Data, models, and applications [R]. University of Otago (April 2006)
14. F. Barpi, S. Valente, Creep and fracture in concrete a fractional order rate approach. Eng. Fract. Mech. **70**, 611–623 (2003)
15. M.M. Meerschaert, C. Tadjeran, Finite difference approximations for two-sided space-fractional partial differential equations. Appl. Numer. Math. **56**, 80–90 (2006)
16. L.Ts. Adzhemyan, N.V. Antonov, M.V. Kompaniets, A.N. Vasil'ev. Renormalization group in the statistical theory of turbulence: two-loop approximation. arXiv:nlin/0205046v1 [nlin.CD] 20 May 2002
17. M.F. Shlesinger, B.J. West, J. Klafter, Lévy dynamics of enhanced diffusion: application to turbulence. Phys. Rev. Lett. **58**(11), 1100–1103 (1987)
18. A. La Porta, G.A. Voth, A.M. Crawford, J. Alexander, E. Bodenschatz, Fluid particle accelerations in fully developed turbulence. Nature **409**, 1017–1019 (2001)
19. N. Mordant, A.M. Crawford, E. Bodenschatz, Experimental Lagrangian acceleration probability density function measurement. Physica D **193**, 245–251 (2004)
20. A.K. Aringazin, M.I. Mazhitov. Gaussian factor in the distribution arising from the nonextensive statistics approach to fully developed turbulence. arXiv:cond-mat/0301040v3. 19 Nov 2003
21. C. Beck. Superstatistics in hydrodynamic turbulence. Physica D, 2004193:195–207

22. C. Beck, On the small-scale statistics of Lagrangian turbulence. Phys. Lett. A **287**, 240–244 (2001)
23. C. Beck, Generalized statistical mechanics and fully developed turbulence. Physica A **306**, 189–198 (2002)
24. C. Beck, Non-extensive statistical mechanics approach to fully developed hydrodynamic turbulence. Chaos, Solitons Fractals **13**, 499–506 (2002)
25. T. Arimitsu, N. Arimitsu, Multifractal analysis of fluid particle accelerations in turbulence. Physica D **193**, 218–230 (2004)
26. M. Sokolov, J. Klafter, A. Blumen, Ballistic versus diffusive pair dispersion in the Richardson regime. Phys. Rev. E **61**(3), 2717–2722 (2000)
27. W. Chen. A speculative study of 2/3-order fractional Laplacian modeling of turbulence: Some thoughts and conjectures. Chaos 16:023126 (2006)
28. T. Arimitsu, N. Arimitsu, Tsallis statistics and fully developed turbulence. J. Phys. A: Math. Gen. **33**, L235–L241 (2000)
29. W. Chen, Time–space fabric underlying anomalous diffusion. Chaos, Solitons Fractals **28**, 923–929 (2006)
30. A.J. Majda, P.R. Kramer, Simplified models for turbulent diffusion: theory, numerical modelling, and physical phenomena. Phys. Rep. **314**, 237–574 (1999)
31. R. Kanno, Representation of random walk in fractal space-time. Physica A **248**, 165–175 (1998)
32. Z. Warhaft, Passive scalars in turbulent flows. Annu. Rev. Fluid Mech. **32**, 203–240 (2000)
33. J.P. Laval, B. Dubrulle, S.V. Nazarenko, Fast numerical simulations of 2D turbulence using a dynamic model for subfilter motions. J. Comput. Phys. **196**, 184–207 (2004)
34. L. Chevillard, S.G. Roux, E. Leveque, N. Mordant, J.F. Pinton, A. Arneodo. Lagrangian velocity statistics in turbulent flows: effects of dissipation. arXiv:cond-mat/0310105v1 6 Oct 2003
35. D. del-Castillo-Negrete, B.A. Carreras, V. E. Lynch. Fractional diffusion in plasma turbulence. Phys. Plasmas 11(8):3854–3864 (2004)
36. X.S. Xu, L.L. Ma, Y.B. Chen, The empirical tests of Lévy distribution on China's stock market. Sci. Technol. Progr. Policy **8**, 103–105 (2005)
37. G.S. Zhao, The fractal simulation and interpolation and Lévy distribution. Petrol. Geophys. Transl. **2**, 55–69 (1992)
38. W. Feller, *An introduction to probability theory and its applications*, 2nd edn. (John Wiley & Sons, Inc., New York, 1971)
39. M. Guarnieri, P. Biancardi, D. D'Aria, F. Rocca. Accurate and robust baseline estimation. In Second International Workshop on ERS SAR Interferometry, 'FRINGE99', Liege, Belgium, 10–12, Nov 1999, ESA (1999)
40. N.E. Sanjana. Lecture 22: Lévy distributions [R]. Department of Brain and Cognitive Sciences, MIT. April 26, 2005
41. http://en.wikipedia.org/wiki/Levy_distribution
42. http://en.wikipedia.org/wiki/Stable_distribution
43. J.P. Nolan. Stable distributions-models for heavy tailed data [Z]. Math/Stat Department American University, Processed April 28, 2004
44. C. Tsallis, Nonextensive statistics: Theoretical, experimental and computational evidences and connections. Braz. J. Phys. **29**(1), 1–35 (1999)
45. C. Tsallis, E. Brigatti, Nonextensive statistical mechanics: a brief introduction. Continuum Mech. Thermodyn. **16**(3), 223–235 (2004)
46. L. Guang, M.P. Qian. Applied Stochastic Processes tutorial-in algorithms and intelligent computing. Tsinghua University Press (2004)
47. Q. Xue, J. Hui. Stochastic process [M]. Hefei University Press (2006)
48. R. Gorenflo, F. Mainardi. Continuous time random walk, Mittag-Leffler waiting time and fractional diffusion: mathematical aspects. arXiv:0705.0797v2 [cond-mat.stat-mech]
49. N. Krepysheva, L.D. Pietro, M.C. Néel, Fractional diffusion and reflective boundary condition. Physica A **368**, 355–361 (2006)
50. R. Gorenflo, F. Mainardi. Approximation of Lévy–Feller diffusion by random walk. J. Anal. Appl. 231–246 (1999)

51. R. Gorenflo, G.D. Fabritiis, F. Mainardi, Discrete random walk models for symmetric Lévy -Feller diffusion processes. Physica A **269**, 79–89 (1999)
52. R. Gorenflo, F. Mainardi, A. Vivoli, Continuous-time random walk and parametric subordination in fractional diffusion. Chaos, Solitons Fractals **34**, 87–103 (2007)
53. R. Gorenflo, F. Mainardi, Random walk models for space-fractional diffusion processes. Frac. Calc. Appl. Anal. **1**, 167–191 (1998)
54. J.W. Hanneken et al. A random walk simulation of fractional diffusion. J. Mol. Liq. **114**, 153–157 (2004)
55. B. Baeumer et al., Advection and dispersion in time and space. Physica A **350**, 245–262 (2005)
56. E.A. Abdel-Rehim, R. Gorenflo, Simulation of the continuous time random walk of the space-fractional diffusion equations. J. Comput. Appl. Math. **222**(2), 274–283 (2008)
57. W. Chen, Lévy stable distribution and [0,2] power law dependence of acoustic absorption on frequency in various lossy media. Chin. Phys. Lett. **22**(10), 2601–2603 (2005)
58. A. Saichev, G.M. Zaslavsky, Fractional kinetic equations: solutions and applications. Chaos **7**(4), 753 (1997)
59. P. Lévy. Theorie de l'addition des variables aleatoires. 2nd Ed. (Gauthier-Villars, 1954)

# Chapter 5
# Typical Applications of Fractional Differential Equations

Fractional differential equations can be used to describe the some complex physical and mechanics processes, memory property in the temporal domain, as well as path dependence and global relevance in the spatial domain, and establish constitutive models for complex mechanics behaviors. These physical and mechanics models are summarized as the following four aspects:

1. Soft matter: It belongs neither to ideal solid nor to Newtonian fluid. Typical examples include polymers, foam, underwater sediments, organisms, oil, gasoline, etc. The classical integer-order differential equation models used to describe ideal solids and Newtonian fluids cannot well describe the mechanics behaviors of soft matter, for example, the frequency dependence of the energy dissipation. Fractional derivative is a powerful mathematical tool to characterize these complex mechanics behaviors.

2. Power-law phenomena: The empirical formula of the complex mechanical process often obeys the power-law function, of which the mechanics constitutive relation does not satisfy the standard "gradient" law (the Darcy law, the Fourier heat conduction law, Newtonian viscosity law, Fick diffusion law, etc.). Fractional derivatives can be used to characterize such properties known as memory and path dependence.

3. Differential description of fractal: It provides accurate differential equation models to describe various physical and mechanics behaviors of the medium which contains fractal structure.

4. Differential equation models for unconventional statistical and stochastic processes: The physical and mechanics models with fractional derivatives have a statistical mechanics background of Lévy stable distribution and fractional Brownian motion.

The background of applications mentioned above can be shown by the following figure (Fig. 5.1).

© Science Press 2022
W. Chen et al., *Fractional Derivative Modeling in Mechanics and Engineering*,
https://doi.org/10.1007/978-981-16-8802-7_5

**Fig. 5.1** Schematic diagram of the physical and mechanics background for fractional differential equations

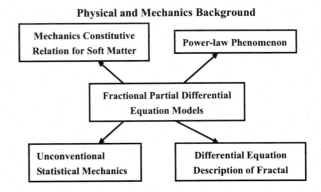

## 5.1   Power-Law Phenomena and Non-Gradient Constitutive Relation

### 5.1.1   Power-Law Distribution Phenomena and Fractal Dimension

The power-law distribution phenomenon which widely exists in nature and human society is first discovered by the economist Pareto. He put forward the famous 80/20 rule, that is, 20% of the population accounted for 80% of the social wealth [1]. And he believed that the probability of the income exceeding a given value $x$ follows a power function. That is known as power-law distribution and shown as the following equation [2]

$$P[X \geq x] = ax^{-b}. \tag{5.1.1}$$

Zipf, a linguist, found that the variation of the usage frequency of a word with the magnitude of the frequency also falls into a similar power-law relationship [2]. That is the famous Zipf's law. Since then, many scholars did a great deal of research on the power-law phenomena and discovered many power-law distribution phenomena in different areas. Christian et al. [3] also proposed the power-law relationship between the population of the European cities and the number of post offices, restaurants, gas stations and so on. A large number of studies have found that the power-law distribution is widespread in many areas, such as physics, biology, ecology and demography [1–3] (Fig. 5.2).

Different from the Gaussian distribution and the Poisson distribution, power-law distribution possesses a bell shape, but the tail does not quickly attenuate to zero. Power-law distribution features the so-called "long tail" phenomenon [1]. The random variables of the Gaussian distribution and the Poisson distribution are limited within a certain scale range, and if outside the range, the probability for the occurring of the random variables rapidly decays to zero. However, there is no such scale range

**Fig. 5.2** Christian et al. made a statistical study of the relationship between the populations of different Germany cities (the abscissa) and the power consumption in the corresponding cities (the ordinate), and the logarithmic values of these two quantities are given, from which one can see a line approximation [3]

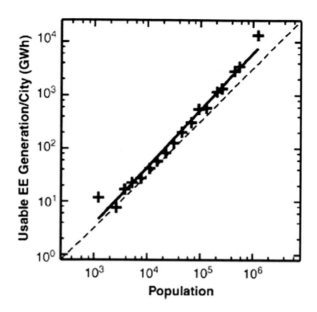

for the power-law distribution, resulting in great difference between the statistical individual. That is the so-called scale-free phenomenon with no characteristic scale [1].

Meanwhile, the scale-free characteristics of the power-law distribution phenomenon is intimatedly tied to the concept of fractal that Mandelbrot proposed, which essentially describes the phenomenon of power-law distribution [4, 5]. The so-called fractal refers to a system containing intensive mesostructures which have a similar structure or other characteristics with the system. This characteristic is also known as self-similarity [4]. Unlike that in mathematical theory which has strict geometrical self-similarity, fractal phenomena in nature are almost statistically self-similar. That is to say they have similar statistical characteristics. This self-similarity makes the system have no characteristic scale, which is the same as the scale-free property of the power-law distribution [1]. Meanwhile, Mandelbrot argued that the dimension corresponding to fractal should be a fractional dimension, rather than an integer-order dimension. In mathematics, it can be written as $y = ax^b$ ($b$ is a positive real number), which corresponds to power-law phenomenon itself [4]. Thereby, fractal is essentially a self-similar statistical phenomenon of power-law distribution.

### 5.1.2   The Mechanism of Power-Law Distribution Phenomenon

There are a large number of power-law phenomena both in nature and social sciences. Question may occur that why these phenomena obey a power-law distribution. So far, some scholars have proposed many different theories aiming at explaining the mechanism of power-law phenomena. This section will briefly introduce four related theories [1–10].

## 1.    Theory of growth and preferential attachment

With the deeper understanding of nature and human society, people gradually began to abandon the idealistic assumptions in some classical theories, and study system models which are more complex and more realistic. Many research fields including biology and physics are facing the need to study complex systems. Many scholars insist that complex system is equivalent to a complex network connecting various individuals, and the linear and nonlinear effects between these individuals are conducted through this complex network [6]. Barabasi and Albert [1, 7] proposed the theory of growth and preferential attachment for the power-law distribution phenomenon of scale-free complex network. They believed that power-law distribution is the consequence of the following two mechanisms: (1) Network growth, which refers to that networks expand continuously by the addition of new vertices, (2) the priority connection, which means that new vertices attach preferentially to sites that are already well connected. That's the same with surfing on the Internet. People will give priority to those websites which contain more links and a large amount of information. In accordance with this theory, the continuous growth of the network vertices make the well-connected vertices increase and the others decline relatively. However, the assumptions of the theory don't take into account that the additional vertices may not define or connect the well-connected vertices [1].

## 2.    Self-organized criticality theory

It has been widely accepted that self-organized criticality theory is the kinetic reason of the power-law distribution phenomenon, and the power-law distribution shows that the system is on the self-organized critical state [1, 8]. Self-organized criticality theory believes that self-organized critical state is the result of individual movements. Even small perturbations will make the system change dramatically [1]. The experiments of sandpile model [8, 9] found that the size that sand collapses and its frequency obeys the power-law relationship, revealing that self-organized critical system is on the edge of chaos. In the works "Das Chaos und seine Ordnung", Stenfan Greschik compared particles' movements to that of pedestrians on different occasions, and believed that power-law distribution if the sign to distinguish the self-organized state and the chaos state [4].

## 3.    HOT theory

Highly optimized tolerance (HOT) theory divides the influence factors of system into two types: high tolerance and high sensitivity, and presents that global optimization of complex systems for the environment can lead to the power-law distribution [1, 10].

## 4.  Maximum complexity principle [5]

The maximum complexity principle [5] concerns that any random system tends to be the most complicated one. Xuewen Zhang [5] mentioned in his works "The Constitution Theory" that it is most probable for a system to become most complicated, and many components in the system will raise the probability of being a complex system to a value close to 1. This implies that the system always tends to stay in the most complex state. He also believed that the system will obey different distributions if it meets different constraints during the process of ensuring the maximum degree of complexity. The following highlights the generation mechanism of the most complicated principle and the power-law distribution.

The most complicated principle is established on the basis of the principle of maximum entropy, and the so-called complexity is a new physical interpretation of entropy. Xuewen Zhang insists that the entropy essentially is the degree of complexity of the system, and makes use of the concept to unify thermodynamic entropy and information entropy. In his works, he takes the system composed of $N$ individuals as an example. All the random characteristic values of the system as well as the individuals are defined as a generalized collection, and its probability density function is $f(x)$. Then, the system's degree of complexity is $C$ [5]

$$C = - \int_a^b N f(x) \log(f(x)) dx. \tag{5.1.2}$$

According to Eq. (5.1.2), the most complicated principle refers to the theory that the probability density function $f(x)$ of the random variables makes the system possess the maximum degree of complexity $C$, thus it simplifies to solve a variational formulation of the functional $C$. Xuewen Zhang supposes that the constraints are that the geometric mean is a constant and the probability density is normalized [5]

$$\ln m = \int_a^b f(x) \ln x dx,$$

$$1 = \int_a^b f(x) dx. \tag{5.1.3}$$

Use the Lagrange multiplier method and substitute the constraints to obtain [5]

$$F = - \int f(x) \ln f(x) dx + C_1 \left[ \int f(x) dx - 1 \right] + C_2 \left[ \int f(x) \ln x dx - \ln m \right]. \tag{5.1.4}$$

The extreme value of the above equation is

$$f(x) = \exp(-1 + C_1)x^{C_2}.\tag{5.1.5}$$

Substituting the solution into the constraints, then the power-law distribution is obtained

$$f(x) = ax^b.\tag{5.1.6}$$

### 5.1.3   Power-Law Phenomena and Non-gradient Constitutive Relation

The so-called constitutive relation is a mathematical model that describes the macroscopic physical and mechanics properties of materials, and its specific mathematical expression is called the constitutive equation. To describe the response of the objects under the external force, we should consider the field equations such as mass conservation, momentum conservation and energy conservation, as well as the unique physical properties. Constitutive relations that are related to the material mechanics characteristics mainly describe the relationship between temperature, time, stress and strain, such as Hooke's law for ideal elastic solid, the Newtonian shear law for Newtonian viscous fluid.

1. **Power-law phenomena and complex viscoelastic constitutive**

Viscoelastic materials whose mechanics properties are between the ideal elastomer and Newtonian fluid have the properties of elasticity and viscosity at the same time, and their stress–strain response depends on the time, strain rate and the history of load and deformation. Therefore, stress and strain of the viscoelastic material possesses the memory property [11]. Classical viscoelastic mechanics model is usually composed by linear elastic elements (spring) and pure viscous elements (dashpot) in series or parallel. These models are simple and intuitive to describe the viscoelastic materials. Among them, the simplest model is Maxwell model with a spring and dashpot in series and Kelvin model with a spring and dashpot in parallel. In the process of relaxation and creep that mechanics model predicts, stress and strain are to meet the natural exponential relationship. However, a large number of experiments found that in most relaxation and creep processes for viscoelastic materials, the stress and strain often obeys a time power function [12]. Therefore, Ferry and Nutting made use of the power function to express the stress–strain response of the viscoelastic body. In order to combine power function and mechanics model together, Gemant recommends the introduction of fractional time derivative into the constitutive relationship of the viscoelastic material [12]. Fractional viscoelastic model with less material parameters can accurately describe the dynamic properties within a wide frequency range for a large number of complex viscoelastic materials, such as power-law frequency-dependent damping [13], while traditional integer-order derivative models need more derivative terms and material parameters.

## 2. Power-law phenomena and Non-Newtonian fluid constitutive relation

The feature of non-Newtonian fluid is that shear stress and shear strain rate does not satisfy the linear relationship. There are diverse non-Newtonian fluids in natural and engineering fields, such as oil, animal blood, mud and oil and the non-Newtonian fluid mechanics are widely used in areas such as chemical industry, plastic industry, oil industry, light industry [14, 15]. The following will describe the power-law constitutive relation of several non-Newtonian fluids.

### (1) Bingham fluid

Bingham fluid is also known as the ideal plastic fluid, i.e. like the plastic property of solid material, it will flow only when the stress exceeds a certain value [14, 15]. If Bingham fluid obeys Newtonian viscous shear law in the state of flow, its constitutive relation is as follows:

$$\sigma = \sigma_s + \eta \dot{\varepsilon}. \tag{5.1.7}$$

It's ordinary Bingham fluid [16]. If flow characteristics of Bingham fluid do not meet the Newtonian viscosity shear rate, it's called nonlinear Bingham fluid. A class of nonlinear Bingham fluid follows the power-law constitutive relation.

$$\sigma = \sigma_s + \eta \dot{\varepsilon}^n. \tag{5.1.8}$$

Such fluid is called Herschel–Bulkley fluid [14, 15].

### (2) Pseudo-plastic fluid

Pseudo-plastic fluid is a fluid which satisfies the Newtonian viscous shear law when the shear strain rate is small, and whose viscosity decreases when the shear strain rate is large [14]. It obeys power-law empirical formula [15]

$$\sigma = \eta \dot{\varepsilon}^n. \tag{5.1.9}$$

Here, $n$ is called flow index or non-Newtonian parameters, and smaller than 1. In order to describe pseudo-plastic fluid more accurately, some scholars have proposed the following power-law constitutive models [15]

$$\text{Carreau equation} \quad \sigma = \frac{a\dot{\varepsilon}}{(1 + b\dot{\varepsilon})^c}, \tag{5.1.10}$$

$$\text{Cross equation} \quad \sigma = \left[ \eta_\infty + \frac{\eta_0 - \eta_\infty}{1 + \eta \dot{\varepsilon}^m} \right] \dot{\varepsilon}. \tag{5.1.11}$$

### (3) Inflation fluid

Inflation fluid has a property contrast to pseudo-plastic fluid, i.e. it meets Newtonian viscous shear law if the shear strain is small, and it has an increasing viscosity and a slight expansion of volume when the shear strain rate is large [14]. There are only a few inflation fluids in nature, and such fluid obeys the same power-law empirical formula as the pseudo-plastic fluid [15]

$$\sigma = \eta \dot{\varepsilon}^n. \tag{5.1.12}$$

Flow index $n$ of inflation fluid is larger than 1.

### 3.  Power-law phenomena and the frequency-dependent acoustic attenuation

When propagating in a dissipative medium, sound waves often exhibit the characteristics of frequency-dependent energy dissipation [17, 18]

$$P(x + \Delta x) = P(x)e^{-\alpha(\omega)\Delta x}, \tag{5.1.13}$$

$$\alpha = \alpha_0 |\omega|^\eta. \tag{5.1.14}$$

Here, $\omega$ is angular frequency, $P$ is sound pressure, $\alpha_0$ and $\eta$ are dissipation coefficients, $x$ is the direction of propagation, $\Delta x$ is wave propagation distance. In order to describe such characteristics, Szabo proposed the Szabo acoustic wave equation [17]. However, Szabo acoustic wave equation contains a hyper-singular integration, which makes the equation difficult to solve numerically. To remedy this problem, Chen and Holm [18] presented the positive fractional derivative and developed the modified Szabo acoustic equation

$$\frac{1}{c_0^2}\frac{\partial^2 P}{\partial t^2} + \frac{2\alpha_0}{c_0} Q_y(P) = \Delta P, \tag{5.1.15}$$

here

$$Q_y(P) = \begin{cases} \partial P/\partial t & \eta = 0 \\ \frac{\partial^{|\eta|+1} P}{\partial t^{|\eta|+1}} & 0 < \eta < 2 \\ \partial^3 P/\partial t^3 & \eta = 2 \end{cases},$$

where $P$ is sound pressure, $\Delta$ is Laplace operator, $\alpha_0$ and $\eta$ are dissipation coefficients ($\eta > 0$) ($\eta$ is the order of fractional derivative as well as dissipation coefficient), $c_0$ is ultrasonic velocity, $\Gamma(\eta)$ is Gamma function, and the definition of the positive fractional derivative [18] is

$$\frac{d^{|\eta|} P}{dt^{|\eta|}} = \frac{1}{q(\eta)} \int_0^t \frac{P}{(t - \tau)^{\eta+1}} d\tau. \tag{5.1.16}$$

$q$ is a constant,

$$q(\eta) = \frac{\pi}{2\Gamma(\eta + 1)\cos[(\eta + 1)\pi/2]}.$$

Different from that of fractional derivative, the Fourier transform of the definite fractional derivative holds the property of positive definiteness. The modified Szabo acoustic wave equation meets the characteristics of the frequency dependence of acoustic attenuation in the frequency domain [18].

## 5.2 Fractional Langevin Equation

### 5.2.1 Langevin Equation

In 1827, Brown, a botanist, found that the pollen did an irregular movement in the water by using a microscope. Thus, the random motion of the particles in a liquid is called Brownian motion. In the nineteenth century, Del Sole proposed that the phenomenon of Brownian motion was caused under the condition of imbalance collision of liquid molecules. When it comes to the twentieth century, based on Del Sole's conjecture, many scientists, such as Albert Einstein and Langevin, proposed several theories for Brownian motion respectively and made them confirmed by the experiments [19]. Brownian motion theory is not only the foundation for the establishment of material atomic theory, but also the most representative fluctuation phenomena [19–21].

Classical Brownian motion theory considers that particles, also known as the Brownian particles, are affected by two different forces coming from the external field and the liquid molecules. The latter force $F$ is caused by the collision between the liquid molecules and the Brownian particles, and it's not completely sure to define. Then, it can be considered to include the definite force—the buoyancy generated by the liquid and resistance force $F_0$, and the uncertain fluctuation effect $F$-$F_0$ generated by the random motion of the molecules [19]

$$F \equiv F_0 + (F - F_0). \tag{5.2.1}$$

If we only take into account the effect of gravity and the force induced by the Brownian motion in the horizontal direction, the motion equation of the Brownian particle with unit mass can be defined as

$$du/dt = -\gamma u + \Gamma(t), \tag{5.2.2}$$

where $\gamma = a/m$ is the damping coefficient of unit mass, $u$ is the velocity in $x$ direction, the fluctuate force of unit mass $\Gamma(t) = X(t)/m$ is called the Langevin force [19].

As the effect on particles from the liquid is random, $\Gamma(t)$ is a random force and Eq. (5.2.2) is called stochastic differential equation. Therefore, the Langevin force $\Gamma(t)$ can only be described through the statistical methods.

The classical theory supposes that fluctuation behavior generally neutralizes on average, that's to say the statistical average value $< \Gamma(t) > = 0$. Brownian motion is also considered to be a Markov process with no aftereffect, which means $\Gamma(t)$ is independent at different time. In other words, $\Gamma(t)$ doesn't have time correlation and it's time correlation moment is $< \Gamma(t)\Gamma(t\prime) > = 2D\delta(t - t\prime)$, here $t$ and $t\prime$ represent different time, $\delta$ represents Dirac-Delta function [19, 21].

According to the statistical properties of the Langevin force, such as $< \Gamma(t) > = 0$ and $< \Gamma(t)\Gamma(t\prime) > = 2D\delta(t - t\prime)$, we can derive the power spectrum by applying Fourier transform to the time correlation moment of $\Gamma(t)$ [19, 21]

$$S(\omega) = \int_{-\infty}^{+\infty} \exp(-i\omega\tau)2D\delta(\tau)d\tau = 2D. \tag{5.2.3}$$

By using Fourier transform, original function can be transferred into the frequency domain and then obtain a frequency-related property for variable. According to the function of inverse Fourier transform, $\exp(i\omega\tau)$ is frequency component corresponding to the function, and the integration can be viewed as the accumulation of these frequency components. Spectrum function is the weight value that each frequency component contributes. The white spectrum is that the power spectrum $S(\omega)$ has no relationship with $\omega$ according to Eq. (5.2.3), that's to say weight function contributed by each frequency component is $2D$. The Langevin force whose time correlation function is $\delta$ function is called white noise [21].

Since it is difficult to measure irregular motion of Brownian particles in a liquid, the mean square displacement is usually taken as equation variables. If we assume that $x$ is the velocity, then Langevin equation can be written as [19]

$$x'' + \gamma x' = \Gamma(t). \tag{5.2.4}$$

By multiplying $x$ on both sides, the equation can be changed to

$$\frac{1}{2}(x^2)'' - (x')^2 = -\frac{\gamma}{2}(x^2)\prime + x\Gamma(t). \tag{5.2.5}$$

To get the statistical average value of the equation, the Langevin equation in the form of mean square displacement is as the following [1, 3]

$$\frac{1}{2}\frac{d^2}{dt^2} < x^2 > - < x\prime >^2 = -\frac{\gamma}{2}\frac{d}{dt} < x^2 > + < x\Gamma(t) > . \tag{5.2.6}$$

Average kinetic energy is $< E > = \frac{1}{2}m < x\prime(t)^2 > = \frac{1}{2}kT$, mean square speed is $< x\prime >^2 = kT/m$, by using the property of Langevin force, we can get the random force as $< x\Gamma(t) >=< x >< \Gamma(t) >= 0$. Then Eq. (5.2.6) can be changed into the following form

$$\frac{d^2}{dt^2} < x^2 > +\gamma \frac{d}{dt} < x^2 > -2kT/m = 0. \tag{5.2.7}$$

If we set $\gamma = \tau^{-1}$, where $\tau$ can be taken as the average impact time of particles, the mean square displacement of Brownian particles is [19]

$$< x^2 >= \frac{2kT\tau}{m} t + C_1 \exp(-t/\tau) + C_2. \tag{5.2.8}$$

If initial velocity and displacement are all equal to 0, and $t >> \tau$, the relationship between mean square displacement and time is

$$< x^2 >= 2Dt, D = \frac{kT}{\gamma m}. \tag{5.2.9}$$

That's the famous Einstein relation [19, 21].

The analysis above does not take the effect on Brownian particles from external force into account. If we do, the external force item should be added to the Langevin equation [19]

$$x'' + \gamma x' = f(x) + \Gamma(t). \tag{5.2.10}$$

Here $f(x)$ is the external force applied to the Brownian particles of unit mass. If the damping force plays a major role and the inertia force can be ignored, Langevin equation can be simplified as

$$\gamma x' = f(x) + \Gamma(t). \tag{5.2.11}$$

As to the equation above, if external force $f(x)$ is a nonlinear function, the equation is a nonlinear Langevin equation [22].

### 5.2.2  Fractional Langevin Equation

The issue discussed above is based on the hypothesis that Brownian motion is a Markov process, that is to say, at different time $\Gamma(t)$ is independent and do not have the property of time correlation and memory. But if Brownian particles move in fluid with dense viscous or internal degrees of freedom and in turbulence, the time correlation moment of random fluctuation force is a power function of $t$, rather than $\delta$ function [23]. That means Langevin force $\Gamma(t)$ has the property of time correlation, and Brownian motion depends on historic process. It's a non-Markov time-dependent process. Under this circumstance, the statistical average value of random fluctuation force equals 0, but its time correlated moment is a power function [23]

$$< \Gamma(t) >= 0,$$

$$< \Gamma(0)\Gamma(t) >= \Gamma_0(\alpha)t^{-\alpha}. \tag{5.2.12}$$

Damping force in Brownian motion also has a memory, represented by time convolution integral which maintains a memory core. Then, Langevin equation is rewritten as [23]

$$x''(t) + \int_0^t B(t - \tau)x'(\tau)d\tau = \Gamma(t). \tag{5.2.13}$$

This is the generalized Langevin equation.

In microscopic fractal structure of complex substances, the diffusion processes usually meet the relation $< x^2 >= 2Dt^{\alpha}$ instead of the classical Einstein relation $< x^2 >= 2Dt$. This diffusion process is commonly referred to anomalous diffusion. The random walk process is known as fractional Brownian motion, and it is a non-Markov process. The spatial distribution of the particles of the standard diffusion process corresponds to a Gaussian distribution, but when it comes to anomalous diffusion process, the spatial distribution corresponds to the non-Gaussian distribution [23]. Anomalous diffusion usually occurs in seepage of the porous media, capillary surface wave, two-dimensional rotating flow, etc. [24]. In order to describe the fractional Brownian motion in anomalous diffusion which has a memory property, time fractional differential operator is employed to describe memorial damping force in the movement of Brownian particles. Thus we obtain the fractional Langevin equation [24, 25]. The following will give a brief introduction to fractional Langevin equation according to Lutz and Burov's work [24, 25].

$$x'' + \gamma \frac{d^{\alpha-1}}{dt^{\alpha-1}}x' = \Gamma(t). \tag{5.2.14}$$

If $0 < \alpha < 1$, the above equation corresponds to the sub-diffusion, if $1 < \alpha < 2$, equation corresponds to the super-diffusion. By employing the application of Laplace transform method, the solution of fractional Langevin equation can be obtained as

$$x(t) = x(0) + x'(0)B(t) + \int_0^t B(t - \tau)\Gamma(\tau)d\tau. \tag{5.2.15}$$

Here, $x(0)$ and $x'(0)$ represents initial displacement and initial velocity, respectively, $B(t) = \int_0^t C_v(\tau)d\tau$ ($C_v(t)$ is the autocorrelation function of the speed, $C_v(t) =$

$\frac{<x'(t)x'(0)>}{<x'(0)x'(0)>}$). Applying Laplace transform to $C_v(t)$, we can obtain $C_v(s) = \frac{1}{s+\gamma s^{\alpha-1}}$, which can be represented by Mittag–Leffler function, i.e. $C_v(s) = E_{2-\alpha,1}(-\gamma t^{2-\alpha})$. When $t$ is small, the limitation of Mittag–Leffler function is $\exp(-at^{\alpha})/\Gamma(\alpha+1)$, otherwise it approaches $-(t\Gamma(\beta-\alpha))^{-1}$. Therefore, if $t$ is small, autocorrelation function exhibits the characteristics of the extended index

$$C_v(t) \exp \frac{-\gamma t^{2-\alpha}}{\Gamma(3-\alpha)}, t << 1/(\gamma)^{1/\alpha}. \tag{5.2.16}$$

Otherwise, in the long process, it shows a tailing phenomenon with inverse power law

$$C_v(t) \frac{t^{\alpha-2}}{\gamma\Gamma(\alpha-1)}, t >> 1/(\gamma)^{1/\alpha}. \tag{5.2.17}$$

Integrating the autocorrelation function $C_v(t)$ with respect to the speed, we can get $B(t) = tE_{2-\alpha,2}(-\gamma t^{2-\alpha})$. In the long process, memory kernel function is expressed as

$$B(t) \frac{t^{\alpha-1}}{\gamma\Gamma(\alpha)}(t \to \infty). \tag{5.2.18}$$

Taking the statistical average value of fractional Langevin equation's solution, the average displacement of the particles is as follows

$$\langle x \rangle = x(0) + x'(0)tE_{2-\alpha,2}(-\gamma t^{2-\alpha}) \frac{x'(0)t^{\alpha-1}}{\gamma\Gamma(\alpha)}(t \to \infty). \tag{5.2.19}$$

And the mean square displacement of the particles is

$$\langle x^2 \rangle = \frac{2kT}{M}t^2 E_{2-\alpha,3}(-\gamma t^{2-\alpha}) \frac{2kTt^{\alpha}}{\gamma M\Gamma(1+\alpha)}(t \to \infty). \tag{5.2.20}$$

Then, we derive the following relationship

$$\langle x^2 \rangle t^{\alpha}. \tag{5.2.21}$$

This is the generalized Einstein relation for fractional Langevin motion.

Giuseppe et al. [26] also obtained the analogous fractional Langevin equation by using $B(t) = \gamma\prime/[\Gamma(1/2)t^{1/2}]$ as memory kernel function to replace the original item in the generalized Langevin equation. In this equation, $\gamma\prime = 1/\tau^{3/2}$, and $\tau$ was the average impact time of particles. They further proposed that there are different governing equations corresponding to different time scales:

$$x'' + \gamma x' = \Gamma(t), t << \tau;$$

$$x'' + \gamma \frac{d^{1/2}x'}{dt^{1/2}} = \Gamma(t), t \gg \tau. \tag{5.2.22}$$

Mainardi and Tampieri [27] proposed fractional Langevin equation by using Basset history force to represent the long time damping effect of Brownian particles, and expressed damping force as

$$F = -A\gamma \frac{1}{\sqrt{\pi}} \int_a^t \frac{x''(\tau)}{\sqrt{t-\tau}} d\tau. \tag{5.2.23}$$

Using the definition of the Caputo fractional derivative, Basset force is expressed as follows (taking $a = 0$ in the above equation)

$$F = -A\gamma \frac{d^{1/2}u(t)}{dt^{1/2}}. \tag{5.2.24}$$

Substitute the damping force into the Langevin equation and get the fractional Langevin equation

$$x'' + \frac{1}{\gamma}\left(x' + \frac{d^{1/2}x'}{dt^{1/2}}\right) = \Gamma(t). \tag{5.2.25}$$

Kobelev et al. [28, 29] employed the fractional derivative operator and got the following fractional Langevin equation [30, 31]

$$\frac{d^{\alpha}x'}{dt^{\alpha}} + \gamma x' = \Gamma(t).$$

## 5.3  The Complex Damped Vibration

### 5.3.1  The Model for Oscillator

#### 1.  The vibration principle of harmonic oscillator

Harmonic oscillator is a dynamic system that the oscillator is subject to an elastic restoring force which is proportional to the displacement $x$ when it leaves the equilibrium position.

$$F = -kx, \tag{5.3.1}$$

**Fig. 5.3** Single degree of
freedom vibration

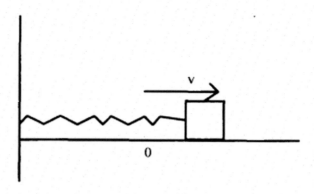

0

where k is the constant which represents the elastic coefficient. When the force in this system is only $F$, this system is called Simple Harmonic Oscillator [32]. The harmonic vibration is shown in Fig. 5.3.

Through Newton's second law, that the acceleration is proportional to the second derivative of displacement $x$, the equilibrium of force is expressed as

$$m\ddot{x} = -kx. \tag{5.3.2}$$

Define $\omega_0^2 = k/m$, and Eq. (5.3.2) can be rewritten as

$$\ddot{x} + \omega_0^2 x = 0. \tag{5.3.3}$$

The general solution of Eq. (5.3.3)

$$x = A\cos(\omega_0 t + \phi), \tag{5.3.4}$$

where the amplitude $A$ and the phase $\phi$ are decided by the initial displacement and the initial velocity. The frequency of the vibration

$$f = \frac{\omega_0}{2\pi}. \tag{5.3.5}$$

The kinetic energy of the vibration

$$T = \frac{1}{2}m(\dot{x})^2 = \frac{1}{2}kA^2\sin^2(\omega_0 t + \phi). \tag{5.3.6}$$

The potential energy of the vibration

$$U = \frac{1}{2}kx^2 = \frac{1}{2}kA^2\cos^2(\omega_0 t + \phi). \tag{5.3.7}$$

The total energy of the vibration system is stated as

$$E = T + U = \frac{1}{2}kA^2, \tag{5.3.8}$$

which means the total energy of the oscillator is constant when it is harmonic vibration [32].

## 2.  Damping vibration

The harmonic oscillator discussed in last section is the ideal oscillator which ignores the friction and damping. Actually, in the mechanical system, the energy attenuation is always existed [32]. When the damping is proportional to the velocity, the dynamic equation is expressed as

$$-kx - c\dot{x} = m\ddot{x}, \tag{5.3.9}$$

or

$$m\ddot{x} + c\dot{x} + kx = 0. \tag{5.3.10}$$

This equation is Constant coefficient homogeneous differential equation, whose characteristic equation is

$$mq^2 + cq + k = 0, \tag{5.3.11}$$

and the characteristic roots are

$$q_{1,2} = \frac{-c \pm \sqrt{c^2 - 4mk}}{2m}. \tag{5.3.12}$$

Three cases of the solution are as follows:

(1)   $c^2 > 4mk$, (which is called Over Damping),
(2)   $c^2 = 4mk$, (which is called Critical Damping),
(3)   $c^2 < 4mk$, (which is called Under Damping).

For the case of Over Damping, assume the two roots are $\gamma_1$, $\gamma_2$, the general solution of Eq. (5.3.10) is

$$x = A_1 e^{-\gamma_1 t} + A_2 e^{-\gamma_2 t}. \tag{5.3.13}$$

We can find that the attenuation is the form of exponent function which attenuates rapidly. In this case, it is no long vibration.

For the case of Critical Damping, Eq. (5.3.11) only has one root $-\gamma$, where $\gamma = \frac{c}{2m}$. The general solution of dynamic equation is

$$x = e^{-\gamma t}(A_1 t + A_2). \tag{5.3.14}$$

This solution illustrates that there is also no vibration in this case, and the displacement will approach to 0 with time.

For the case of Under Damping, the two roots of Eq. (5.3.11) are conjugate complex, i.e., $-\gamma \pm i\omega_1$, where $\gamma = \frac{c}{2m}$, $\omega_1 = \sqrt{\frac{k}{m} - \frac{c^2}{4m^2}}$, and the solution is

$$x = Ae^{-\gamma t}\cos(\omega_1 t + \phi), \tag{5.3.15}$$

where $\gamma = c/2\,m$. Equation (5.3.15) illustrates that the dynamical system is periodic vibration, but the amplitude $Ae^{-\gamma t}$ decays in the exponential behavior.

The damping vibration can be maintained in several periods. The mechanical energy of the total system is

$$E = \frac{1}{2}m\dot{x}^2 + \frac{1}{2}kx^2. \tag{5.3.16}$$

The rate of change of the total energy over the time is

$$\frac{dE}{dt} = -c\dot{x}^2. \tag{5.3.17}$$

Equation (5.3.17) is always negative, which denotes the dissipated mechanical energy of system in unit time. The dissipation is related to the damping coefficient and the motion velocity. Therefore, if we want to keep the vibration in practical problems, the external force should be acted on the system continuously.

### 5.3.2 Fractional Oscillator

1. **Harmonic oscillator**

In reality, many factors affect the harmonic oscillator and make it damping vibrate. The factors are divided into two classes, internal friction and external friction. The dynamic behavior of oscillator with internal friction can be described by fractional calculus, which is already used by many mathematicians and physicists [33–35]. In the following, we will introduce the existing research result of the fractional oscillator.

Considering the simple fractional harmonic vibration equation [34]

$$\begin{cases} m\frac{d^\alpha x}{dt^\alpha} + kx = 1, & 1 < \alpha \leq 2, \\ x(0) = c_0, \dot{x}(0) = c_1, \end{cases} \tag{5.3.18}$$

where $\frac{d^\alpha x}{dt^\alpha}$ is Caputo fractional derivative. Here, $M$, $T$ and $L$, respectively, denote the dimensions of mass, time and length. In Eq. (5.3.18), the dimension of $\frac{d^\alpha x}{dt^\alpha}$ is $\frac{L}{T^\alpha}$ and the dimension of elastic coefficient $k$ is $\frac{M}{T^2}$; therefore, the dimension of coefficient $m$ in the first item is deduced as $[m] = MT^{\alpha-2}$ by the equilibrium $[m]\frac{L}{T^\alpha} = \frac{M}{T^2}L$.

Applying Laplace transform for Eq. (5.3.18), we have

$$L({}_0D_t^\alpha x(t)) = s^\alpha X(s) - s^{\alpha-1}x(0) - s^{\alpha-2}x'(0), \tag{5.3.19}$$

and then

$$X(s) = \frac{1}{m}\frac{c_0 s^{\alpha-1} + c_1 s^{\alpha-2}}{s^\alpha + k/m}. \tag{5.3.20}$$

By the Laplace transform of Mittag–Leffler function,

$$\int_0^\infty e^{-st}t^{\alpha k+\beta-1}E_{\alpha,\beta}^{(k)}(\pm at^\alpha)\mathrm{d}t = \frac{k!s^{\alpha-\beta}}{(s^\alpha \mp a)^{k+1}}, \tag{5.3.21}$$

the analytical solution of Eq. (5.3.18) is obtained as

$$x(t) = \frac{1}{m}\left[c_0 E_{\alpha,1}\left(-\frac{k}{m}t^\alpha\right) + c_1 E_{\alpha,2}\left(-\frac{k}{m}t^\alpha\right)\right]. \tag{5.3.22}$$

Simply, let $k/m = 1, c_0 = 1, c_1 = 0$, and the numerical solution is shown in Fig. 5.4 [36–40].

In Fig. 5.4, we find that the damping vibration occurs when $\alpha \in (1, 2)$, and the attenuation is larger with the decrease of $\alpha$. And when $\alpha = 2$, the vibration is undamped simple harmonic vibration.

**Fig. 5.4** The vibration of single degree freedom oscillator with fractional derivative damping

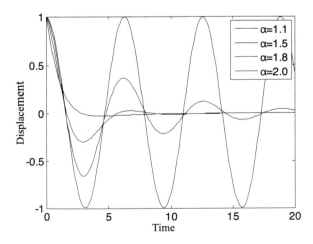

2.  **Force vibration equation**

The force vibration equation with integer-order derivative

$$mu''(t) + cu'(t) + ku(t) = f(t). \tag{5.3.23}$$

This equation with $cu'(t)$, which is corresponding to the ideal Rayleigh proportion damping, is suitable for the pure viscous material, like water. However, it cannot be applied to describe the complex behavior of viscoelastic damping material. In order to apply the integer-order derivative model to simulate the viscoelastic damping, some empirical parameters are added to construct the nonlinear differential equation. But in this nonlinear model, the parameters without exact physical meaning are difficult to obtain in application. After applying the fractional differential operator, the damping vibration of viscoelastic media can be described by simple fractional differential equation. The fractional Bagley–Torvik equation is [38]

$$my''(t) + c_0 D_t^\alpha y(t) + ky(t) = f(t), \ t > 0, \tag{5.3.24}$$

where $m$ denotes mass, $my''(t)$ denotes the inertial force, $c_0 D_t^\alpha y(t)$ denotes the damping force which is dissipation item, $k$ denotes the stiffness, $ky(t)$ denote the elastic force, $f(t)$ is the external force. Let

$$m = 1, c = 0.5, k = 0.5, f(t) = \begin{cases} 8, (0 \le t \le 1) \\ 0, (t > 1) \end{cases}, \alpha = 1.5, y(0) = 0, y'(0) = 0.$$

The numerical solution of fractional Bagley–Torvik equation is shown in Fig. 5.5, where we can find the movement situation of fractional force vibration [38].

**Fig. 5.5** Single freedom degree damping vibration with fractional Bagley–Torvik equation

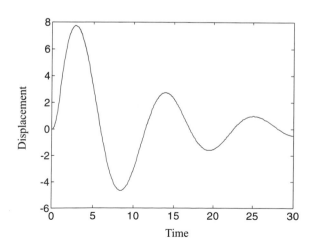

**Fig. 5.6** Single freedom
degree damping vibration
with fractional
Bagley–Torvik equation for
different orders

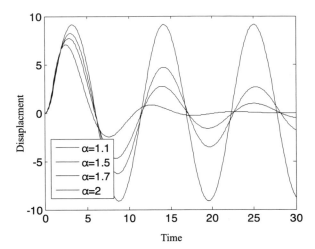

Four cases of vibration with order $\alpha = 1.1, 1.5, 1.7$ and $2.0$, respectively, are shown
in Fig. 5.6. The relationship between the movement of oscillator and the fractional
order illustrates that the displacement changes to larger with the increase of order $\alpha$.
With a smaller order $\alpha$, the memory is stronger. Figure 5.7 gives the solutions of the
integer order and fractional order damping vibration equations.

### 3.   The fractional model for viscoelastic damping

The vibration of the high-speed devices, such as high-speed CD-ROM drive, will
affect their normal work, heat the devices and reduce the service life. To keep the
normal work and avoid the adverse effect of vibration, the vibration isolation and
damping are needed for these high-speed devices. Linchao Liu et al. [41] mentioned
that the viscoelastic material has the dissipation and damping characteristics, and

**Fig. 5.7** The damping
vibrations of integer and
fractional order equations

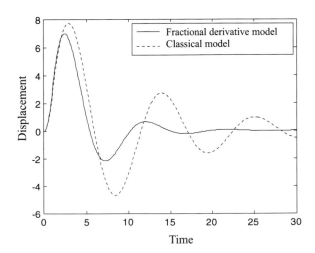

it's able to get very good damping effect to the high-speed CD-ROM drive frame structure by viscoelastic dampers. The mechanics characteristics of the viscoelastic damper can be described by Fractional Kelvin solid model [41],

$$\begin{cases} S_{ij}(t) = 2G\left(1 + \tau^r \frac{\partial^r}{\partial t^r}\right)e_{ij}(t), \ 0 < r < 1, \\ \sigma_{11}(t) = 3K\varepsilon_{11}(t), \end{cases} \tag{5.3.25}$$

where $S_{ij}$ is the stress deviator, $e_{ij}$ the strain deviator, $\sigma_{11}$ the stress tensor, $\varepsilon_{11}$ the strain tensor. $G, K, r, \tau$ are material parameters, $\frac{d^r}{dt^r}$ is Riemann–Liouville fractional differential operator. The corresponding relaxation function is

$$\begin{cases} G(t) = G\left[1 + \frac{\tau^r}{\Gamma(1-r)}t^r\right), \\ K(t) = K. \end{cases} \tag{5.3.26}$$

The finite element form of the fractional Kelvin viscoelastic damping force $P_{ve}(t)$ is stated as [41]

$$P_{ve}(t) = M_{ve}\ddot{a}_{ve}(t) + C_{ve}\frac{\partial^r}{\partial t^r}a_{ve}(t) + K_{ve}a_{ve}(t). \tag{5.3.27}$$

Thus, the finite element equation for the fractional viscoelatic damping dynamics of high-speed CD-ROM drive frame structure is

$$M_e\ddot{a}(t) + K_e a(t) = P_e(t) - P_{ve}(t), \tag{5.3.28}$$

$$(M_e + M_{ve})\ddot{a}(t) + CD^r a(t) + Ka(t) = P(t). \tag{5.3.29}$$

## 5.4 Viscoelastic and Rheological Models

It is usually supposed that when solid materials are subjected to external forces and deformation, they always reserve the elasticity before the stress of material reach or exceed the yield limit. Namely, in this case, the deformation of solid material under external force is restorable. And when the solid material is at the linear elasticity stage, its stress and strain meets Hooke's law of elasticity which means the stress is directly proportional to strain. On the other hand, the fluid can only sustain the normal stress but has no resistance to shear stress. More precisely, the fluid just performs a specific flow, rather than a limit deformation under the implementation of the shear force.

As the elastomer produces only recoverable deformation, when the external load disappears, the material restores the original shape. Therefore, elastic deformation only stores the energy in the form of elastic potential energy without any consuming energy.

Viscous fluid is generally considered as the Newtonian fluid which meets the Newtonian viscous shear rate, namely, shear stress is proportional to the shear strain rate. The so-called viscosity refers to the frictional resistance of the relative motion of fluid particles. Thus, the viscous fluid only dissipates energy, rather than storing energy.

However, in practical engineering fields, plastics, rubber, paint, resin, concrete, rock, soil, oil, bones and blood and other materials involved in the chemical, petroleum, biology, medicine, civil engineering and other fields, often at the same time show elasticity and viscosity both, usually called viscoelasticity [11]. Materials which exhibit viscoelastic properties to a certain extent are called viscoelastic materials, and most viscoelastic materials are polymers [11]. Because of manifesting the nature mechanics properties of both elastic and viscous material, the viscoelastic material under a certain stress state displays both a finite deformation and a finite or infinite flow, mainly in a certain degree of relaxation, creep and hysteresis phenomenon. Relaxation is the phenomenon that stress decreases gradually when materials are subjected to a constant strain condition; creep refers to the phenomenon that the deformation of material keeps continuous increase under the role of constant stress. Relaxation and creep phenomena are rheological phenomena and the viscoelasticity theory is to study the rheological properties of viscoelastic material [42]. For elastic solids, the stress and strain have a linear relationship; moreover, for viscous bodies, stress and strain rate has a linear relationship. The stress and strain responses of the viscoelastic bodies depend on time and strain rate, relate to the loading and deformation history, which leads to that stress and strain variation of the viscoelastic material has memory [42].

The viscoelastic material is divided into viscoelastic solids (such as concrete, biological tissues and organs, polymers) and viscoelastic fluids (pharmaceutical preparation, blood, tissue fluid, mud, oil, etc.). The so-called viscoelastic solid is the matter, response of which tends to be limited deformation when an external force is loaded; but the response of viscoelastic fluid under an external force tends to steady flow. The traditional modeling methods of viscoelastic materials are component models, relaxation modulus and creep compliance, as well as complex relaxation modulus and complex creep compliance which are used to describe the dynamic characteristics of viscoelastic materials [11, 42].

## 5.4.1  Component Model

Component model is the most simple and intuitive viscoelastic material constitutive model. The so-called component model considers that the viscoelastic material has linear viscoelastic characteristics, and can use superimposed linear elastic and ideal viscous to describe the viscoelastic behavior of the material [42], usually make combination of the elastic components and viscous components in series or parallel

**Fig. 5.8**  Spring element

$$E$$

**Fig. 5.9**  Viscous component

$$\eta$$

to have the model show elasticity and viscosity at the same time. So it exhibits elastic properties and viscous nature at the same time, similar to the combination of electrical components in the electrics. The elastic component here is represented by the spring element, i.e. (Fig. 5.8).

$$\sigma = E\varepsilon, \tag{5.4.1}$$

where $E$ is Young's modulus which is equal to spring coefficient. Viscous component is represented by the Newtonian dashpot model, namely, the stress and strain rate have linear relation (Fig. 5.9).

$$\sigma = \eta\dot{\varepsilon}. \tag{5.4.2}$$

Here are the two most basic components of the viscoelastic material model [11, 42]: Maxwell model and Kelvin–Voigt model.

1.   **Maxwell Model:**

Maxwell model is composed by an elastic component and a viscous component in series, and is used to describe the stress relaxation of viscoelastic fluid, as shown in Fig. 5.10.

Because the Maxwell model is obtained by components in series, its total stress is equal to the stress of each component and its total strain is equal to the summation of strain of elastic component and viscous component

$$\sigma = \sigma_e = \sigma_v,$$
$$\varepsilon = \varepsilon_e + \varepsilon_v, \tag{5.4.3}$$

where the subscript $e$ means the elastic component, the subscript $v$ denotes the viscous component. Hence, the constitutive equation of the Maxwell model can be stated as follows:

**Fig. 5.10**  Maxwell Model

$$E \qquad \eta$$

$$\dot{\sigma} + \frac{1}{\tau}\sigma = E\dot{\varepsilon}, \quad \tau = \frac{\eta}{E}, \tag{5.4.4}$$

where $\tau$ is the relaxation time. Under the constant strain $\varepsilon_0$, the constitutive equation of Maxwell model is

$$\dot{\sigma} + \frac{1}{\tau}\sigma = 0. \tag{5.4.5}$$

Then, the stress relaxation can be expressed as

$$\sigma(t) = \sigma_0 e^{-t/\tau} = E\varepsilon_0 e^{-t/\tau} = G(t)\varepsilon_0, \tag{5.4.6}$$

where $\sigma_0$ is the initial stress, $G(t)$ is the relaxation modulus. Thus, the relaxation modulus of Maxwell model is

$$G(t) = E e^{-t/\tau}. \tag{5.4.7}$$

When $t$ is small, the response of Maxwell model approaches to the elastic body; moreover, when $t$ is much larger than $\tau$, the response of Maxwell model approaches to the Newtonian fluid and the Maxwell model tends to the steady viscous flow. In fact, $\tau$ represents that when time $t$ is far more than the relaxation time, the viscoelastic material tends to the Newtonian fluid. Meanwhile, if the rate of stress decay keeps the constant which is equal to the initial rate, the stress decays to zero at time $\tau$. Therefore, $\tau$ is called relaxation time [11, 42]. Figure 5.11 is the stress relaxation curve of the Maxwell model.

To study the creep behavior of material described by the Maxwell model, the stress should be assumed as constant $\sigma_0$, and the corresponding constitutive equation is stated as follows:

$$\dot{\varepsilon}(t) = \frac{\sigma_0}{\eta}, \tag{5.4.8}$$

and the creep behavior can be expressed as

**Fig. 5.11**  Stress relaxation
curve of the Maxwell model

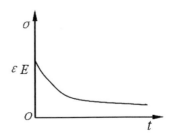

**Fig. 5.12** Creep curve of the Maxwell model

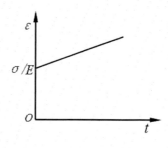

**Fig. 5.13** Kelvin–Voigt model

$$\varepsilon(t) = \frac{\sigma_0}{\eta}t + \frac{\sigma_0}{E}. \tag{5.4.9}$$

Figure 5.12 is the creep curve of the Maxwell model.

From Fig. 5.12 and the constitutive equation, the creep of Maxwell model manifests a straight line. This illustrates that Maxwell model produces instantaneous elasticity and steady flow. But almost no creep curve is in this linear form, which make few scholars use this model to reflect the creep phenomenon. Hence, Maxwell model is called relaxation model [11, 42], and mostly is applied to describe the relaxation behaviors.

2.  **Kelvin–Voigt Model:**

Kelvin–Voigt model is composed by an elastic component and a viscous component in parallel, and are mainly used to describe the creep behavior of viscoelastic solid, as shown in Fig. 5.13.

Because the Kelvin–Voigt model is combined by components in parallel, its total strain is equal to the strain of each component and its total stress is equal to the summation of the strain of the elastic component and viscous component:

$$\varepsilon = \varepsilon_e = \varepsilon_v,$$
$$\sigma = \sigma_e + \sigma_v, \tag{5.4.10}$$

where the subscript $e$ means the elastic component, the subscript $v$ denotes the viscous component. Hence, the constitutive equation of the Kelvin–Voigt model can be stated as follows:

$$\dot{\varepsilon} + \frac{1}{\tau}\varepsilon = \frac{\sigma}{\eta}, \tau = \frac{\eta}{E}, \tag{5.4.11}$$

**Fig. 5.14** Creep curve of the
Kelvin–Voigt model

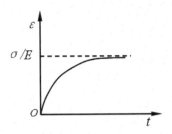

where $\tau$ is the retardation time. Under the constant strain $\sigma_0$, the constitutive equation
of Kelvin–Voigt model is

$$\dot{\varepsilon} + \frac{1}{\tau}\varepsilon = \frac{\sigma_0}{\eta}.  \tag{5.4.12}$$

Then, the stress relaxation can be expressed as

$$\varepsilon(t) = [1 - e^{-t/\tau} H(t)]\frac{\sigma_0}{E} = J(t)\sigma_0,  \tag{5.4.13}$$

where $J(t)$ is the creep compliance, $H(t)$ means the Heaviside function. The creep
compliance of Kelvin–Voigt model is

$$J(t) = [1 - e^{-t/\tau} H(t)]\frac{1}{E}.  \tag{5.4.14}$$

As the instant of time $t$ is much larger than $\tau$, the response of Kelvin–Voigt model
tends to an elastic body. $\tau$ represents that when time $t$ is far more than the retardation
time, the viscoelastic material tends to the elastic body. Meanwhile, if the strain rate
keeps the constant which is equal to the initial rate, the strain reaches to the elastic
strain at time $\tau$. Therefore, $\tau$ is called retardation time [11, 42]. Figure 5.14 is the
creep curve of the Kelvin–Voigt model.

To study the stress relaxation behavior of material described by the Kelvin–Voigt
model, the strain should be assumed as constant $\varepsilon_0$, and the corresponding constitutive
equation is stated as follows:

$$\sigma(t) = E\varepsilon_0.  \tag{5.4.15}$$

Based on the constitutive relationship, the Kelvin–Voigt model reduces to the
linear elastic model, therefore this model cannot describe the stress relaxation
behaviors.

The above-mentioned two viscoelastic component models are the most simple
ones and one-dimensional models. Actually, the viscoelastic materials are more
complex. Hence, a few Maxwell models and Kelvin–Voigt models are combined
in series or in parallel to describe the actual viscoelastic material.

Component models describe the viscoelastic behavior by simply making combination of elastic components and viscous components in series or in parallel, and are only finite discrete models. But many actual viscoelastic materials are very complex and are equivalent to an infinite number of combinations of elastic components and viscous components. In the next subsection, the continuous viscoelastic material model will be introduced.

### 5.4.2  Relaxation Modulus and Creep Compliance

The viscoelastic constitutive models mentioned in this subsection are not obtained by combining the elastic components with viscous components, but by considering how to describe the memory of the viscoelastic materials. For viscoelastic materials, convolution integral is commonly used to describe their memory. For example, the creep phenomenon of strain can be expressed by [11, 42]

$$\varepsilon(t) = \sigma_0 J(t) + \int_0^t J(t - \tau) \frac{d\sigma(\tau)}{d\tau} d\tau, \tag{5.4.16}$$

where $J$ is the creep compliance. With integration by parts, Eq. (5.4.16) can be written as

$$\varepsilon(t) = J(0)\sigma(t) - \int_0^t \sigma(\tau) \frac{dJ(t - \tau)}{d\tau} d\tau. \tag{5.4.17}$$

Relatively, for stress relaxation

$$\sigma(t) = \varepsilon_0 G(t) + \int_0^t G(t - \tau) \frac{d\varepsilon(\tau)}{d\tau} d\tau, \tag{5.4.18}$$

where $G$ is the relaxation modulus. Applying the same method, Eq. (5.4.18) can be written as

$$\sigma(t) = \varepsilon(t)G(0) - \int_0^t \varepsilon(\tau) \frac{dG(t - \tau)}{d\tau} d\tau. \tag{5.4.19}$$

Here, the convolution integral is called hereditary integral.

### 5.4.3   Complex Relaxation Modulus and Complex Creep Compliance

Complex relaxation modulus and complex creep compliance are usually used to study the dynamic response of viscoelastic materials [11, 42]. And many materials show viscoelasticity when dynamic force is loaded. Considering a general stress–strain relation [42]

$$P(\sigma(t)) = Q(\varepsilon(t)). \tag{5.4.20}$$

Using Taylor series expansion of functions P and Q, it is easy to obtain

$$\sum_{r=0}^{n} p_r \frac{d^r \sigma(t)}{dt^r} = \sum_{r=0}^{n} q_r \frac{d^r \varepsilon(t)}{dt^r}, \tag{5.4.21}$$

where $p_r$ and $q_r$ are material parameters. For elastic bodies, according to Hooke's law, stress is in direct proportion to strain, thus only considering the linear term. But, for complex materials, more terms are considered, the constitutive equation is much closer to the real condition. The Maxwell model and the Kelvin–Voigt model introduced in subsection 5.4.1 can also be unified with Eq. (5.4.21). When $p_0 = 1/\tau$, $p_1 = 1$, $q_0 = 0$, $q_1 = E$, Eq. (5.4.21) is the Maxwell model; when $p_0 = 1/\eta$, $p_1 = 0$, $q_0 = 1/\tau$, $q_1 = 1$, the Kelvin–Voigt can be obtained. By Fourier transform, the stress–strain relation can be transformed to frequency domain, shown as following

$$\sum_{r=0}^{n} p_r (i\omega)^r \overline{\sigma} = \sum_{r=0}^{n} q_r (i\omega)^r \overline{\varepsilon}. \tag{5.4.22}$$

Let $P(\omega) = \sum_{r=0}^{\infty} p_r (i\omega)^r$, $Q(\omega) = \sum_{r=0}^{\infty} q_r (i\omega)^r$, then the complex relaxation modulus is $\overline{G}(\omega) = Q(\omega)/P(\omega)$, the complex creep compliance is $\overline{J}(\omega) = P(\omega)/Q(\omega)$. $\overline{G}(\omega)$ can be written as $\overline{G}_1(\omega) + i\overline{G}_2(\omega)$, where $\overline{G}_1(\omega)$ is the storage modulus which means energy storage due to the elasticity, $\overline{G}_2(\omega)$ is the loss modulus which denotes the energy dissipation due to viscosity; $\overline{J}(\omega)$ can also be written as $\overline{J}_1(\omega) + i\overline{J}_2(\omega)$, $\overline{J}_1(\omega)$ is the storage compliance which means energy storage due to the elasticity, $\overline{J}_2(\omega)$ is the loss compliance which denotes the energy dissipation due to viscosity [11, 42].

Therefore, the relation between the complex relaxation modulus, the complex creep compliance and the integral model can be got [11, 42]. The creep integral can be rewritten as [11, 42]

$$\sigma(t) = G(\infty) \int_{0}^{t} \frac{d\varepsilon(\tau)}{d\tau} d\tau + \int_{0}^{t} [G(t-\tau) - G(\infty)] \frac{d\varepsilon(\tau)}{d\tau} d\tau, \tag{5.4.23}$$

where $G(\infty)$ is the elastic modulus of delay which denotes the value of $G(t)$ when time $t$ tends to infinity. For viscoelastic fluid, $G(\infty)$ is equal to zero; for viscoelastic solid, $G(\infty)$ is equal to zero; for viscoelastic solid, $G(\infty)$ is a constant. By Fourier transform of Eq. (5.4.23), the complex modulus can be obtained

$$\overline{G}(\omega) = \frac{\overline{\sigma}}{\overline{\varepsilon}} = G(\infty) + i\omega \int_0^\infty [G(t) - G(\infty)] e^{-i\omega t} \, dt. \qquad (5.4.24)$$

Respectively, the storage modulus and the loss modulus are

$$\overline{G}_1(\omega) = G(\infty) + \omega \int_0^\infty [G(t) - G(\infty)] \sin \omega t \, dt, \overline{G}_2(\omega)$$

$$= G(\infty) + i\omega \int_0^\infty [G(t) - G(\infty)] \cos \omega t \, dt.$$

Relatively, the complex creep compliance is $\overline{J}(\omega) = \frac{\overline{\varepsilon}}{\overline{\sigma}} = 1 / \left[ G(\infty) + i\omega \int_0^\infty [G(t) - G(\infty)] e^{-i\omega t} \, dt \right]$.

### 5.4.4 Complex Relaxation Modulus and Complex Creep Compliance of Fractional Derivative Viscoelastic Models

The initial relaxation and creep curve of traditional integer-order derivative viscoelastic models cannot agree well with the experimental data. Therefore, fractional derivative viscoelastic models attract more and more attentions. The viscoelastic models based on the standard integral-order derivative cannot describe the viscoelastic behaviors well, such as the stress relaxation of viscosity, super slow relaxation of high molecular polymer and so on. These phenomena of viscoelastic materials show history dependency and memory. Therefore, the fractional derivative which has memory is applied to model the viscoelastic constitutive relation. As shown in Eq. (5.4.21), the general form of the traditional constitutive relation is stated as [11, 42]

$$\sum_{r=0}^n p_r \frac{d^r \sigma(t)}{dt^r} = \sum_{r=0}^n q_r \frac{d^r \varepsilon(t)}{dt^r}.$$

For traditional viscoelastic theory, $r$ is an integer and the above equation contains only integer-order derivative. The general fractional derivative stress–strain relation can be got by introducing the fractional derivative operator, where the derivative order of the left of equation is replaced with $\alpha_r$, the derivative order of the right of equation is replaced with $\beta_r$, $\alpha_r$ and $\beta_r$ are arbitrary real number.

$$\sum_{r=0}^{n} p_r \frac{d^{\alpha_r} \sigma(t)}{dt^{\alpha_r}} = \sum_{r=0}^{n} q_r \frac{d^{\beta_r} \varepsilon(t)}{dt^{\beta_r}}. \tag{5.4.25}$$

Using Fourier transform on Eq. (5.4.25)

$$\sum_{r=0}^{n} p_r (i\omega)^{\alpha_r} \overline{\sigma}(\omega) = \sum_{r=0}^{n} q_r (i\omega)^{\beta_r} \overline{\varepsilon}(\omega). \tag{5.4.26}$$

The complex relaxation modulus is

$$\overline{G}(\omega) = \frac{\sum\limits_{r=0}^{n} q_r (i\omega)^{\beta_r}}{\sum\limits_{r=0}^{n} p_r (i\omega)^{\alpha_r}}, \tag{5.4.27}$$

the complex creep compliance is

$$\overline{J}(\omega) = \frac{\sum\limits_{r=0}^{n} p_r (i\omega)^{\alpha_r}}{\sum\limits_{r=0}^{n} q_r (i\omega)^{\beta_r}} \tag{5.4.28}$$

Same with the integer-order derivative models, $\overline{G}(\omega) = \overline{G}_1(\omega) + i\overline{G}_2(\omega)$, $\overline{J}(\omega) = \overline{J}_1(\omega) + i\overline{J}_2(\omega)$, real part and imaginary part represent storage modulus/compliance and loss modulus/compliance, respectively [43].

## 5.4.5  Fractional Derivative Viscoelastic Models

The fractional derivative viscoelastic models can be obtained with the same manner as above-mentioned. With special parameters $p_r$ and $q_r$, the corresponding mechanics models can be obtained. Hooke's law of elastic solid is $\sigma(t) \, d^0 \varepsilon(t)/dt^0$ and the constitutive relation of Newtonian viscous fluid is $\sigma(t) \, d^1 \varepsilon(t)/dt^1$. Hence, some scholars argued that since the mechanics properties of the viscoelastic body was between the elastic solids and viscous fluids, the viscoelastic constitutive relation should be $\sigma(t) \sim d^\beta \varepsilon(t)/dt^\beta \, (0 < \beta < 1)$. Viscosity and elasticity are two limit

**Fig. 5.15** Abel dashpot

state of viscoelasticity. So far, The most common method is to replace the original traditional integer-order derivative Newton dashpot by the fractional derivative viscoelastic Abel dashpot model $\sigma = \eta d^{\beta} \varepsilon(t)/dt^{\beta}$ [44] (Fig. 5.15).

When $\beta = 1$, the Abel dashpot reduces to the Newtonian dashpot and means the Newtonian viscous fluid; when $\beta = 0$, the Abel dashpot is a spring component and denotes elastic body. When the stress is constant, namely $\sigma = \sigma_0 H(t)$, the Abel dashpot can describe the creep behavior. Based on the Riemann–Liouville fractional derivative definition, fractional integral on both sides of $\sigma(t) \sim d^{\beta} \varepsilon(t)/dx^{\beta} (0 < \beta < 1)$ is

$$\varepsilon(t) = \frac{\sigma_0}{\eta} \frac{t^{\beta}}{\Gamma(1 + \beta)}, \quad (0 \le \alpha \le 1). \tag{5.4.29}$$

Taking different $\beta$, Eq. (5.4.29) will plot a series of curves, as shown in 5.16. It is clear to see from Fig. 5.16, under constant stress, the strain of viscoelastic materials increases slowly, does not like the linear increase of Newtonian fluid, and does not keep constant strain as linear elastic body. Conclusively, the Abel dashpot is able to reflect the nonlinear gradual process of strain [45].

Similarly, when the strain is constant, the Abel dashpot can describe the stress relaxation behavior

$$\sigma(t) = \eta\varepsilon\frac{t^{-\beta}}{\Gamma(1 - \beta)}, \quad (0 \le \beta \le 1). \tag{5.4.30}$$

Figure 5.17 shows a series of curves by Eq. (5.4.30) with different $\beta$. It can be seen that when strain remains unchanged, the stress of Abel dashpot displays a slow decrease, unlike the stress of Newtonian fluid which rapidly becomes zero.

**Fig. 5.16** Creep curve of Abel dashpot [45]

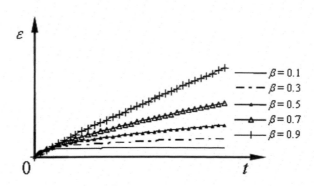

**Fig. 5.17** Stress relaxation
of Abel dashpot [45]

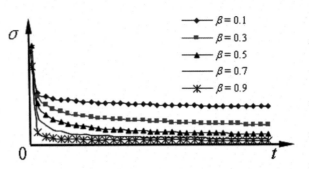

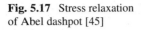

From Figs. 5.16 and 5.17, it is obvious to see that the Abel dashpot elements can describe the various rheological behaviors of materials between the elastic solid and the Newtonian fluid. When $\beta$ tends to 1, the Abel dashpot closes to the elastic body, while when $\beta$ tends to 0, the Abel dashpot closes to the Newtonian fluid.

Therefore, the Abel dashpot is a component including an elastic component and a damping component. The Abel dashpot actually has two parameters $\eta$ and $\beta$. Compared with the classical integer-order derivative damping component, the Abel dashpot is able to control the rate of deformation, but also is able to control the stress (strain). But the damping component can only control the rate of deformation by the viscous coefficient. From this point of view, the Abel dashpot is better to reflect the nonlinear gradual processes of rheological problems.

However, for plastic materials, the traditional model and the fractional derivative model may both need to add the friction component to construct the viscoelastic plastic model. The friction component is shown in Fig. 5.18. It consists of two plates which contact with each other and have cohesive force and frictional force on the contact surface, and reflects the rigid plasticity of materials. When the stress $\sigma$ is smaller than the flow limit $\sigma^0$, the friction component has no deformation; when the stress $\sigma$ is larger than the flow limit $\sigma^0$ and achieve the yield limit, the deformation of friction component will grow unlimited.

The discrete fractional derivative viscoelastic constitutive models can be constructed by components in series or in parallel. The traditional integer-order derivative model cannot agree well with the experimental data in early creep and relaxation. Such as the geotechnical rheology, the creep curve of Maxwell model cannot be used to well fit the experimental curve of soil [46]. And the stress relaxation curve of the Maxwell model rapidly decays from the initial stress to 0 disadvantages. On the other hand, the stress relaxation curve of the Kelvin–Voigt model is a horizontal line and cannot be used to fit the experimental curve of soil. For creep, when $t = 0$, all stress is acted on the damping component, but the elastic component and the damping component do not have strain at time $t = 0$. As time continues, the elastic component and the damping component become to produce strain, but the

**Fig. 5.18** Friction
component

**Fig. 5.19** Fractional Maxwell model

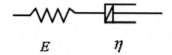

$$E \qquad \eta$$

stress of the damping element keeps decreasing and is converted to the elastic component (because the summation of the stress of the elastic component and the damping component is constant). Finally, the elastic element bears all stress and the strain closes to $\sigma/E$. Of course, since the damping component corresponds to the Newtonian viscous fluid, the strain of which varies from the initial strain to the final strain will be completed within a very short time. Thus, the initial creep curve suddenly increases, and then the curve rapidly becomes a horizontal line. The whole process is very fast, which does not agree with most actual processes. When integer-order derivative mechanics models are used to simulate the viscoelastic material models, lots of components are necessary to fit the experimental data well.

After introducing the fractional derivative Abel dashpot, models constructed by combination of just a few components in series or in parallel can be well consistent with the experimental data, can well simulate the storage modulus and loss modulus of materials within a very broad range of frequency at the same time, and maintain the advantage of clear physics concepts of the traditional theory [47]. The fractional derivative viscoelastic models constructed by the above-mentioned method will be described below [48].

1. **Fractional Maxwell model** [44, 48]

Fractional Maxwell model is composed by an elastic component and an Abel dashpot component in series, and is used to describe the viscoelastic solid, as shown in Fig. 5.19

Based on physical analysis, this model will be more in line with the actual situation. For the creep behavior, namely when stress is a constant, the function of the spring component is used to produce the initial strain which is equal to $\sigma_0/E$, and the Abel dashpot describes the variation of strain with time. Therefore, compared with the Maxwell model, the fractional Maxwell model agrees better with the real situation.

$$\sigma_e = E\varepsilon_e, \sigma_v = \eta \mathrm{d}^\alpha \varepsilon_v / \mathrm{d}t^\alpha;$$

$$\varepsilon = \varepsilon_e + \varepsilon_v, \sigma = \sigma_e = \sigma_v, \tag{5.4.31}$$

where the subscript $e$ means the elastic component, the subscript $v$ denotes the viscous component. The constitutive equation of the fractional Maxwell model is

$$\frac{\mathrm{d}^\alpha \sigma}{\mathrm{d}t^\alpha} + \frac{1}{\tau^\alpha}\sigma = E\frac{\mathrm{d}^\alpha \varepsilon}{\mathrm{d}t^\alpha}, \tag{5.4.32}$$

where $\tau$ means the relaxation time. The Laplace transform of Eq. (5.4.32) is

$$s^\alpha \overline{\varepsilon} = \frac{1}{E} s^\alpha \overline{\sigma} + \frac{1}{\eta} \overline{\sigma}. \tag{5.4.33}$$

Then, Eq. (5.4.33) can be rewritten as

$$J(s) = \frac{\overline{\varepsilon}}{s\overline{\sigma}} = \frac{1}{sE} + \frac{s^{-\alpha-1}}{\eta}. \tag{5.4.34}$$

By the inverse Laplace transform, the creep compliance is

$$J(t) = \frac{1}{E} + \frac{1}{\eta} \frac{t^\alpha}{\Gamma(1+\alpha)}. \tag{5.4.35}$$

Similarly, Eq. (5.4.33) can be rewritten as

$$G(s) = \frac{\overline{\sigma}}{s\overline{\varepsilon}} = \frac{s^{\alpha-1}}{\frac{s^\alpha}{E} + \frac{1}{\eta}}. \tag{5.4.36}$$

By the inverse Laplace transform, the relaxation modulus is

$$G(t) = L^{-1} \left[ E \sum_{k=0}^{\infty} (-1)^k \left( \frac{\eta}{E} \right)^{-k} s^{-\alpha k-1} \right] = E H_{1,2}^{1,1} \left[ \frac{E t^\alpha}{\eta} \middle| \begin{array}{l} [0,1] \\ [0,1]; \ (0,1) \end{array} \right], \tag{5.4.37}$$

where $H_{1,2}^{1,1}(x)$ is H-Fox function (See Appendix VI). The deducing process of the above equation applies the following property of H-Fox function

$$\sum_{n=0}^{\infty} \frac{(-z)^n \prod_{i=1}^{P} \Gamma(a_i + A_i n)}{n! \prod_{i=1}^{q} \Gamma(b_i + B_i n)} = H_{P,Q+1}^{1,P} \left[ z \middle| \begin{array}{l} [1 - a_p, A_p] \\ [0,1]; \quad (1 - b_q, B_q) \end{array} \right]. \tag{5.4.38}$$

## 2.   Fractional Kelvin–Voigt Model [48]

Fractional Kelvin–Voigt model is composed by an elastic component and an Abel dashpot component in parallel, and is used to describe the viscoelastic solid, as shown in Fig. 5.20.

Based on the physical analysis, for creep behavior, when the Abel dashpot is acted by stress, it cannot respond suddenly as Newtonian fluid. Therefore, the strain will

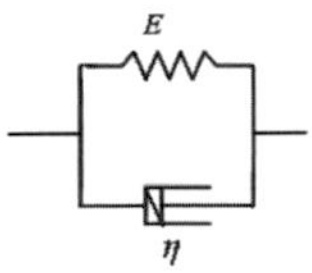

**Fig. 5.20** Fractional
Kelvin–Voigt model

not develop from the initial strain 0 to the final strain $\sigma/E_0$ in a very short time and the creep curve will not show the sudden change, which is good at fitting well with the experimental data. Of course, the actual effects need further inspection of actual instances, rather than just physical analysis.

$$\sigma_e = E\varepsilon_e, \sigma_v = \eta \mathrm{d}^\alpha \varepsilon_v / \mathrm{d}t^\alpha,$$

$$\varepsilon = \varepsilon_e = \varepsilon_v, \sigma = \sigma_e + \sigma_v, \tag{5.4.39}$$

where the subscript $e$ means the elastic component, the subscript $v$ denotes the viscous component. The constitutive equation of fractional Kelvin–Voigt model is

$$\frac{\mathrm{d}^\alpha \varepsilon}{\mathrm{d}t^\alpha} + \frac{1}{\tau^\alpha}\varepsilon = \frac{\sigma}{\eta}. \tag{5.4.40}$$

And the creep compliance of the fractional Kelvin–Voigt model can be obtained as follows:

$$J(t) = L^{-1}\left[\eta^{-1}\sum_{k=0}^{\infty}(-1)^k\left(\frac{\eta}{E}\right)^k s^{-\alpha k-\alpha-1}\right]$$

$$= \eta^{-1}\sum_{k=0}^{\infty}\frac{(-1)^k}{\Gamma(\alpha k+\alpha+1)}\left(\eta\frac{t^\alpha}{E}\right)^{k+1} = \frac{t^\alpha}{E}H_{1,2}^{1,1}\left[\eta\frac{t^\alpha}{E}\bigg|\begin{matrix}[0,1]\\ [0,1];\ (-\alpha,\alpha)\end{matrix}\right]. \tag{5.4.41}$$

The relaxation modulus is

$$G(t) = E + \eta\frac{t^{-\alpha}}{\Gamma(1-\alpha)}. \tag{5.4.42}$$

### 3. Fractional Zener Model [48]

Fractional Zener model is composed by an elastic component and a fractional Kelvin–Voigt model in series, and is used to describe the viscoelastic solid.

$$\varepsilon = \varepsilon_e + \varepsilon_K, \sigma = \sigma_e = \sigma_K, \tag{5.4.43}$$

where the subscript $e$ means the elastic component, the subscript $K$ denotes the fractional Kelvin–Voigt model. The constitutive equation of the fractional Zener model is

$$\left(E_e + E_K + \eta_K\frac{\partial^{\alpha_K}}{\partial t^{\alpha_K}}\right)\sigma(t) = \left(E_e E_K + E_e\eta_K\frac{\partial^{\alpha_K}}{\partial t^{\alpha_K}}\right)\varepsilon(t). \tag{5.4.44}$$

4.   **Fractional Burgers Model** [48]

Fractional Burgers model is composed by a fractional Maxwell model and a fractional Kelvin–Voigt model in series, and is used to describe the viscoelastic solid with a viscous flow.

$$\varepsilon = \varepsilon_M + \varepsilon_K, \ \sigma = \sigma_M = \sigma_K, \tag{5.4.45}$$

where the subscript $M$ means the fractional Maxwell model, the subscript $K$ denotes the fractional Kelvin–Voigt model. The constitutive equation of the fractional Burgers model isofmassmeans change ofmass ins

$$\left( E_K + \eta_K \frac{\partial^{\alpha_K}}{\partial t^{\alpha_K}} + \frac{E_M \eta_M (\partial^{\alpha_M}/\partial t^{\alpha_M})}{E_M + \eta_M (\partial^{\alpha_M}/\partial t^{\alpha_M})} \right) \sigma(t)$$

$$= \frac{(E_K + \eta_K (\partial^{\alpha_K}/\partial t^{\alpha_K})) E_M \eta_M (\partial^{\alpha_M}/\partial t^{\alpha_M})}{E_M + \eta_M (\partial^{\alpha_M}/\partial t^{\alpha_M})} \varepsilon(t). \tag{5.4.46}$$

5.   **Generalized Zener model and generalized Poynting–Thomson model**

Because many complex viscoelastic materials show self-similar properties, some scholars combined the fractional viscoelastic models in series and in parallel to describe the viscoelastic constitutive relation and proposed the generalized Zener model and the generalized Poynting–Thomson model [49].

(1)   Generalized Zener model

Generalized Zener model is composed by a combination of two tandem Fractional Abel dashpots and a Fractional Abel dashpot in parallel (Fig. 5.21).

$$\sigma_L = \left[ E_1 \tau_1^{\alpha} \frac{\partial^{\alpha}}{\partial t^{\alpha}} E_2 \tau_2^{\beta} \frac{\partial^{\beta}}{\partial t^{\beta}} \Big/ \left( E_1 \tau_1^{\alpha} \frac{\partial^{\alpha}}{\partial t^{\alpha}} + E_2 \tau_2^{\beta} \frac{\partial^{\beta}}{\partial t^{\beta}} \right) + E_3 \tau_3^{\lambda} \frac{\partial^{\lambda}}{\partial t^{\lambda}} \right] \varepsilon;$$

$$\sigma_R = E_3 \tau_3^{\lambda} \frac{\partial^{\lambda} \varepsilon}{\partial t^{\lambda}}, \tag{5.4.47}$$

where the subscript $L$ means the Left model, the subscript $R$ denotes the right model. The constitutive equation of the generalized Zener model is

$$\sigma_L = \left[ E_1 \tau_1^{\alpha} \frac{\partial^{\alpha}}{\partial t^{\alpha}} E_2 \tau_2^{\beta} \frac{\partial^{\beta}}{\partial t^{\beta}} \Big/ \left( E_1 \tau_1^{\alpha} \frac{\partial^{\alpha}}{\partial t^{\alpha}} + E_2 \tau_2^{\beta} \frac{\partial^{\beta}}{\partial t^{\beta}} \right) + E_3 \tau_3^{\lambda} \frac{\partial^{\lambda}}{\partial t^{\lambda}} \right] \varepsilon. \tag{5.4.48}$$

(2)   Generalized Poynting–Thomson model

Generalized Poynting–Thomson model is composed by a combination of two parallel Fractional Abel dashpots and a Fractional Abel dashpot in series [50] (Fig. 5.22).
    The constitutive equation of the generalized Poynting–Thomson model is [49]

**Fig. 5.21** Generalized Zener model [49]

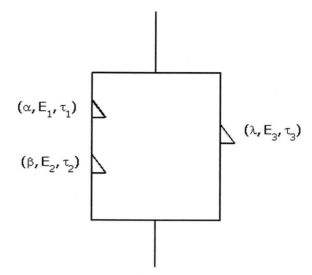

**Fig. 5.22** Generalized Poynting–Thomson model [49]

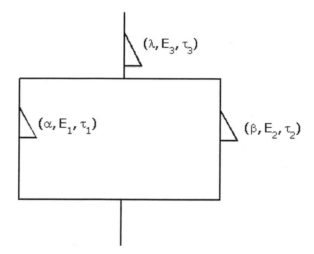

$$\sigma(t) + \frac{E_0}{E}\tau^{\alpha-\lambda}{}_0D_t^{\alpha-\lambda}\sigma(t) + \frac{E_0}{E}\tau^{\beta-\lambda}{}_0D_t^{\beta-\lambda}\sigma(t)$$
$$= E_0\tau^{\alpha}{}_0D_t^{\alpha}\varepsilon(t) + E_0\tau^{\beta}{}_0D_t^{\beta}\varepsilon(t). \tag{5.4.49}$$

6.   **Fractional creep compliance and fractional relaxation modulus**

Zhang [48] introduced the fractional derivative operator and proposed the fractional creep compliance and fractional relaxation modulus:

$$J(t) = J_0\{1 + (J_\infty/J_0 - 1)[1 - \exp(-\beta_1[(\gamma + \alpha)t]^{1-\alpha})]\};$$

$$G(t) = G_0\{1 - (1 - G_\infty/G_0)[1 - \exp(-\beta[(\gamma + \alpha)t]^{1-\alpha})]\}, \tag{5.4.50}$$

where $J_0$ and $G_0$ are the initial creep compliance and the initial relaxation modulus, respectively; $J_\infty$ and $G_\infty$ are balancing creep compliance and balancing relaxation modulus, respectively; $\alpha$, $\beta_1$, $\beta$ and $\gamma$ are material parameters which can be obtained by experiments. This model can agree well with experimental data.

In addition to the above model, Pan and Tan apply the fractional Maxwell model $\sigma + \lambda^\alpha d^\alpha \sigma/dt^\alpha = G\lambda^\beta d^\beta \varepsilon/dt^\beta$ to study unsteady flow between two parallel plates [51]; Su and Xu use the fractional calculus to describe the gel layer of otolith organs (viscoelastic solid) and endolymph (viscoelastic fluid) [52]; Zhang et al. [53] use the fractional Zener model to describe the creep and relaxation of aging concrete.

### 7.  Fractal Tree Model

Elastic components and viscous components, respectively, are described by Hooke's Law and the Newton law of viscosity, and the two basic components, Maxwell model and Kelvin–Voigt model, of viscoelastic materials model are constructed by elastic component and viscous component in series or in parallel. This subsection will introduce a fractal tree model proposed by Zhu and Hu [54] which is constructed by elastic components and viscous components in fractal sequences, as shown in Fig. 5.23.

From Fig. 5.23, it is seen that the fractal tree model has self-similar feature and each branch is similar to the whole model. Therefore, the fractal tree model is equal to (Fig. 5.24).

Based on series and parallel formulas and the principle of self-similarity, we can get

$$T = \frac{TT_1}{T + T_1} + \frac{TT_2}{T + T_2}, \tag{5.4.51}$$

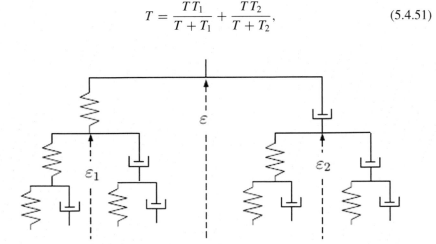

**Fig. 5.23**  Fractal tree model [54]

**Fig. 5.24** Equivalent form
of the fractal tree model [54]

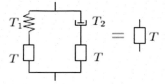

where $T = 1$, $T_2 = \frac{\eta}{G}D$, then

$$T = \sqrt{T_1 T_2} = \lambda^{1/2} D^{1/2}, \tag{5.4.52}$$

where $D^{1/2} = \frac{d^{1/2}}{dt^{1/2}}$ means 1/2 order derivative. Finally, the stress–strain relation of viscoelastic materials described by the fractal tree model can be stated as

$$\sigma = G\lambda^{1/2}\frac{d^{1/2}\varepsilon}{dt^{1/2}}. \tag{5.4.53}$$

If the viscous component of Fig. 5.23 is replaced by the fractal tree, then the double fractal network can be obtained and the constitutive operator is

$$T = \lambda^{1/4} D^{1/4}; \tag{5.4.54}$$

If the viscous component of Fig. 5.23 is replaced by the fractal tree, then another double fractal network can be obtained and the constitutive operator is

$$T = \lambda^{3/4} D^{3/4}; \tag{5.4.55}$$

Generally, if the elastic component and the viscous component of Fig. 5.23 are replaced by two kinds of fractal tree components $T_3 = (\lambda D)^{\alpha}$, $T_4 = (\lambda D)^{\beta}$, then the constitutive operator of the fractal tree model is

$$T = \sqrt{T_3 T_4} = (\lambda D)^{\frac{\alpha+\beta}{2}}. \tag{5.4.56}$$

### 5.4.6 Data Fitting and Analysis of Rheological Experiments of Soil

Most rheological experiments of soil are creep experiments, therefore, the experimental data from Ref. [55] is fitted by the fractional Maxwell model, Merchant model [46] and the five-parameter Kelvin model proposed in [55]. The fitting result is compared and shown in Fig. 5.25; the parameters of models are listed in Table 5.1.

**Fig. 5.25** Creep curves

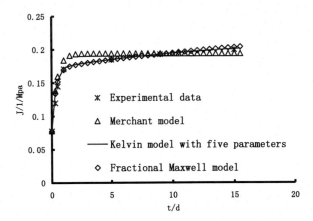

**Table 5.1** Parameters of models

| Parameters | Fractional Maxwell model | Merchant model | Five-Parameter Kelvin model |
|---|---|---|---|
| $E_0$/MPa | 12.804 | 12.804 | 12.804 |
| $E_1$/MPa | | 10.625 | 10.625 |
| $\kappa$ or $\xi$/d·Mpa$^{-1}$ | 11.8 | 3.536 | 3.536 |
| $E_2$/MPa | | | 22.058 |
| $\kappa_1$/d·Mpa$^{-1}$ | | | 325.218 |
| $\beta$ | 0.118 | | |

The fractional Maxwell model and the Merchant model are both three parameters models. From Fig. 5.23, it is seen that the fractional Maxwell model can fit with the experimental data much better than the Merchant model. Compared with the five-parameter Kelvin model, it is found that the fractional Maxwell model can fit with the experimental data as well as the five-parameter Kelvin model, which illustrates that the model combined by the soft body component and the spring component in series are more effective to portray the rheological properties of soil.

In the case of the same effect, the application of Abel dashpot can achieve the purpose of decreasing the number of parameters. It can be seen that the expression of the above-mentioned fractional derivative models is relatively simple and have clear physical meaning. However, they also have some shortcomings, such as when the series model is applied to describe the creep phenomenon, its curve is a monotonically increasing curve but does not have asymptotic behavior, which is unfavorable to describe the rheological problems.

### 5.4.7 Exact Solution of the Fractional Derivative Model of Stokes Second Problem[1] [56]

Stokes second problem is a kind of important flow problems, and its exact solution is accessible. The exact solution can be used to describe the basic flow which deserves a further investigation, and helps in benchmarking numerical solutions. It can be seen that the exact solution of such equations is of great importance.

Stokes second problem is an unstable shear flow of viscous fluid near plates with harmonic vibration, which has a wide range of applications in chemistry, medicine, biology, microscopy and nanotechnology. Here are solving process of fractional Stokes second problem of viscoelastic fluids [56].

First, the control equation of Stokes second problem of viscoelastic fluid is introduced [56]. The constitutive equation is the constitutive equation of fractional derivative component

$$\sigma_{yx} = G\lambda^\beta \frac{\partial^\beta \varepsilon_{yx}}{\partial t^\beta} = G\lambda^\beta \frac{\partial^{\beta-1}}{\partial t^{\beta-1}} \left( \frac{\partial u}{\partial y} \right). \tag{5.4.57}$$

The momentum equation is

$$\rho \frac{\partial u}{\partial t} = \frac{\partial \sigma_{yx}}{\partial y}, \tag{5.4.58}$$

where $\rho$ is the density of fluid. The constitutive equation is substituted into the momentum equation, then

$$\rho \frac{\partial u}{\partial t} = G\lambda^\beta \frac{\partial^{\beta-1}}{\partial t^{\beta-1}} \left( \frac{\partial^2 u}{\partial y^2} \right). \tag{5.4.59}$$

If the vibration has frequency $\omega$ and amplitude $U$, the boundary conditions and initial conditions can be written as

$$u(0, t) = U \exp(i\omega t), \quad t > 0, \tag{5.4.60}$$

$$u(\infty, t) = 0, \tag{5.4.61}$$

$$u(y, 0) = 0, \quad y > 0. \tag{5.4.62}$$

The analytic solution of equation can be got by applying the Laplace transform [56]. Non-dimensional-normalized the equation is

---

[1] Part of this subsection selects from and refers the reference [56].

$$u^* = \frac{u}{U}, \, y^* = \frac{yUp}{\mu}, \, t^* = \frac{tU^2\rho}{\mu}, \, \eta = \frac{tU^2\rho}{G}. \tag{5.4.63}$$

The master equation, boundary condition and initial condition can be written as

$$\frac{\partial u}{\partial t} = \eta^{\beta-1} \frac{\partial^{\beta-1}}{\partial t^{\beta-1}} \left( \frac{\partial^2 u}{\partial y^2} \right), \quad 0 < \beta < 1 \tag{5.4.64}$$

$$u(0, t) = \exp(i\omega t), \quad t > 0, \tag{5.4.65}$$

$$u(\infty, t) = 0, \tag{5.4.66}$$

$$u(y, 0) = 0, \quad y > 0. \tag{5.4.67}$$

The Laplace transform of the velocity is

$$\overline{u}(y, s) = L\{u(y, t), s\} = \int_0^\infty e^{-st} u(y, t) \mathrm{d}t. \tag{5.4.68}$$

The Laplace transform of Eqs. (5.4.64), (5.4.65) and (5.4.66) is

$$\frac{\partial^2 \overline{u}}{\partial y^2} = \frac{1}{\eta^{\beta-1} s^{\beta-2}} \overline{u}, \tag{5.4.69}$$

$$\overline{u}(\infty, s) = 0, \tag{5.4.70}$$

$$\overline{u}(0, s) = \frac{1}{s - i\omega}. \tag{5.4.71}$$

The solution of the above equation can be got

$$\overline{u}(y, s) = \frac{1}{s - i\omega} \cdot \exp\left(-\left(\frac{1}{\eta^{\beta-1} s^{\beta-2}}\right)^{\frac{1}{2}} y\right) \tag{5.4.72}$$

The Taylor series expansion of above equation is

$$\overline{u}(y, s) = \sum_{n=0}^\infty \frac{(-y)^n}{\eta^{\frac{n(\beta-1)}{2}} n!} \cdot \sum_{k=0}^\infty (i\omega)^k \frac{1}{s^{\frac{n(\beta-2)}{2}+1-k}}. \tag{5.4.73}$$

The analytic solution can be got by the inverse Laplace transform

$$u(y, t) = \sum_{n=0}^{\infty} \frac{(-y)^n}{\eta^{\frac{n(\beta-1)}{2}} n!} \cdot \sum_{k=0}^{\infty} (i\omega)^k \frac{t^{\frac{n(\beta-2)}{2}+k}}{\Gamma\left(\frac{n(\beta-1)}{2} + 1 + k\right)}. \tag{5.4.74}$$

## 5.5 The Power-Law Frequency Dependent Acoustic Dissipation

Sound is a very common phenomenon in the daily life. Human can hear a variety of voices and use sound to communicate with others. People have long realized that sound is actually caused by the vibration of medium. People are able to hear sound due to the vibration in the air which causes the vibration of eardrum. The nature of sound is a kind of mechanical wave propagation in continuous media; the so-called wave means the perturbation of energy and movement form propagates in media and is the dynamic response of medium [57, 58]. Only in the presence of the media, sound can only propagate in media and the space with sound is called sound field [58].

### 5.5.1 Wave Propagation in Elastic Solids

The mechanical wave in solid media is usually called elastic wave (or plastic wave/viscous wave) and the sound wave means the mechanical wave propagation in fluid. Before introducing sound wave, the first to introduce is the basic theory of wave propagation in solids. In fact, wave propagation in media is the dynamic response of media under the external disturbance. Considering the dynamic balance caused by the inertia force (without body force) [59]

$$\sigma_{ij,i} = \rho u_{i,tt}, \tag{5.5.1}$$

where $\sigma_{ij}$ is the stress tensor, $u$ is the displacement. Substituting the elastic constitutive relation and geometric compatibility condition

$$\sigma_{ij} = \lambda \delta_{ij} \Theta + \mu(\varepsilon_{ij} + \varepsilon_{ji}), \tag{5.5.2}$$

$$\sigma_{ij,i} = \mu \nabla^2 u + (\lambda + \mu)\Theta_{,i}, \quad \Theta = \nabla \cdot \mathbf{u}. \tag{5.5.3}$$

where $\sigma_{ij}$ is the stress tensor, $\varepsilon_{ij}$ means the strain tensor, $\Theta$ is the bulk strain, $\lambda$ and $\mu$ are the Lame coefficients, thus the motion equation can be got [59]

$$\mu\nabla^2\mathbf{u} + (\lambda + \mu)\Theta_i = \rho\mathbf{u}_{tt}. \tag{5.5.4}$$

The motion equation can be solved by transforming this time-domain equation to frequency-domain equation with Fourier transformation. Without loss of generality, it is assumed that wave propagates in $x$ axial [58]

$$\begin{pmatrix} (\lambda + 2\mu)k_1^2 - \rho\omega^2 & 0 & 0 \\ 0 & \mu k_1^2 - \rho\omega^2 & 0 \\ 0 & 0 & \mu k_1^2 - \rho\omega^2 \end{pmatrix} \begin{pmatrix} U_{01} \\ U_{02} \\ U_{03} \end{pmatrix} = 0, \tag{5.5.5}$$

where $k$ is the wave number, $U_{0i}$ is the displacement of medium in $i$ direction. Therefore, three kinds of waves can be solved as follows:

(1)   Frequency is $\omega_1 = c_L k_1$, wave velocity is $c_L = \sqrt{\lambda + 2\mu/\rho}$, displacement is $U_1 = U_{01}\exp(ik_1 x - i\omega_1 t)$ which is in the same direction with wave propagation and means longitudinal wave;
(2)   Frequency is $\omega_2 = \omega_3 = c_T k_1$, wave velocity is $c_L = \sqrt{\mu/\rho}$, displacement is $U_2 = U_{02}\exp(ik_1 x - i\omega_2 t)$ and $U_3 = U_{03}\exp(ik_1 x - i\omega_3 t)$ whose direction is perpendicular to the wave propagation direction and means transverse wave.

In addition, the displacement vector can be decomposed as [57]

$$\mathbf{U} = \nabla\Phi + \nabla \times \Psi, \tag{5.5.6}$$

where $\Phi$ is the displacement of scalar potential and $\Psi$ means vector displacement potential. Therefore, because the curl of gradient of scalar and divergence of curl of vector are both equal to zero, the displacement part represented by scalar potential is irrotational and the displacement part represented by vector potential does not cause a change in volume [57]. Substituting Eq. (5.5.6) into the equation of equilibrium, then we can get

$$\nabla[(\lambda + 2\mu)\nabla^2\Phi + \rho\Phi_{tt}] + \nabla \times (\mu\nabla^2\Psi + \rho\Psi_{tt}) = 0, \tag{5.5.7}$$

thus, the wave equation described by displacement of scalar potential $\Phi$ and vector displacement potential $\Psi$ are, respectively [58]

$$\nabla^2\Phi + \frac{\Phi_{tt}}{c_L^2} = 0 \quad \text{(scalar wave)}, \tag{5.5.8}$$

and

$$\nabla^2\Psi + \frac{\Psi_{tt}}{c_T^2} = 0 \quad \text{(vector wave)}. \tag{5.5.9}$$

For scalar waves, displacement and wave propagation has the same direction and no vortex, and displacement causes the change in volume, hence, the scalar wave is called longitudinal waves or compression waves; for vector waves, displacement direction is perpendicular to the wave propagation direction, and there is vortex but no change in volume, hence, the vector wave is called transversal wave, also known as equivoluminal wave or shear wave [58].

### 5.5.2 Classical Models of Dissipative Wave in Viscoelastic Medium

The energy dissipation of wave propagation is not considered in the last subsection. However, in the actual process, wave propagation must be accompanied by energy dissipation. The reason is that the medium in which wave propagates is not ideal medium, but viscoelastic medium which can both store and dissipate the energy. For example, the experiments of seismic wave verify that the earth, in fact, is a kind of viscoelastic materials [57]. When wave propagates in medium, there is not only the transformation between the kinetic and potential energy, but also the mechanical energy dissipation which makes part of energy are transformed to heat due to damping of medium [43]. From Sect. 5.4, the general model of viscoelastic constitutive relation is

$$P(\sigma(t)) = Q(\varepsilon(t)).$$

When describing dissipative wave propagation in viscoelastic medium, the viscoelastic constitutive models are written as equivalent elastic constitutive model by using the corresponding principle

$$\sigma_{ij} = \bar{\lambda}\delta_{ij}\Theta + 2\bar{\mu}\varepsilon_{ij}, \ \bar{\lambda} = \frac{P}{Q}\lambda, \ \bar{\mu} = \frac{P}{Q}\mu. \tag{5.5.10}$$

Then, substituting the motion Eq. (5.5.10), the corresponding dissipative wave equation can be got [58, 60]. Taking the Maxwell model and the Kelvin–Voigt model as examples [11, 57]

The Maxwell model

$$\bar{\lambda} = \frac{\tau D}{1 + \tau D}\lambda, \ \bar{\mu} = \frac{\tau D}{1 + \tau D}\mu; \tag{5.5.11}$$

The Kelvin–Voigt model

$$\bar{\lambda} = (1 + \tau D)\lambda, \ \bar{\mu} = (1 + \tau D)\mu, \tag{5.5.12}$$

where $D$ is the first derivative. Substituting this relation into the Navier equation, then we can get

$$\overline{\mu}\nabla^2\mathbf{u} + (\overline{\lambda} + \overline{\mu})\Theta_i = \rho\mathbf{u}_{tt}. \tag{5.5.13}$$

With the method mentioned in the last subsection, the longitudinal wave in the viscoelastic medium can be stated as [57]

$$\nabla^2\Phi = \frac{\rho}{(\overline{\lambda} + 2\overline{\mu})}\Phi_{tt}. \tag{5.5.14}$$

Similarly, the transversal wave equation is

$$\nabla^2\Psi = \frac{\rho}{\overline{\mu}}\Psi_{tt}. \tag{5.5.15}$$

From this, the wave speed of longitudinal wave and transversal wave are, respectively,

$$\overline{c}_L = \sqrt{\frac{\overline{\lambda} + 2\overline{\mu}}{\rho}}, \ \overline{c}_T = \sqrt{\frac{\overline{\mu}}{\rho}}. \tag{5.5.16}$$

The following takes the one-dimensional longitudinal wave equation in viscoelastic media as example, and will give the wave equations based on the Maxwell model and the Kelvin–Voigt model. Substituting the equivalent Lame coefficients of viscoelastic materials, the corresponding wave equation in viscoelastic medium can be got. Without loss of generality, the one-dimensional longitudinal wave equation is [11, 57, 60]

$$\Phi_{xx} = \frac{\rho}{(\overline{\lambda} + 2\overline{\mu})}\Phi_{tt}. \tag{5.5.17}$$

**The Maxwell Model**

$$\overline{\lambda} = \frac{\tau D}{1 + \tau D}\lambda, \ \overline{\mu} = \frac{\tau D}{1 + \tau D}\mu, \tag{5.5.18}$$

then

$$\overline{c}_L^2 = \frac{\tau D}{1 + \tau D}c_L^2, \tag{5.5.19}$$

and the wave equation can be written as

$$c_L^2 \Phi_{xxt} = \Phi_{ttt} + \frac{1}{\tau}\Phi_{tt}. \tag{5.5.20}$$

Integrating on both sides of the equation, thus the equation is equal to

$$c_L^2 \Phi_{xx} = \Phi_{tt} + \frac{1}{\tau}\Phi_t + g(x), \tag{5.5.21}$$

where $g(x)$ relates to the initial condition and is the dissipative term. If considering the displacement $u_L$ represented by the scalar potential, thus the longitudinal wave equation is

$$c_L^2 u_{Lxxt} = u_{Lttt} + \frac{1}{\tau}u_{Ltt} \text{ or } c_L^2 u_{Lxx} = u_{Ltt} + \frac{1}{\tau}u_{Lt} + g(x); \tag{5.5.22}$$

**The Kelvin–Voigt Model**

$$\bar{\lambda} = (1 + \tau D)\lambda, \quad \bar{\mu} = (1 + \tau D)\mu, \tag{5.5.23}$$

therefore

$$\bar{c}_L^2 = (1 + \tau D)c_L^2, \tag{5.5.24}$$

then the wave equation is

$$\Phi_{xx} + \tau\Phi_{xxt} = \frac{1}{c_L^2}\Phi_{tt}, \tag{5.5.25}$$

where $\tau u_{xxt}$ is the dissipative term. If considering the displacement $u_L$ represented by the scalar potential, thus the longitudinal wave equation is

$$u_{Lxx} + \tau u_{Lxxt} = \frac{1}{c_L^2}u_{Ltt}. \tag{5.5.26}$$

### 5.5.3 Fractional Derivative Model of Wave in Viscoelastic Medium

The last subsection introduces the master equation of wave in viscoelastic medium by elastic wave equation and the viscoelastic Maxwell model and the Kelvin–Voigt model. However, the traditional viscoelastic models use the integer-order derivatives, and cannot effectively describe the memory relaxation and creep behaviors and power-law frequency dependent damping. This subsection uses fractional viscoelastic models introduced in Sect. 5.4, and introduces the fractional derivative wave equation to describe the dissipative wave in viscoelastic medium [11, 33, 43, 57–64]. Without loss of generality, this subsection uses the fractional Maxwell model and the Kelvin–Voigt model to describe the one-dimensional longitudinal wave in viscoelastic medium [11, 57, 60].

1.  **Fractional Maxwell model**

Based on Eq. (5.5.10), the equivalent Lame coefficients of viscoelastic medium are

$$\bar{\lambda} = \frac{\tau \partial^\alpha / \partial t^\alpha}{1 + \tau \partial^\alpha / \partial t^\alpha} \lambda, \quad \bar{\mu} = \frac{\tau \partial^\alpha / \partial t^\alpha}{1 + \tau \partial^\alpha / \partial t^\alpha} \mu. \tag{5.5.27}$$

The wave speed of longitudinal wave in viscoelastic medium is

$$\bar{c}_L^2 = \frac{\tau \partial^\alpha / \partial t^\alpha}{1 + \tau \partial^\alpha / \partial t^\alpha} c_L^2, \tag{5.5.28}$$

and the wave equation is

$$c_L^2 \frac{\partial^\alpha \Phi_{xx}}{\partial t^\alpha} = \frac{\partial^\alpha \Phi_{tt}}{\partial t^\alpha} + \frac{1}{\tau} \Phi_{tt}. \tag{5.5.29}$$

After integrating on both sides of the equation, the equation can be written as

$$c_L^2 \Phi_{xx} = \Phi_{tt} + \frac{1}{\tau} \frac{\partial^{2-\alpha} \Phi}{\partial t^{2-\alpha}} + g(x), \tag{5.5.30}$$

where $g(x)$ relates to the initial condition, $\frac{1}{\tau} \frac{\partial^{2-\alpha} \Phi}{\partial t^{2-\alpha}}$ is the dissipative term. If considering the displacement $u_L$ represented by the scalar potential, thus, the longitudinal wave equation is

$$c_L^2 \frac{\partial^\alpha u_{Lxx}}{\partial t^\alpha} = \frac{\partial^\alpha u_{Ltt}}{\partial t^\alpha} + \frac{1}{\tau} u_{Ltt} \text{ or } c_L^2 u_{Lxx} = u_{Ltt} + \frac{1}{\tau} \frac{\partial^{2-\alpha} u_L}{\partial t^{2-\alpha}} + g(x). \tag{5.5.31}$$

2.  **The Fractional Kelvin–Voigt Model**

Based on Eq. (5.5.10), the equivalent Lame coefficients of viscoelastic medium are

$$\bar{\lambda} = \left(1 + \tau \frac{\partial^\alpha}{\partial t^\alpha}\right) \lambda, \quad \bar{\mu} = \left(1 + \tau \frac{\partial^\alpha}{\partial t^\alpha}\right) \mu. \tag{5.5.32}$$

The wave speed of longitudinal wave in viscoelastic medium is

$$\bar{c}_L^2 = \left(1 + \tau \frac{\partial^\alpha}{\partial t^\alpha}\right) c_L^2, \tag{5.5.33}$$

and the wave equation is

$$\Phi_{xx} + \tau \frac{\partial^\alpha \Phi_{xx}}{\partial t^\alpha} = \frac{1}{c_L^2} \Phi_{tt}, \tag{5.5.34}$$

where $\frac{\partial^\alpha \Phi_{xx}}{\partial t^\alpha}$ is the dissipative term. If considering the displacement $u_L$ represented by the scalar potential, then the longitudinal wave equation is

$$u_{Lxx} + \tau \frac{\partial^\alpha u_{Lxx}}{\partial t^\alpha} = \frac{1}{c_L^2} u_{Ltt}. \tag{5.5.35}$$

### 5.5.4 Acoustic Wave in Fluids

This subsection introduces the differential equation of acoustic wave in fluid medium. In describing the mechanical wave propagation in fluid medium, we mainly consider variety of pressure, density and velocity of acoustic wave. This subsection first considers a simple condition, namely acoustic wave propagation in uniform, static and adiabatic ideal fluids. This subsection will discuss the effects of damping on acoustic wave. Different from the elastic wave in solids, three conversation laws are needed to be considered in describing acoustic wave in fluid [58]:

(1) Conservation of mass means change of mass inside the volume is equal to mass flowing into the volume. The differential form of continuity equation is [65]

$$\frac{\partial \rho}{\partial t} = -\nabla \cdot (\rho \boldsymbol{v}); \tag{5.5.36}$$

(2) Momentum conservation means change of momentum is equal to active force. The differential form of motion equation is [65]

$$\frac{\partial}{\partial t}(\rho \boldsymbol{v}) + \nabla \cdot (\rho \boldsymbol{v} \boldsymbol{v}) + \nabla p = 0. \tag{5.5.37}$$

Substituting continuity equation into motion equation, then the equation can be stated as

$$\rho \frac{\partial \boldsymbol{v}}{\partial t} + \rho \boldsymbol{v} \cdot \nabla \boldsymbol{v} + \nabla p = 0. \tag{5.5.38}$$

Considering linear problems and neglecting second-order small quantities in the equation, the linear motion equation can be stated as [58]

$$\rho_0 \frac{\partial v}{\partial t} + \nabla p = 0. \tag{5.5.39}$$

$\rho_0$ is the initial density, $\rho$ the change of density, $p$ denotes the sound pressure.

(3)   Constitutive equation of pressure and density can be got by energy conservation [65]. For adiabatic process, sound pressure is the function of density [58]

$$p = c_0^2 \rho, \tag{5.5.40}$$

where $c_0$ is wave speed. If considering linear problems and neglecting second-order small quantities, the linear motion equation can be obtained [58]

$$\frac{\partial \rho}{\partial t} = -\rho_0 \nabla \cdot v. \tag{5.5.41}$$

Differentiating both sides of the equation and substituting into the motion equation and constitutive equation, then the acoustic wave equation of pressure and density are, respectively [65]

$$\frac{\ddot{p}}{c_0^2} = \nabla^2 p \text{ or } \frac{\ddot{\rho}}{c_0^2} = \nabla^2 \rho. \tag{5.5.42}$$

Similarly, the acoustic wave equation of velocity can be obtained. The velocity vector of sound field can be expressed as [58]

$$\mathbf{v} = -\frac{\nabla \left( \int p \mathrm{d}t \right)}{\rho_0}. \tag{5.5.43}$$

Based on the above equation, velocity of acoustic wave is the gradient of scalar potential. Because fluid cannot bear shear force, only longitudinal wave, rather than transversal wave, can propagate in fluid. Hence, acoustic wave is longitudinal wave, can only cause change of volume. The acoustic wave equation of velocity potential is [58]

$$\frac{\ddot{\phi}}{c_0^2} = \nabla^2 \phi, \tag{5.5.44}$$

which illustrates that acoustic wave in fluid is scalar wave.

### 5.5.5   Acoustic Absorption and Frequency Dependent Dissipation

The acoustic wave equation, Eq. (5.5.44), introduced in the last subsection, is only available to the acoustic wave propagation in non-viscous ideal fluid. However, actual fluid medium is usually viscous fluid, even viscoelastic fluid or multi-phase medium. Acoustic wave in these media will produce dissipation of mechanical energy due to viscosity of medium which leads to decrease in amplitude of acoustic wave. This subsection mainly introduces theories and mechanical models of describing acoustic attenuation.

There are three reasons which lead to decrease of amplitude of acoustic wave, namely acoustic diffusion, acoustic scattering and acoustic absorption. Acoustic diffusion means that when acoustic wave propagates, such as sound radiation, the wave front will keep expanding, which makes the energy density and amplitude of sound decrease. This phenomenon is called radiation damping in some references. Acoustic scattering means that when sound propagates, sound will be reflected by obstacles, thus, the direction of sound propagation is changed and part of acoustic energy losses. Acoustic absorption means when sound propagates, mechanical energy is transformed to heat by viscoelastic damping of media, also called acoustic attenuation [58].

This subsection discusses acoustic attenuation. Classical fluid mechanics considers fluids with Newtonian viscosity, shear stress of which is proportional to gradient of velocity

$$\tau = \eta \frac{\partial}{\partial x} \frac{\partial u}{\partial t}. \tag{5.5.45}$$

Based on balance of force, the viscous effects can be introduced into the following constitutive equation

$$p = c_0^2 \rho + \tau = c_0^2 \rho + \frac{\eta}{\rho} \rho \frac{\partial}{\partial x} \frac{\partial u}{\partial t}. \tag{5.5.46}$$

With the method in Sect. 5.5.4, substituting the above equation into the motion Eq. (5.5.37), then, the viscous acoustic wave equation with energy dissipation can be got [58]

$$\rho u'' = \rho c_0^2 \frac{\partial^2 u}{\partial x^2} + \eta \frac{\partial^2}{\partial x^2} \frac{\partial u}{\partial t}. \tag{5.5.47}$$

The complex wave number of dissipative acoustic wave $k = \omega \sqrt{\frac{\rho}{\rho c_0^2 - i\omega\eta}}$ can be obtained by Fourier transformation. If the viscosity of medium is small, then the complex wave number can be simplified as $k = \frac{\omega}{c_0} + i \frac{\omega^2 \eta}{2\rho c_0^3}$, and its imaginary part means energy dissipation which is proportional to square of frequency, namely acoustic dissipation in classical Newtonian fluid is proportional to square of frequency [58].

Dissipative acoustic wave Eq. (5.5.47) can only describe the frequency-squared dependent acoustic dissipation in Newtonian viscous fluid. Many materials in nature and engineering field have complex viscoelastic properties and rheological behaviors, and represent more complex dissipative properties, such as biological tissues, rock and soil, gel, foam, underwater sediment and so on. Moreover, nonuniformity of medium will result in acoustic scattering which plays an important role in attenuation of acoustic wave [58]. Lots of papers [17, 18, 58] report experiments on acoustic absorption in biological tissues and soil show acoustic attenuation are power-law frequency dependent, rather than frequency-squared dependent in most cases. Numerous experimental and field measurements find that the acoustic attenuation coefficient $\alpha(\omega)$ of a wide range of human tissues and other soft matters can be expressed as the following empirical power law with respect to frequency

$$P(x + \Delta x) = P(x)e^{-\alpha(\omega)\Delta x}, \tag{5.5.48}$$

$$\alpha = \alpha_0 |\omega|^{\eta}, \tag{5.5.49}$$

where $x$ is the direction of wave propagation, $\Delta x$ denotes the wave propagation distance, $\omega$ is the angular frequency, $P$ the pressure, $\alpha_0$ and $\alpha$ the attenuation coefficients, $\eta$ exponent of frequency power law which are obtained by fitting measurement data. It is obvious that the higher frequency is, acoustic wave has more dissipation. For general ideal solid or fluid, $\eta$ is equal to 0 or 2, namely energy dissipation is not frequency-independent or frequency-squared dependent, such as, $\eta$ of ultrasound in water is equal to 2. But the exponent of frequency power law $\eta$ of complex media (such as human tissues, polymer, liquid crystal, soil, gel, particulate matter, porous rock, foam, fabric, oil and other soft condensed state matters) is in the range of 0 to 2. For example, $\eta$ of layered porous rock and underwater sediment is near to 1. $\eta$ of kinds of human tissues are between 1.1 and 1.7, such as $\eta$ of fatty tissue is 1.7, $\eta$ of tumor is 1.3, $\eta$ of connective tissue is 1.5 [3–6]. Therefore, we can differentiate the cancer tissue with the normal tissue.

Figure 5.26 shows the variety of acoustic absorption of breast tissue, muscle and blood versus frequency from 0 to 10 MHz. From this figure, it is clear to see that linear proportional relationship of power-law function, and the slop of these curves means the exponent of frequency power law. Because the breast tissue consists of many different tissues (skin tissue, fatty tissue, glandular tissue and vessel), its function of dissipation-frequency relationship is the combination of power-law relation of these tissues and then displays as a curve. This kind of power-law relation of frequency-dependent energy dissipation is not limited to human tissues. For example, Fig. 5.27 shows power-law attenuation data of some typical soft and crystal materials, such as different porous rocks and liver tissue, in log–log plot, where "shear" and "long" means shear wave and longitudinal wave, respectively. The unit dB means denary logarithm and is used to measure the relative size of the acoustic intensity. The

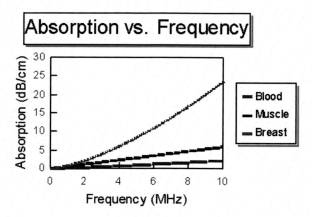

**Fig. 5.26** Frequency-dependent energy dissipation of human tissues (supplied by Prof. Szabo of Boston University)

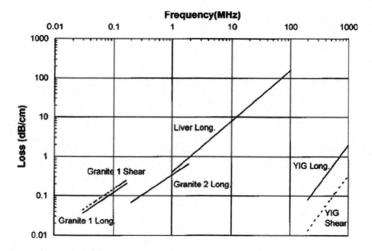

**Fig. 5.27** Data for shear and longitudinal wave loss which show power-law dependence over four decades of frequency (From the reference [66])

slop of straight line is exponent of frequency power law of energy dissipation. For example, exponent of frequency power law $\eta$ of longitudinal wave in bovine liver is equal to 1.3, in the range of frequency 1–100 MHz. YIG (Indium yttrium Garnet) is a kind of ideal crystal material and exponent of frequency power law $\eta$ of shear wave and longitudinal wave in YIG is equal to 2, even with a very high frequency. Granite 1 and granite 2 are two kinds of granite. Exponent of frequency power law $\eta$ of longitudinal wave in granite 1 is equal to 1, in the range of frequency 140 Hz to 2.2 MHz. These examples clearly illustrate the power-law frequency-dependent attenuation of ultrasound in complex media.

The following classical damped wave Eq. (5.5.50) can only describe the dissipative wave with frequency-independent dissipation, namely $\eta = 0$ [17, 67]

$$\nabla^2 p = \frac{1}{c_0^2}\frac{\partial^2 p}{\partial t^2} + \frac{2\alpha_0}{c_0}\frac{\partial p}{\partial t},$$  (5.5.50)

where $p$ means the sound pressure, $c_0$ denotes sound speed. The standard thermo-viscous wave Eq. (5.5.51) can only describe the dissipative wave equation with frequency-squared dependent dissipation, namely $\eta = 2$ [17, 67]

$$\nabla^2 p = \frac{1}{c_0^2}\frac{\partial^2 p}{\partial t^2} + 2\alpha_0 c_0 \frac{\partial}{\partial t}\left(-\nabla^2 p\right).$$  (5.5.51)

The second terms on the right side of Eqs. (5.5.50) and (5.5.51) are the dissipative terms. Hence, the standard dissipative wave equations can only describe the frequency-independent or frequency-squared dependent acoustic dissipation, but cannot describe the anomalous acoustic dissipation in complex media, such as human tissues, with $0 < \eta < 2$. The power-law frequency-dependent energy dissipation in complex media are called anomalous dissipation, and are also called non-exponential relaxation, hysteresis, non-elastic damping, and Constant-Q attenuation in different academic disciplines [68]. The anomalous acoustic dissipation is related to the anomalous diffusion in many papers [17], which will not be discussed here.

Due to the requests of researches on geophysical exploration, new materials, biomechanics, medical engineering and other academic disciplines, in recent years, physical modeling of these anomalous dissipation processes has aroused widespread attention at home and abroad. Anomalous dissipation, in fact, is the power-law frequency-dependent acoustic dissipation. Therefore, some scholars try to model this acoustic attenuation behavior in frequency domain [69], but these models cannot be applied to nonlinear and complex geometric domain. On the other hand, some scholars use nonlinear models to describe the power-law frequency-dependent acoustic attenuation. But nonlinear models are very complex and expensive in numerical calculation, and very difficult in model analysis. The representative models constructed by using standard integer-order partial differential equations in time–space domain are the multi-relaxation model [70] and the adaptation Rayleigh ratio damping model [71]. Although these two models are successful to apply in many cases, but the former needs many parameters which are not easy to get by experiments, are very complex, has lots of variables and expensive computational cost; and the latter is not available to describe the acoustic dissipation of broadband signal and pulse signal [72]. These two models are empirical or semi-empirical models, essentially cannot accurately describe the anomalous energy dissipation processes.

Since Caputo et al. [73] did the pioneering works, fractional derivative is found as a powerful method in describing this kind of physical and mechanics problems [74, 75]. In recent years, the number of papers related to the fractional derivative increases quickly. Fractional derivative models only request a few parameters which can be obtained by fitting experimental data and have clear physical meanings. And fractional derivative models have simple form and can satisfy the causality relation when describing the dissipative acoustic wave and vibration.

To achieve accurate numerical simulation of ultrasonic medical imaging, an acoustic wave model is indispensable to represent the power-law frequency-dependent viscous dissipation. One of the main difficulties of ultrasonic medical imaging is human tissues are very complex and multi-phase media composed by gas, liquid and solid matters. But recent models in describing acoustic wave in complex media are most empirical or semi-empirical models with lots of parameters without clear physical meanings, and classical dissipative wave equations which are used to describe acoustic wave propagation in ideal solid and fluid do not agree with the experimental data. For example, because human body has much water, thus, human body is considered water in traditional ultrasonic medical imaging, however, the calculated and experimental results cannot match well. Taking ultrasonic medical imaging of human body as example, sound speed and energy dissipation have strong relation with physical and mechanics properties of human tissues. For example, sound speed and acoustic dissipation properties of tumors and normal tissues are much different. The standard integer-order derivative models cannot accurately describe the frequency-dependent acoustic attenuation [13]. So far, fractional calculus has become one of the main modeling methods of this kind of complex mechanics problems [13, 43, 76, 77], but the standard fractional derivative is not positive. The following Eq. (5.5.52) is the time-fractional derivative dissipative acoustic wave equation with power-law frequency-dependent attenuation:

$$\nabla^2 p = \frac{1}{c_0^2} \frac{\partial^2 p}{\partial t^2} + \gamma \frac{\partial}{\partial t} \frac{\partial^y p}{\partial t^y} \qquad (5.5.52)$$

where coefficient $\gamma$ is decided by wave speed $c_0$, exponent of frequency power law $\eta$ and attenuation coefficient $\alpha_0$. $\eta$ of Eq. (5.5.52) can be arbitrary real number between 0 and 2.

On the other hand, Professor Szabo in US [17] published the causal time convolutional integral wave equation which can describe the power-law frequency-dependent acoustic attenuation. Although the time convolutional integral wave equation by Szabo is widely used, this equation includes the hypersingular improper integral leading to troubles in numerical computation. Chen and Holm [18] noted the similarity of the attenuation term of Szabo's time convolutional integral wave equation and fractional derivative. To remedy the hypersingular improper integral of the time convolution integral wave equation by Szabo, Chen and Holm [18] introduced the positive fractional derivative and thus proposed the modified Szabo's wave equation:

$$\frac{1}{c_0^2} \frac{\partial^2 P}{\partial t^2} + \frac{2\alpha_0}{c_0} Q_y(P) = \Delta P, \qquad (5.5.53)$$

where

$$Q_y(P) = \begin{cases} \partial P/\partial t & \eta = 0 \\ \frac{\partial^{|\eta|+1} P}{\partial t^{|\eta|+1}} & 0 < \eta < 2 \\ \partial^3 P/\partial t^3 & \eta = 2 \end{cases},$$

where $P$ is sound pressure, $\Delta$ means the Laplace operator, $\alpha_0$ denotes the attenuation coefficient and $\eta$ is the exponent of power law ($\eta > 0$, here $\eta$ means the derivative order of positive fractional derivative), $c_0$ is the wave speed of ultrasound, $\Gamma(\eta)$ is the Gamma function, and the definition of positive fractional derivative is [18]

$$\frac{d^{|\eta|} P(t)}{dt^{|\eta|}} = \begin{cases} \frac{-1}{\eta q(\eta)} \int_0^t \frac{P'(\tau)}{(t-\tau)^\eta} d\tau & 0 < \eta \le 1 \\ \frac{1}{\eta(\eta-1)q(\eta)} \int_0^t \frac{P''(\tau)}{(t-\tau)^{\eta-1}} d\tau & 1 < \eta < 2 \end{cases} \tag{5.5.54}$$

where the constant $q$ is

$$q(\eta) = \frac{\pi}{2\Gamma(\eta+1)\cos[(\eta+1)\pi/2]}.$$

Compared with the fractional derivative, the Fourier transformation of positive fractional derivative is equal to ($|\omega|^\eta P$) and is positive, thus, the modified Szabo's wave equation agrees with the power-law frequency-dependent acoustic attenuation, as shown in the above-mentioned Eqs. (5.5.48) and (5.5.49) [18].

In this subsection, the finite difference method (FDM) is used to numerically solve the modified Szabo's wave equation and simulate the CARI of breast tumors. The FDM discretization of the modified Szabo's wave equation viscous attenuation term with the positive fractional derivative is written as follows [78, 79],

(1)  When $0 < \eta < 1$,

$$\frac{d^{|\eta|+1} P}{dt^{|\eta|+1}} = \frac{-1}{\eta q(\eta)} \int_0^t \frac{P''(\tau)}{(t-\tau)^\eta} d\tau$$

$$\approx \frac{-1}{\eta q(\eta)} \sum_{k=0}^{n-1} \frac{P^{k+2} - 2P^{k+1} + P^k}{\Delta t^2} \int_{k\Delta t}^{(k+1)\Delta t} \frac{d\tau}{(t_n - \tau)^\eta}$$

$$= \frac{-1}{\eta q(\eta)} \sum_{k=0}^{n-1} \frac{P^{n-k+1} - 2P^{n-k} + P^{n-k-1}}{\Delta t^2} \int_{k\Delta t}^{(k+1)\Delta t} \frac{dr}{r^\eta}$$

$$= \frac{-1}{\eta(1-\eta)q(\eta)} \sum_{k=0}^{n-1} \frac{P^{n-k+1} - 2P^{n-k} + P^{n-k-1}}{\Delta t^{\eta+1}}[(k+1)^{1-\eta} - k^{1-\eta}],$$

$$(5.5.55)$$

(2)　when $1 < \eta < 2$,

$$\frac{d^{|\eta|+1}P}{dt^{|\eta|+1}} = \frac{1}{\eta(\eta-1)q(\eta)} \int_0^t \frac{P'''(\tau)}{(t-\tau)^{\eta-1}} d\tau$$

$$\approx \frac{1}{\eta(\eta-1)q(\eta)} \sum_{k=0}^{n-3} \frac{P^{n-k+1} - 3P^{n-k} + 3P^{n-k-1} - P^{n-k-2}}{\Delta t^3} \int_{k\Delta t}^{(k+1)\Delta t} \frac{dr}{r^{\eta-1}}$$

$$+ \frac{1}{\eta(\eta-1)q(\eta)} \frac{P^3 - 3P^2 + 3P^1 - P^0}{\Delta t^3} \int_{(n-2)\Delta t}^{n\Delta t} \frac{dr}{r^{\eta-1}}$$

$$= \frac{1}{(2-\eta)\eta(\eta-1)q(\eta)}$$

$$\sum_{k=0}^{n-3} \frac{P^{n-k+1} - 3P^{n-k} + 3P^{n-k-1} - P^{n-k-2}}{\Delta t^{\eta+1}}[(k+1)^{2-\eta} - k^{2-\eta}]$$

$$+ \frac{1}{(2-\eta)\eta(\eta-1)q(\eta)} \frac{P^3 - 3P^2 + 3P^1 - P^0}{\Delta t^{\eta+1}}[n^{2-\eta} - (n-2)^{2-\eta}].$$

$$(5.5.56)$$

Considering the initial and boundary conditions according to actual ultrasonic medical imaging [80, 81], as shown in Fig. 5.28.

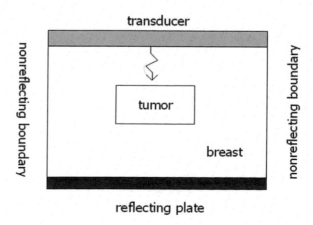

**Fig. 5.28** Two-dimensional configuration of CARI of breast tumors [81]

As tissue is exposed to air during the ultrasonography examination, the initial conditions are stated as

$$P(\mathbf{x}, 0) = P_{atm}, \ \partial P(\mathbf{x}, 0)/\partial t = 0; \tag{5.5.57}$$

and the pressure on the transducer is equal to the pressure $P_{tran}$ of the input ultrasound

$$P(\mathbf{x}_{tran}, t) = P_{tran}(\mathbf{x}, t); \tag{5.5.58}$$

where $P_{tran}$ is inputted ultrasound. The lower plate is the reflecting plate subjected to the reflecting boundary condition

$$\partial P(\mathbf{x}, t)/\partial n = 0; \tag{5.5.59}$$

and the others are non-reflecting boundaries

$$\frac{\partial P(\mathbf{x}, t)}{\partial n} = -\frac{1}{c_0}\frac{\partial P(\mathbf{x}, t)}{\partial t}. \tag{5.5.60}$$

To ensure accuracy and stability of the numerical computation, the following condition should be satisfied [80, 81]

$$\Delta t \le \frac{1}{c_0}\left(\frac{1}{\Delta x^2} + \frac{1}{\Delta y^2}\right)^{-1/2}. \tag{5.5.61}$$

The modified Szabo wave equation is the positive fractional derivative equation. Same with the fractional derivative equation, the computational cost of positive fractional equation is very large. Hence, here we consider a small computational domain. We consider the 20 mm × 20 mm homogeneous breast tissue and another one with a 2 mm × 4 mm tumor in its center. The 3.75 Hz ultrasound is transformed into the tissue. The mean wave speeds and attenuation parameters in the fatty tissue and the tumor, respectively, are $c_{0fat} = 1475$m/s, $\eta_{fat} = 1.7$, $\alpha_{0fat} = 15.8/(2\pi)^{1.7}$dB/m/MHz$^{1.7}$ and $c_{0tum}=1527$m/s, $\eta_{tum} =1.3$, $\alpha_{0tum} = 57.0/(2\pi)^{1.3}$dB/m/MHz$^{1.3}$, which were got from the medicinal examination [82, 83]. Wave length $\lambda = c_0/f \approx 0.4$ mm, the time before reflection is about 13.33μs. In this subsection, the time step is $\Delta t = 2.6657 \times 10^{-2}$μs, the space grid sizes are $\Delta x = 0.05$ mm, $\Delta y = 0.1$ mm ($x$ is the direction of wave propagation, $y$ is the direction of reflecting plate). The inputted ultrasound is [81]

$$P_{tran}(\mathbf{x}, t) = \begin{cases} P_{atm}\left[1 + \frac{\cos\omega t}{2}\left(1 + \cos\frac{\omega t}{4}\right)\right] & 0 \le t \le 1.2\mu s, \\ P_{atm} & t \ge 1.2\mu s, \end{cases} \tag{5.5.62}$$

where $P_{tran}$ is the inputted ultrasound, $P_{atm}$ is the standard atmospheric pressure, $\omega_1$ $= 2\pi f$, $\omega_2 = \omega_1/4$.

By numerical computation, Figs. 5.29 and 5.30 show numerical results of normal tissues and tissue with a tumor in the center.

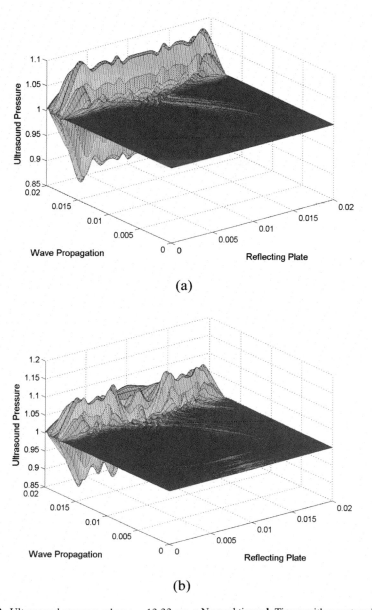

(a)

(b)

**Fig. 5.29**   Ultrasound pressure when $t = 13.33\ \mu s$. **a** Normal tissue **b** Tissue with a centered tumor

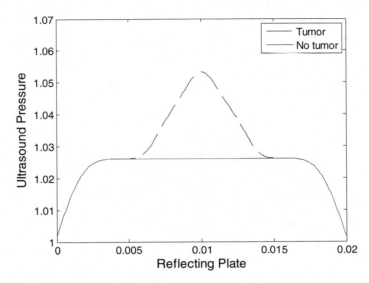

**Fig. 5.30** Ultrasound pressure along reflecting plate ($x = 20$ mm) when $t = 13.33$ μs. Dash line means the pressure of normal tissue; solid line denotes the pressure of normal tissue

From the numerical results shown in Fig. 5.29, the profile of normalized ultrasound pressure displays fluctuation when ultrasound travels through the region of tumor, which differs greatly from that of the normal tissue. Additionally, the reflecting line of ultrasound pressure of the normal tissue is straight. Otherwise, if there is a tumor in the breast tissue, as shown in Fig. 5.30, the reflecting line is enhanced, which is a recognizable signature of tumor existence inside the fatty tissue. This finding coincides with clinical observations. Our above numerical results suggest that the modified Szabo's wave equation can well describe the frequency-dependent power law of acoustic dissipation in human tissues.

**Power-law Frequency Dependency and Fractal Characters of Media**

Chen [84] rewrites the power-law dissipative formula (5.5.49) as

$$\eta = \frac{\ln \alpha(\omega)/\alpha_0}{\ln|\omega|}, \qquad (5.5.63)$$

which clearly reveals the self-similarity of power-law frequency-dependent dissipation. Fractal can portray the similarity of medium. Exponent of power law $\eta$ can be explained as fractal dimension of medium. Mandelbrot [85] and Sato [86] noted the relation between the Lévy stable distribution and fractal. Section 2.8.3 points that $\eta$, in fact, is exponent of the Lévy distribution. Hence, $\eta$ is the fractal dimension.

Exponent of power law $\eta$ is keeps constant in the wide range of frequency, which illustrates that $\eta$ is decided by spacial structures and denotes fractal dimension of diffusion processes. For example, exponent of power law $\eta$ of different human tissues is different which means the fractal dimension $\eta$ portrays statistical geometric characteristics of biological macromolecules and determines the physical and mechanics behavior of the media.

Herrchen [87] pointed self-similarity of real physical problems only exists in certain scales. This agrees with many experimental results. Exponent of power law $\eta$ of power-law frequency-dependent dissipation keeps constant in a certain range of frequency.

### 5.5.6 Fractional Laplacian Equation of Power-Law Frequency-Dependent Dissipative Acoustic Wave

The time-fractional derivative models introduced in subsection 5.5.5 simplified the time–space energy dissipation process (as shown in the second term in right side of Eq. (5.5.51)) to a time process described by time-fractional derivative operator (as shown in Eqs. (5.5.52) and (5.5.53)). Pure-time modeling of attenuation considers only the correlation between time and energy dissipation. Energy dissipation is not only a time-dependent process, but also depends on space. Ochmann and Makarov [88] point that this simplification is only possible if the interaction between two oppositely traveling sound waves can be neglected and the thermoviscous term is relatively small. From a mathematical point of view, when the anomalous energy dissipation with exponent $\eta$ greater than 1, master equation has time derivative with order more than 2 and needs not only the initial displacement and speed conditions, but also requires initial acceleration conditions. But for many actual engineering problems, initial acceleration is unknown. In addition, Pure-time modeling with time derivative cannot describe the spacial anisotropy of medium, and only represents the time average results. In one word, time-fractional derivative models have limitations. Hence, applying spacial fractional derivative models to describe the anomalous acoustic dissipation is very meaningful.

Spatial fractional derivative is the fractional Laplace operator based on the fractional Riesz potential [89] (also called the Riesz fractional derivative, refer to Sect. 2.1.4). The spacial fractional derivative is used to model various physical and mechanics problems in recent ten years, such as the anomalous diffusion equation. This operator is defined by Fourier transformation. Cheng and Chen did some path-breaking works. However, the definition based on Fourier transformation is very useful in mathematical analysis, but are not available for applications in engineering fields. Then, the definition of spatial fractional derivative based on the approximated difference operator is proposed and widely used. But this kind of definition has hyper-singularity, which is applicable only for simply boundary conditions [90, 91], and are not available in engineering modeling and numerical computation. Fractional Laplacian is far inferior to the time-fractional derivative in research on theory, application and numerical algorithms.

Based on Caputo's fractional derivative, Chen and Holm [90] proposed the analytic definition of fractional Laplace operator. This definition does not use the Fourier transformation and remedy the hypersingularity of the definition based on differential operator.

Thus, Chen and Holm [90] deduced the fractional Laplacian wave equation which is causal and can describe the power-law frequency-dependent acoustic dissipation

$$\nabla^2 p = \frac{1}{c_0^2} \frac{\partial^2 p}{\partial t^2} + \frac{2\alpha_0}{c_0^{1-y}} \frac{\partial}{\partial t} (-\nabla^2)^{\eta/2} p, \tag{5.5.64}$$

where $c_0$ is the wave speed, $(-\nabla^2)^{y/2}$ is the fractional Laplacian operator, $0 \le \eta \le 2$. When $\eta = 0$ or 2, Eq. (5.5.64) reduces to the classical damped wave Eq. (5.5.50) and the thermoviscous wave Eq. (5.5.51).

Because the traditional definition of fractional Laplacian is defective, researches of numerical computation on the fractional Laplacian are only on one-dimensional problems [92–94]. Chen and Holm [90] proposed an analytic definition of fractional Laplacian which can be used to develop numerical method of two or three-dimensional fractional Laplacian wave equations. Chen and Zhang [95] have made the first attempt to develop a numerical approach for the two or three-dimensional fractional Laplacian wave equation. The proposed method introduces a numerical integral technique of boundary element method which employs triangular subelements with local coordinate systems to obtain a weak singular integral. Consequently, the numerical discretization of the fractional Laplacian wave equation is established by combining the above-mentioned numerical integral method and the Galerkin finite element formulation. In addition, we numerically simulate the ultrasound propagation through human tissues in ultrasonic medical imaging of breast tumors by the clinical amplitude-velocity reconstruction imaging (CARI). Two-dimensional numerical results verify the effectiveness of this new scheme and show that the fractional Laplacian wave equation describes well the power-law frequency-dependent acoustic attenuation. The spatial fractional Laplacian, in fact, is a global convolution operator. Its computing cost is very expensive. The development of the fast algorithm is an important issue.

With the numerical technique for singular integral, the finite element formulation of fractional Laplacian wave equation is stated as follows:

$$\sum_e \int \left[ (\nabla \mathbf{N})^T (\nabla \mathbf{N}) \mathbf{P}^e + \mathbf{N}^T \frac{1}{c_0^2} \frac{\partial^2}{\partial t^2} \mathbf{N} \mathbf{P}^e + \mathbf{N}^T \frac{2\alpha_0}{c_0^{1-\eta}} \frac{\partial}{\partial t} \mathbf{Q}^e \mathbf{P} \right] d\Omega$$
$$- \sum_e \int \mathbf{N}^T \frac{\partial \mathbf{N}}{\partial n} \mathbf{P}^e dS' + \sum_e \int \mathbf{N}^T \frac{1}{c_0} \frac{\partial}{\partial t} \mathbf{N} \mathbf{P}^e dS_2 = 0 \tag{5.5.65}$$

$$Q^e = \begin{bmatrix} q_{1k_1}^e & q_{1k_2}^e & \cdots & q_{1k_n}^e \\ q_{2k_1}^e & q_{2k_2}^e & \cdots & q_{2k_n}^e \\ \vdots & \vdots & \ddots & \vdots \\ q_{Nk_1}^e & q_{Nk_2}^e & \cdots & q_{Nk_n}^e \end{bmatrix}_{N \times n}$$

where $\eta$ is the order of the fractional Laplacian. Overall coefficient matrix is assembled with element coefficient matrix and the finite element discretization can be written as

$$\frac{\partial^2}{\partial t^2}\mathbf{MP} + \frac{\partial}{\partial t}\mathbf{BP} + \mathbf{KP} = 0, \qquad (5.5.66)$$

where

$$\mathbf{M} = \mathbf{C} = \frac{1}{c_0^2}\sum_e \int \mathbf{N}^T \mathbf{N} d\Omega$$
$$\mathbf{B} = \sum_e \int \mathbf{N}^T \frac{2\alpha_0}{c_0^{1-\eta}} \mathbf{Q}^e d\Omega + \sum_e \int \mathbf{N}^T \frac{1}{c_0} \mathbf{N} dS_2 \quad .$$
$$\mathbf{K} = \sum_e \int (\nabla \mathbf{N})^T (\nabla \mathbf{N}) d\Omega - \sum_e \int \mathbf{N}^T \frac{\partial \mathbf{N}}{\partial n} dS'$$

Using the above finite difference method in time domain, we can numerically calculate the fractional Laplacian wave equation.

We employ 800 8-node quadrilateral isoparametric structural elements, and consider 4 mm × 2 mm tissue and 0.4 mm × 0.4 mm centered tumor here. The element size is 0.1 mm × 0.1 mm, and time step is $6.5490 \times 10^{-3}$ ms. Numerical results in Fig. 5.31 show that the sound pressure profile along reflecting line is enhanced which indicates the suspicious existence of the tumor, same as clinical observations. The accuracy of this numerical method needs to be further improved.

One of the main research goals of soft matter mechanics is to construct the mechanics phenomenological constitutive equation of soft matters. Acoustic wave propagation and damped vibration can reflect the mechanics properties of soft matters and related models and researches are meaningful in many engineering fields. For example, the damping property of dampers which are used to shock absorption of buildings do not like that of viscous fluid, but show power-law frequency-dependent energy dissipation; anomalous dissipation of seismic waves in porous rocks and oil in seismic exploration; biological bodies (such as cell, blood, protein, DNA and so on) are typical of soft matters. To achieve accurate numerical simulation of ultrasonic medical imaging, an acoustic wave model is indispensable to represent the power-law frequency-dependent viscous dissipation.

### 5.5.7  Fractional Derivative Models of Nonlinear Acoustic Wave

Most nonlinear acoustic wave models are only available to describe the frequency-independent and frequency-squared dependent acoustic dissipation [96]. Thus, their actual scope of application has been largely restricted. This subsection refers to reference [18] by Chen and Holm. The fractional Laplacian wave equation introduced in Sect. 5.7.6 is generalized to describe the nonlinear dissipative acoustic

**Fig. 5.31** Normalized ultrasound pressure when $t$ = 1.3 ms, **a** pressure in the normal breast tissue, **b** pressure in the tissue with a centered tumor. **c** Normalized ultrasound pressure along the reflecting line ($x$ = 2 mm) when $t$ = 1.3 ms. The solid line denotes the sound pressure profile of the normal tissue, while the dashed line means the sound pressure profile of the tissue with a centered tumor

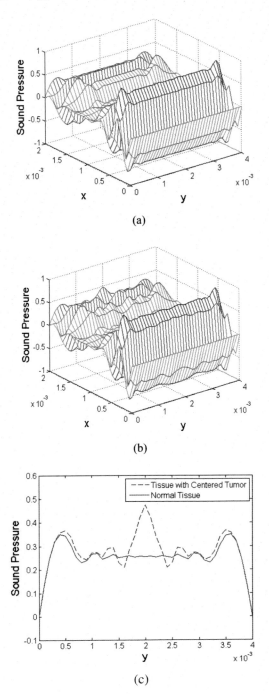

wave with power-law frequency dependent dissipation. It is pointed that the dissipative term of the nonlinear dissipative wave equation has the same form with that of the linear dissipative wave equation. Blackstock [97] proposed a simple method in constructing nonlinear anomalous diffusion equation which replaces the dissipative term of nonlinear models with the corresponding dissipative terms of linear models. Blackstock [97] used perturbation methods to verify this method. Szabo [17] applied this method to generalize the time convolutional integral wave equation to nonlinear acoustic wave equation, such as Burgers, KZK and Westervelt equations. Chen and Holm [18] used the same method to generalize the fractional Laplacian wave equation to the nonlinear acoustic wave equation.

One-dimensional plane wave expression of the Burgers equation, one of the most simple nonlinear acoustic wave equation:

$$\frac{\partial p}{\partial t} + Bp\frac{\partial p}{\partial z} - \varepsilon\frac{\partial^2 p}{\partial z^2} = 0, \tag{5.5.67}$$

where $B$ is the nonlinear coefficient [98], $\varepsilon$ denotes a constant which is proportional to viscous and thermal-conductivity coefficient [97]. It is known that the integer-order derivative Burgers equation [96, 97] describes the frequency-squared dependent acoustic dissipation. In this subsection, we neglect the detailed deducing process and directly give the following fractional derivative Burgers dissipative acoustic equation.

$$\frac{\partial p}{\partial t} + Bp\frac{\partial p}{\partial z} + 2\alpha_0 c_0^{1+y}\left(-\frac{\partial^2}{\partial z^2}\right)^{y/2} p = 0. \tag{5.5.68}$$

The above fractional derivative model (5.5.68) belongs to fractal Burgers equation or fractional derivative convection–diffusion equation. Biler et al. [99] introduced this equation in detail. The three-dimensional form of this equation can be written as

$$\frac{\partial p}{\partial t} + Bp \cdot \nabla p + 2\alpha_0 c_0^{1+y}\left(-\nabla^2\right)^{y/2} p = 0, \tag{5.5.69}$$

where $\nabla p$ denotes pressure gradient vector, $< \cdot >$ means the dot product of vectors. Ochmann and Makarov [88] also developed the time-fractional derivative Burgers equation. Similarly, Chen and Holm [18] also developed nonlinear KZK, Westervelt, Boussinesq models with fractional Laplacian which can describe the power-law frequency-dependent dissipation.

### 5.5.8   Fractional Derivative Seismic Wave Model

Lots of researches on seismic wave show that its attenuation is proportional to frequency, namely $Q$ is frequency-independent, thus, the constant $Q$ model is an effective method to describe the attenuation of seismic wave [100]. The so-called $Q$ is the quality factor

$$Q = \frac{\omega m}{c}, \tag{5.5.70}$$

is used to measure the medium damping and energy dissipation of wave propagation, where $\omega$ means the natural frequency of vibration without damping, $c$ is the damping coefficient, $m$ is the mass [58]. The lower $Q$ is, the medium damping and energy dissipation of wave will be larger. Carcione et al. [100] proposed the fractional derivative models with constant $Q$ to describe the seismic wave. The stratum is not ideal elastic body, but viscoelastic body. Kjartansson considered one-dimensional viscoelastic constitutive relation of the stratum as [101]:

$$\sigma(t) = J(t) * \dot{\varepsilon}(t), \quad J(t) = \frac{K_0}{\Gamma(1 - 2\gamma)} \left(\frac{t}{\tau}\right)^{-2\gamma} H(t), \tag{5.5.71}$$

where $J(t)$ is relaxation function, $t$ the time, * means the convolutional integral, $\tau$ is the relaxation time, $H$ is the step function, $K_0$ is the bulk modulus, $\gamma$ is a constant coefficient

$$\gamma = \frac{1}{\pi} \tan^{-1}\left(\frac{1}{Q}\right) (0 < \gamma < 1). \tag{5.5.72}$$

By the Fourier transformation and analysis in frequency domain, the constitutive relation (5.5.72) can be rewritten as [101]

$$\overline{\sigma}(\omega) = i\omega \overline{J}(\omega)\varepsilon(\omega), \quad M(\omega) = i\omega J(\omega) = K_0 \left(\frac{i\omega}{\omega_0}\right)^{2\gamma}. \tag{5.5.73}$$

Based on the frequency domain constitutive relation (5.5.73) and the Fourier transformation of fractional derivative, the time domain constitutive relation is [43, 102]

$$\sigma = \rho D \frac{\partial^{2\gamma}\varepsilon}{\partial t^{2\gamma}}. \tag{5.5.74}$$

Carcione et al. [100] used the constitutive relation in Eq. (5.5.71) to get the following fractional wave equation to describe the seismic wave:

$$\frac{\partial^{\beta}u}{\partial t^{\beta}} = D\Delta u + f, \quad \beta = 2 - 2\gamma, \tag{5.5.75}$$

where $u$ is the displacement, $\beta$ is the time-fractional derivative order, $D$ is the material parameter. Reference [100] uses the fractional derivative constant Q model to numerically simulate seismic wave and compare the numerical results with the experimental data. The numerical results can agree with the experimental data.

## 5.6  The Fractional Variational Principle of Mechanics

### 5.6.1  Variational Principle of Mechanics

Variational principle is one of the basic principles in the theory of classical analytical mechanics, mainly involving Hamilton's principle, Lagrange equations, Hamilton–Jacobi equations, Hamiltonian canonical equations and other important principles of mechanics [57, 103]. Variational principle was first used in the study of mechanical and optical problems, and then gradually expanded to various physical areas. In physics, variational principle is called the principle of least action, that is to say, the real path from one state to another makes one action get the minimum value. Variational principle is a mathematical form of the principle of least action, and Hamilton's principle is an important method to study dynamical systems and quantum mechanics. By utilizing Hamilton's principle, as long as the Hamiltonian of the system is obtained, the characteristics of the phase space of the dynamical system are easy to achieve. Meanwhile, Hamilton's principle is an effective tool to study the chaotic characteristics of nonlinear dynamical systems.

The following is a brief introduction to the variational principle [57, 103]. Assume that $\Pi$ is the function of $L(x)$, and for all functions with the endpoint $A$ and $B$, $L_1(x)$ makes functional $\Pi$ get the minimum value. The function $\Pi$ can be written as $\Pi(L(x)) = \Pi(L_1(x) + af(x))$. Mathematical expression of variation is as follows:

$$\delta\Pi = \left.\frac{\partial\Pi(L_1(x) + af(x))}{\partial a}\right|_{a\to 0} \cdot \delta a. \qquad (5.6.1)$$

If $L = L_1$, its variation value equals 0, i.e. the situation that functional variation is zero corresponds to that functional gets the extreme value. The mathematical expression of the principle of least action is that the variation of the integration of a variable along the path is zero. That's the variational principle. Assuming that in function $L = L(t, q, \dot{q})$, $q$ is generalized coordinate and $\dot{q}$ is generalized velocity, representing arbitrary coordinates and velocity components in arbitrary coordinate respectively. Supposing that the functional $\Pi$ is an integration with respect to $L$ from $t_A$ to $t_B$, then its form is as follows [103]

$$\Pi = \int_{t_A}^{t_B} L(t, q, \dot{q})\mathrm{d}t. \qquad (5.6.2)$$

To get the extreme value, then set the variation as 0

$$\delta \int_{t_A}^{t_B} L(t, q, \dot{q}) dt = \int_{t_A}^{t_B} \frac{\partial L(t, q, \dot{q})}{\partial a}\bigg|_{a=0} dt = 0. \qquad (5.6.3)$$

Expand Eq. (5.6.3) as follows

$$\int_{t_A}^{t_B} \left[ \frac{\partial L(t, q, \dot{q})}{\partial t} \frac{\partial t}{\partial a}\bigg|_{a\to 0} + \frac{\partial L(t, q, \dot{q})}{\partial q} \frac{\partial q}{\partial a}\bigg|_{a\to 0} + \frac{\partial L(t, q, \dot{q})}{\partial \dot{q}} \frac{\partial \dot{q}}{\partial a}\bigg|_{a\to 0} \right] dt = 0. \qquad (5.6.4)$$

As $\frac{\partial L(t,q,\dot{q})}{\partial t} \frac{\partial t}{\partial a}\big|_{a\to 0} = 0$, it is straightforward to show

$$\int_{t_A}^{t_B} \left[ \frac{\partial L(t, q, \dot{q})}{\partial q} \delta q + \frac{\partial L(t, q, \dot{q})}{\partial \dot{q}} \delta \dot{q} \right] dt = 0. \qquad (5.6.5)$$

By integrating the second item by parts, then the following form is obtained

$$\int_{t_A}^{t_B} \left[ \frac{\partial L(t, q, \dot{q})}{\partial q} \delta q - \frac{d}{dt} \frac{\partial L(t, q, \dot{q})}{\partial \dot{q}} \delta q \right] dt + \frac{\partial L(t, q, \dot{q})}{\partial \dot{q}} \delta q \bigg|_{t=t_B}^{t=t_A} = 0. \qquad (5.6.6)$$

As the endpoint is fixed, then $\delta q|_{t=t_A} = \delta q|_{t=t_B} = 0$ and

$$\int_{t_A}^{t_B} \left[ \frac{\partial L(t, q, \dot{q})}{\partial q} - \frac{d}{dt} \frac{\partial L(t, q, \dot{q})}{\partial \dot{q}} \right] \delta q \, dt = 0. \qquad (5.6.7)$$

As the above equation is established for arbitrary start and end points, the following Euler–Lagrange equation is obtained [103]

$$\frac{\partial L(t, q, \dot{q})}{\partial q} - \frac{d}{dt} \frac{\partial L(t, q, \dot{q})}{\partial \dot{q}} = 0. \qquad (5.6.8)$$

For systems with multi-degrees of freedom, by using the variational principle, Euler–Lagrange equations set can be obtained [104]

$$\frac{\partial L}{\partial q_i} - \frac{d}{dt} \frac{\partial L}{\partial \dot{q}_i} = 0 \quad (i = 1, 2, \ldots n). \qquad (5.6.9)$$

In mechanics, Eq. (5.6.9) is called Lagrange equations, here $L = T - V$, $L$ is Lagrange function, $T$ is kinetic energy and $V$ is potential energy. For oscillator with singular degree of freedom, its kinetic energy is $T = m\dot{x}^2/2$, potential energy is $V = kx^2/2$, and the Lagrange function is $L = m\dot{x}^2/2 - kx^2/2$. Substitute the Lagrange function $L$ into Euler–Lagrange equation and get the equation of the motion of oscillator

$$m\ddot{x} = -kx. \tag{5.6.10}$$

This simple example proves that Lagrange equation equals Newton's second law. Similarly, Lagrange equation under the generalized coordinates can also be deduced from Newton's second law [104].

The above variational principle applies to conservative systems, where the external force is a conservative force, and energy exchanging is only between kinetic energy and potential energy. If external force is a kind of non-conservative force containing, for example, the viscous resistance, mechanical energy will transfer into heat energy, then the above Lagrange variational principle is no longer applicable, and the non-conservative force power should be introduced. Set [57]

$$\int_{t_A}^{t_B} \left[ \delta L(t, q, \dot{q}) \mathrm{d}t + \sum_i f \delta q_i \right] = 0, \tag{5.6.11}$$

where $\sum_i f \delta q_i$ is the virtual work of the non-conservative force. Thus, its corresponding variational principle is as follows [57]

$$\frac{\partial L}{\partial q_i} - \frac{\mathrm{d}}{\mathrm{d}t} \frac{\partial L}{\partial \dot{q}_i} = -f. \tag{5.6.12}$$

Considering the variational problems with constraints, if its constraint condition is $\varphi_r(t, q_i, \dot{q}_i) = 0$, then Lagrange multiplier method can be used, and Lagrange function is as follows [103]

$$\overline{L}(t, q_i, \dot{q}_i) = L(t, q_i, \dot{q}_i) + \sum_r \lambda_r \varphi_r(t, q_i, \dot{q}_i). \tag{5.6.13}$$

Thus, Euler–Lagrange equation with constraint condition is obtained

$$\frac{\partial L}{\partial q_i} - \frac{\mathrm{d}}{\mathrm{d}t} \frac{\partial L}{\partial \dot{q}_i} + \sum_r \lambda_r \left( \frac{\partial \varphi_r(t, q_i, \dot{q}_i)}{\partial q_i} - \frac{\mathrm{d}}{\mathrm{d}t} \frac{\partial \varphi_r(t, q_i, \dot{q}_i)}{\partial \dot{q}_i} \right) = 0. \tag{5.6.14}$$

Hamilton further developed the Lagrange principle. He defined the Hamiltonian as [104]

$$H = \sum_i p_i \dot{q}_i - L. \tag{5.6.15}$$

Here $p_i$ corresponds to generalized momentum of the generalized coordinate $q_i$, i.e., $p_i = m\dot{q}_i$. Therefore, the kinetic energy is the double of $\sum_i p_i \dot{q}_i$,

$$H = 2T - L = 2T - (T - V) = T + V, \tag{5.6.16}$$

where $H$ is total mechanical energy. If $H$ is a constant, it corresponds to energy conservative system.

The following will give an introduction to Hamilton canonical equations according to the relationship between Hamiltonian $H$ and Lagrange equations $L$ [57]. If we take Hamiltonian $H$ as a function of generalized coordinates, generalized momentum and time, i.e. $H = H(q_i, p_i, t)$, then the total differential of Hamiltonian $H$ is

$$dH = \sum_i \frac{\partial H}{\partial q_i} dq_i + \sum_i \frac{\partial H}{\partial p_i} dp_i + \frac{\partial H}{\partial t} dt. \tag{5.6.17}$$

By using the relationship between Hamiltonian $H$ and Lagrange equations $L$, the following equation is

$$dH = \sum_i p_i d\dot{q}_i + \sum_i \dot{q}_i dp_i - \sum_i \frac{\partial L}{\partial q_i} dq_i - \sum_i \frac{\partial L}{\partial \dot{q}_i} d\dot{q}_i - \frac{\partial L}{\partial t} dt. \tag{5.6.18}$$

As

$$\frac{\partial L}{\partial \dot{q}_i} = p_i, \quad \frac{\partial L}{\partial q_i} = \dot{p}_i, \tag{5.6.19}$$

Equation (5.6.18) can be written as

$$dH = \sum_i \dot{q}_i dp_i - \sum_i \dot{p}_i dq_i - \frac{\partial L}{\partial t} dt. \tag{5.6.20}$$

Comparing Eqs. (5.6.17) and (5.6.18), Hamilton canonical equations are obtained [57]

$$\frac{\partial H}{\partial q_i} = -\dot{p}_i, \quad \frac{\partial H}{\partial p_i} = \dot{q}_i, \quad \frac{\partial H}{\partial t} = -\frac{\partial L}{\partial t}. \tag{5.6.21}$$

At last, a brief introduction to Hamilton–Jacobi equation is given. Substitute $H = \sum_i p_i \dot{q}_i - L$ and Hamiltonian $K(\mathbf{Q}, \mathbf{P}, t)$ which is related to initial coordinate vector $\mathbf{Q}$ and initial momentum vector $\mathbf{P}$, into Lagrange equations and get [57]

$$\delta \int_{t_A}^{t_B} [\mathbf{p} \cdot \dot{\mathbf{q}} - H(\mathbf{q}, \mathbf{p}, t)]dt = 0 \text{ and } \delta \int_{t_A}^{t_B} [\mathbf{P} \cdot \dot{\mathbf{Q}} - K(\mathbf{Q}, \mathbf{P}, t)]dt = 0. \quad (5.6.22)$$

Then

$$\mathbf{P} \cdot \dot{\mathbf{Q}} - K(\mathbf{Q}, \mathbf{P}, t) = \mathbf{p} \cdot \dot{\mathbf{q}} - H(\mathbf{q}, \mathbf{p}, t) - \frac{dS(\mathbf{q}, \mathbf{P}, t)}{dt}. \quad (5.6.23)$$

Hamilton's principle that formally equals to Lagrange principle is

$$\delta S = \delta \int L = 0 \quad \left( \frac{dS}{dt} = L \right), \quad (5.6.24)$$

where $S$ is the Hamiltonian main function [57]. Hamiltonian main function can be written as [57]

$$S(\mathbf{q}, \mathbf{P}, t) = F(\mathbf{q}, \mathbf{P}, t) - \mathbf{P} \cdot \mathbf{Q}. \quad (5.6.25)$$

Then Lagrangian is equivalent to

$$\frac{dS}{dt} = \frac{\partial S}{\partial t} + \sum_i \frac{\partial F}{\partial q_i} \dot{q}_i + \sum_i \frac{\partial F}{\partial P_i} \dot{P}_i - \sum_i \dot{P}_i Q_i - \sum_i P_i \dot{Q}_i. \quad (5.6.26)$$

Substitute the above equation into Eq. (5.6.23), and set generalized coordinates independent, i.e., make Eq. (5.6.23) contains no $\dot{q}_i$ and $\dot{P}_i$ [57], then

$$\frac{\partial S}{\partial q_i} = \frac{\partial F}{\partial q_i} = p_i, \frac{\partial F}{\partial P_i} = Q_i, K - H = \frac{\partial S}{\partial t}. \quad (5.6.27)$$

Without loss of generality, $K = 0$, then Hamilton–Jacobi equations are obtained [57]

$$H\left(q_i, \frac{\partial S}{\partial q_i}, t\right) + \frac{\partial S}{\partial t} = 0. \quad (5.6.28)$$

## 5.6.2  Fractional Variational Principle

With the development of dynamics, many works have introduced the fractional derivative damping to the equations of motion, and established fractional dynamic equation to describe the non-conservative damping systems. The development of fractional dynamic systems has endued fractional variational principle important practical significance. In this section, two types of fractional variational principles are discussed: one is the variational principle applied to the fractional derivative terms contained by Lagrange function $L$ and Hamiltonian $H$ [105–109], the other is the Lagrange equations of non-conservative derived from using the variation of the fractional integration [110].

## 1    Lagrange function L and Hamiltonian H with fractional derivatives

Here Riemann–Liouville fractional derivative definition is introduced, and Lagrange function contains Caputo fractional derivative [107]

$$L(t, q, {}^C D_L^\alpha q, {}^C D_R^\beta q).$$

The left and right Caputo fractional derivative definition are as follows:

$$
\begin{aligned}
{}^C D_L^\alpha f(t) &= \frac{1}{\Gamma(n-\alpha)} \int_{t_A}^{T} (t - \tau)^{n-\alpha-1} f^{(n)}(\tau) \mathrm{d}\tau \\
{}^C D_R^\alpha f(t) &= \frac{(-1)^n}{\Gamma(n-\alpha)} \int_{t}^{t_B} (t - \tau)^{n-\alpha-1} f^{(n)}(\tau) \mathrm{d}\tau
\end{aligned}
\tag{5.6.29}
$$

In order to facilitate the writing, the left and right Caputo fractional derivative definition are denoted as ${}^C D_L^\alpha$ and ${}^C D_R^\beta$.

The following introduces the Euler–Lagrange equations that contain the fractional Lagrange function. Similar to Eq. (5.6.3), let's set that the functional $\Pi$ is an integration with respect to $L$ from $t_A$ to $t_B$ [107]

$$
\Pi = \delta \int_{t_A}^{t_B} L(t, q, {}^C D_L^\alpha q, {}^C D_R^\beta q) \mathrm{d}t = \delta a \cdot \int_{t_A}^{t_B} \frac{\partial L(t, q, {}^C D_L^\alpha q, {}^C D_R^\beta q)}{\partial a} \Big|_{a=0} \mathrm{d}t = 0.
\tag{5.6.30}
$$

Based on Eq. (5.6.30), we can get

$$
\int_{t_A}^{t_B} \left[ \frac{\partial L(t, q, {}^C D_L^\alpha q, {}^C D_R^\beta q)}{\partial q} \delta q + \frac{\partial L(t, q, {}^C D_L^\alpha q, {}^C D_R^\beta q)}{\partial^C D_L^\alpha q} {}^C D_L^\alpha \delta q \right.
$$
$$
\left. + \frac{\partial L(t, q, {}^C D_L^\alpha q, {}^C D_R^\beta q)}{\partial^C D_R^\beta q} {}^C D_R^\beta \delta q \right] \mathrm{d}t = 0
\tag{5.6.31}
$$

Fractional Euler–Lagrange equations can be obtained by further integration by parts [106]

$$\frac{\partial L}{\partial q} + D_R^\alpha \frac{\partial L}{\partial^C D_L^\alpha q}$$

$$+ D_L^\beta \frac{\partial L}{\partial^C D_R^\beta q} = 0. \tag{5.6.32}$$

Here, $D_R^\alpha$ and $D_L^\beta$ are left and right Riemann–Liouville fractional derivative [105]

$$\frac{d^\alpha f(x)}{dt^\alpha}^L = \frac{1}{\Gamma(n-\alpha)} \frac{d^n}{dt^n} \int_{t_A}^{T} (t-\tau)^{n-\alpha-1} f(\tau) d\tau$$

$$\frac{d^\alpha f(x)}{dt^\alpha}^R = \frac{(-1)^n}{\Gamma(n-\alpha)} \frac{d^n}{dt^n} \int_{t}^{t_B} (t-\tau)^{n-\alpha-1} f(\tau) d\tau$$

Fractional Euler–Lagrange equation has a similar form with the classical Euler–Lagrange equation. The equation reflects the energy dissipation characteristics of non-conservative system by applying fractional derivative to generalized coordinates. Fractional derivative has the property of memory, so the application of fractional derivative can reflect the memory characteristics appearing in the description of the physical and mechanics processes.

When the variational problem with constraints is considered, and the constraint condition is supposed to be $\varphi = 0$, Lagrange multiplier method can also be employed. Lagrange function is as follows [106]

$$\bar{L} = L + \lambda\varphi. \tag{5.6.33}$$

References [105] and [107] proposed fractional Hamiltonian $H$, related fractional Hamiltonian canonical equation and fractional Hamilton–Jacobi equation based on the traditional variational principle. The generalized momentum of fractional Lagrange function is [105]

$$p_\alpha = \frac{\partial L}{\partial^C D_L^\alpha q}, \quad p_\beta = \frac{\partial L}{\partial^C D_R^\beta q}. \tag{5.6.34}$$

Then the relationship between fractional Hamiltonian $H$ and fractional Lagrange multiplier $L$ is [107]

$$H(t, q, p_\alpha, p_\beta) = p_\alpha {}^C D_L^\alpha q + p_\beta {}^C D_R^\beta q - L. \tag{5.6.35}$$

The total differential of fractional Hamiltonian $H$ is

$$dH = \frac{\partial H}{\partial q} dq + \frac{\partial H}{\partial p_\alpha} dp_\alpha + \frac{\partial H}{\partial p_\beta} dp_\beta + \frac{\partial H}{\partial t} dt. \tag{5.6.36}$$

Utilizing Eq. (5.6.35), the total differential of fractional Hamiltonian $H$ can be written as

$$dH = {}^C D_L^\alpha q \mathrm{d}p_\alpha + p_\alpha \mathrm{d}^C D_L^\alpha q + {}^C D_R^\beta q \mathrm{d}p_\beta + p_\beta \mathrm{d}^C D_R^\beta q - \frac{\partial L}{\partial q} \mathrm{d}q$$

$$- \frac{\partial L}{\partial {}^C D_L^\alpha q} \mathrm{d}^C D_L^\alpha q - \frac{\partial L}{\partial {}^C D_R^\alpha q} \mathrm{d}^C D_R^\alpha q - \frac{\partial L}{\partial t} \mathrm{d}t; \qquad (5.6.37)$$

Fractional Hamilton–Jacobi equation is obtained by using (5.6.34) [107]

$$\frac{\partial H}{\partial q} = D_R^\alpha p_\alpha + D_L^\beta p_\beta, \frac{\partial H}{\partial p_\alpha} = {}^C D_L^\alpha q, \frac{\partial H}{\partial p_\beta} = {}^C D_R^\beta q, \frac{\partial H}{\partial t} = -\frac{\partial L}{\partial t}. \quad (5.6.38)$$

The following is an introduction of fractional Hamilton–Jacobi equation that given by references [107–109, 111]. Substitute Eq. (5.6.35) and Hamiltonian $K(Q, P_\alpha, P_\beta, t)$ which is related to the initial coordinates vector $Q$ and initial momentum vector $P_\alpha, P_\beta$, into the Lagrange function and get [107]

$$\delta \int_{t_A}^{t_B} [p_\alpha {}^C D_L^\alpha q + p_\beta {}^C D_R^\beta q - H(t, q, p_\alpha, p_\beta)] \mathrm{d}t = 0$$

$$\delta \int_{t_A}^{t_B} [P_\alpha {}^C D_L^\alpha Q + P_\beta {}^C D_R^\beta Q - K(t, Q, P_\alpha, P_\beta)] \mathrm{d}t = 0 \qquad (5.6.39)$$

According to Eq. (5.6.39), we can get [107–109]

$$p_\alpha {}^C D_L^\alpha q + p_\beta {}^C D_R^\beta q - H = P_\alpha {}^C D_L^\alpha Q + P_\beta {}^C D_R^\beta Q - K + \frac{\mathrm{d}S}{\mathrm{d}t}. \qquad (5.6.40)$$

Hamilton's principle which is formally equivalent to Lagrange principle is

$$\delta S = \delta \int_{t_A}^{t_B} L = 0.$$

It's the same with traditional Hamilton's principle, and $S$ is the Hamiltonian main function. If [57]

$$S({}^C D_L^{\alpha-1} q, {}^C D_R^{\beta-1} q, P_\alpha, P_\beta) = F({}^C D_L^{\alpha-1} q, {}^C D_R^{\beta-1} q, P_\alpha, P_\beta)$$
$$- P_\alpha {}^C D_L^{\alpha-1} Q - P_\beta {}^C D_R^{\beta-1} Q, \qquad (5.6.41)$$

then Lagrangian can be written as [108, 109]

$$\frac{\mathrm{d}S}{\mathrm{d}t} = \frac{\partial S}{\partial t} + \frac{\partial F}{\partial {}^C D_L^{\alpha-1} q} {}^C D_L^\alpha q + \frac{\partial F}{\partial {}^C D_R^{\beta-1} q} {}^C D_R^\beta q + \frac{\partial F}{\partial P_\alpha} \frac{\mathrm{d}P_\alpha}{\mathrm{d}t} + \frac{\partial F}{\partial P_\beta} \frac{\mathrm{d}P_\beta}{\mathrm{d}t}$$

$$- \frac{\mathrm{d}P_\alpha}{\mathrm{d}t} {}^C D_L^{\alpha-1} Q - P_\alpha {}^C D_L^\alpha Q - \frac{\mathrm{d}P_\beta}{\mathrm{d}t} {}^C D_R^{\beta-1} Q - P_\beta {}^C D_R^\beta Q. \qquad (5.6.42)$$

Substitute it into Eq. (5.6.40) and consider that generalized coordinates are independent and Eq. (5.6.34),

$$\frac{\partial F}{\partial P_\alpha} = D_L^{\alpha-1}Q, \frac{\partial F}{\partial P_\beta} = D_R^{\beta-1}Q, p_\alpha = \frac{\partial F}{\partial^C D_L^{\alpha-1}q} = \frac{\partial S}{\partial^C D_L^{\alpha-1}q}, p_\beta = \frac{\partial F}{\partial^C D_R^{\beta-1}q}$$

$$= \frac{\partial S}{\partial^C D_R^{\beta-1}q}. \tag{5.6.43}$$

Without loss of generality, set $K = 0$ in Eq. (5.6.40), then fractional Hamilton–Jacobi equations are obtained [107–109]

$$H\left(t, q, \frac{\partial S}{\partial^C D_L^{\alpha-1}q}, \frac{\partial S}{\partial^C D_R^{\beta-1}q}\right) + \frac{\partial S}{\partial t} = 0. \tag{5.6.44}$$

## 2   Variational principle of fractional integration

Different from the first part, the method in reference [110] is making the variation of the fractional integration of Lagrange equations equal to zero, instead of applying fractional derivative to Lagrange equations

$$\delta \frac{1}{\Gamma(\alpha)} \int_{t_A}^{t_B} L(t, q, \dot{q})(t - \tau)^{\alpha-1} d\tau = 0. \tag{5.6.45}$$

According to the definition of variation principle and Eq. (5.6.45), it's easy to get [110]

$$\frac{1}{\Gamma(\alpha)} \int_{t_A}^{t_B} \left[\frac{\partial L}{\partial q}(t - \tau)^{\alpha-1}\delta q + \frac{\partial L}{\partial \dot{q}}(t - \tau)^{\alpha-1}\delta \dot{q}\right] d\tau = 0. \tag{5.6.46}$$

Integrate the second item by parts

$$\frac{1}{\Gamma(\alpha)} \int_{t_A}^{t_B} \left[\frac{\partial L}{\partial q}(t - \tau)^{\alpha-1} - \frac{d}{dt}\frac{\partial L}{\partial \dot{q}}(t - \tau)^{\alpha-1} - (1 - \alpha)\frac{\partial L}{\partial \dot{q}}(t - \tau)^{\alpha-2}\right]$$

$$\delta q d\tau = 0 = \tag{5.6.47}$$

Then Lagrange equations of the non-conservative system are obtained [110]

$$\frac{\partial L}{\partial q} - \frac{d}{dt}\frac{\partial L}{\partial \dot{q}} = \frac{1 - \alpha}{t - \tau}\frac{\partial L}{\partial \dot{q}}. \tag{5.6.48}$$

Here, $\partial L / \partial q_i$ corresponds to the conservative force $F_i$ in generalized coordinate, $\frac{d}{dt} \frac{\partial L}{\partial \dot{q}_i}$ corresponds to the inertia force $ma$, $\frac{1-\alpha}{t-\tau} \frac{\partial L}{\partial \dot{q}}$ is the non-conservative force that causes the energy dissipation of system.

## 5.7  Fractional Schrödinger Equation

Before twentieth century, classical mechanics theory is thought to be able to explain any physical phenomenon and it has no further room for development. However, with the insight into the microscopic world, it was discovered that the classical Newtonian mechanics cannot describe the movement of microscopic particles. Thus, physicists proposed the quantum mechanics to describe the law of motion of microscopic particles [112, 113]. Wave mechanics and matrix mechanics of quantum mechanics are two equivalent descriptions, and they are founded by Schrödinger and Heisenberg, respectively. Wave dynamics is established on the basis of the Schrödinger equation and the theory of matter waves. De Broglie's theory of matter waves is that the material and the photon exhibit both wave and particle properties, namely, the wave-particle duality. The Schrödinger wave propagation equation is given on this basis, which is Schrödinger equation [112]. This section is an introduction to the Schrödinger equation and fractional Schrödinger equation.

### 5.7.1  Wave-Particle Duality

The wave-particle duality theory is that photons not only have particle property, but also have the wave property; De Broglie generalized the photon wave-particle duality theory proposed by Einstein, believed that any particle has wave-particle duality, namely, the concept of matter waves. The so-called particle property refers to the individual particles, which mainly shows the characteristics of particle motion, while the fluctuation means when a large number of particles appear they mainly exhibit wave property [112, 113]. Classic Huygens wave theory is that wave is the spatial distribution of particles, the fluctuation is the propagation of the particle energy and its movement mode, but the classic particle theory is that particle does not show wave characteristics, so this phenomenon cannot be explained by the classic particle theory or the wave theory.

### 5.7.2  Schrödinger Equation

Schrödinger equation is a fundamental equation of quantum mechanics; it can describe the movement of low-speed microscopic particles in the conservative field, and does not consider the spin of the particles [20]. Based on the matter-wave theory,

Schrödinger used the wave function $\Psi(x, t)$ to describe the fluctuation of the particles [113]. According to Einstein–De Broglie relation, the momentum of a particle and wave number has the following relationship, that is: $\mathbf{p} = \hbar\mathbf{k}$ (Where $\lambda$ is wavelength, $\mathbf{p}$ is the momentum, $\hbar = h/2\pi$, h is Planck's constant, $|\mathbf{k}| = 2\pi/\lambda$ is the modulus of the wave number); at the same time, the energy and frequency of the particles satisfy the relationship: $E = hv = \hbar\omega$[113]. These two formulas link particles' particle property and fluctuation, and demonstrate the wave-particle duality of particles [112]. Thus the form of the complex amplitude of the wave function can be written as

$$\psi = Ae^{i(\mathbf{k}\cdot\mathbf{r}-\omega t)} = Ae^{i(\mathbf{P}\cdot\mathbf{r}-Et)/\hbar}. \tag{5.7.1}$$

Then, we get the following equation [113]:

$$i\hbar\frac{\partial}{\partial t}[Ae^{i(\mathbf{P}\cdot\mathbf{R}-Et)/\hbar}] = E\psi; \tag{5.7.2}$$

$$-\frac{\hbar^2}{2m}\nabla^2[Ae^{i(\mathbf{P}\cdot\mathbf{R}-Et)/\hbar}] = \frac{P^2}{2m}\psi, \tag{5.7.3}$$

where the operator $i\hbar\frac{\partial}{\partial t}$ means the energy of the particles, which are denoted by $\widehat{E}$; $-i\hbar\nabla$ represents the momentum of a particle, can be denoted by $\hat{\mathbf{p}}$. According to the relationship of energy and momentum of low-speed particles in the conservative field $E = T + V = p^2/2\,m + V$, we can obtain the Schrödinger equation

$$i\hbar\frac{\partial\psi}{\partial t} = -\frac{\hbar^2}{2m}\nabla^2\psi + V\psi, \tag{5.7.4}$$

and the operator representation form of Schrödinger equation [113]

$$\widehat{E}\psi = \hat{\mathbf{p}}^2\psi/2m + V\psi. \tag{5.7.5}$$

If we consider Hamilton quantity H in classical mechanics, $H = T + V$, and rewrite H as the operator $\widehat{H}$, we can obtain the wave equation expressed by Hamiltonian [112]

$$\widehat{E}\psi = i\hbar\frac{\partial\psi}{\partial t} = \widehat{H}\psi, \tag{5.7.6}$$

where Hamilton operator $\widehat{H}$ is equivalent to the energy operator $\widehat{E}$. This also shows that formula (5.7.6) describes a conservative system.

Considering the particles' high-speed movement, we need to take into account relativistic effects, and the energy–momentum relationship could be recast as $E^2 = c^2\mathbf{p}^2 + m^2c^4$. Thus, we obtain the famous relativistic wave equation [113] (also called Klein–Gordon equation)

$$\nabla^2 \psi - \frac{1}{c^2} \frac{\partial^2 \psi}{\partial t^2} - \frac{m^2 c^2}{\hbar^2} \psi = 0. \tag{5.7.7}$$

### 5.7.3 The Physical Significance of the Wave Equation

German physicist Bohr proposed the intensity of the matter wave at a particular point is proportionally increasing to the probability of finding the corpuscle at that point. According to Bohr's interpretation, the matter wave is a probability wave, and he put forward the probabilistic description of the wave function. It does not refer to the fluctuation generated by the entity movement, but refers to the fluctuation which indicates the probability that the particles appear somewhere in the space, i.e. $P(x, t)dx = \psi^*(x, t) \psi(x, t)dx = |\psi(x, t)|^2 dx$ represents the probability of occurrence of the particles in the region $(x, x + dx)$, therefore it is called the probability wave, where $\psi^*(x, t)$ is the conjugate function of the wave function $\psi(x, t)$, $\psi^*(x, t) \psi(x, t)$ is probability density [20]. To ensure $\psi^*(x, t) \psi(x, t)$ is the probability density, we should prove that $\psi^*(x, t) \psi(x, t)$ is a normalized function, i.e. that

$$\int\limits_{-\infty}^{+\infty} \psi^* \psi \, \mathrm{d}x = 1. \tag{5.7.8}$$

Hamiltonian operator introduced in Schrödinger Eq. (5.7.6) is Hermitian operator in the case of quantum mechanics that has the following properties

$$\int\limits_{-\infty}^{+\infty} (\widehat{H}g)^* f \, \mathrm{d}x = \int\limits_{-\infty}^{+\infty} g^* (\widehat{H}f) \, \mathrm{d}x, \tag{5.7.9}$$

where $g$ and $f$ are vectors, * represents conjugate [114]. According to this property, Eq. (5.7.8) can be reorganized

$$\int\limits_{-\infty}^{+\infty} (\widehat{H}\psi)^* \psi \, \mathrm{d}x = \int\limits_{-\infty}^{+\infty} \psi^* (\widehat{H}\psi) \, \mathrm{d}x. \tag{5.7.10}$$

Therefore

$$\int_{-\infty}^{+\infty} \left( i\hbar \frac{\partial \psi}{\partial t} \right)^* \psi \, dx = \int_{-\infty}^{+\infty} \psi^* \left( i\hbar \frac{\partial \psi}{\partial t} \right) dx, \tag{5.7.11}$$

$$\int_{-\infty}^{+\infty} -i\hbar \frac{\partial \psi^*}{\partial t} \psi \, dx - i\hbar \int_{-\infty}^{+\infty} \psi^* \frac{\partial \psi}{\partial t} \, dx = -i\hbar \frac{d}{dt} \left( \int_{-\infty}^{+\infty} \psi^* \psi \, dx \right) = 0, \tag{5.7.12}$$

where $\int_{-\infty}^{+\infty} \psi^* \psi \, dx$ is a constant irrelevant with time [114]. Thus as long as we normalize the wave function we can obtain $\int_{-\infty}^{+\infty} \psi^* \psi \, dx = 1$, and $\psi^*(x, t) \, \psi(x, t)$ indicates the probability density that the particles appear at a point in space.

At the same time, the Hamiltonian corresponding to traditional Schrödinger equation is time-independent quantity. That is, for the traditional Schrödinger equation, the wave function only with respect to the time interval, this indicates that the Schrödinger equation remains unchanged under time translation [20].

### 5.7.4 Time Fractional-Order Schrödinger Equation

While using the traditional Schrödinger equation to describe the motion of microscopic particles, we do not suppose that particles movement has memory. In order to consider the influence of the history of particles movement, Naber [115] introduced time-fractional derivative in fractional Schrödinger equation, i.e.

$$(i\hbar)^\alpha \frac{\partial^\alpha \psi}{\partial t^\alpha} = -\frac{\hbar^2}{2m} \nabla^2 \psi + V\psi. \tag{5.7.13}$$

Time fractional derivative contains time convolution integral, which reflects the history dependence of particles motion. Naber [115] analyzed the properties of time fractional Schrödinger equation, and rewrote it as

$$\frac{\partial^\alpha \psi}{\partial t^\alpha} = -\frac{\beta}{i^\alpha} \nabla^2 \psi + \frac{\gamma}{i^\alpha} V\psi, \beta = \frac{\hbar^2}{2m\hbar^\alpha}, \gamma = \frac{1}{\hbar^\alpha}, \tag{5.7.14}$$

if we differentiate both sides of order $1 - \alpha$ with respect of t and multiple by $i\hbar$, according to Caputo fractional derivative definition, we get

$$i\hbar \frac{\partial \psi}{\partial t} = -\frac{\beta\hbar}{i^{\alpha-1}} \nabla^2 \frac{\partial^{1-\alpha} \psi}{\partial t^{1-\alpha}} + \frac{\gamma\hbar}{i^{\alpha-1}} V \frac{\partial^{1-\alpha} \psi}{\partial t^{1-\alpha}} + i\hbar \frac{[\partial^\alpha \psi/\partial t^\alpha]|_{t=0}}{t^{1-\alpha} \Gamma(\alpha)}, \tag{5.7.15}$$

thus we gain the Hamilton $\widehat{H}$ of time fractional order Schrödinger equation

$$\widehat{H} = -\frac{\beta\hbar}{i^{\alpha-1}}\nabla^2\frac{\partial^{1-\alpha}}{\partial t^{1-\alpha}} + \frac{\gamma\hbar}{i^{\alpha-1}}V\frac{\partial^{1-\alpha}}{\partial t^{1-\alpha}} + i\hbar\frac{[\partial^{\alpha}/\partial t^{\alpha}]|_{t=0}}{t^{1-\alpha}\Gamma(\alpha)}. \tag{5.7.16}$$

Therefore, Naber [115] thought that time fractional Schrödinger equation and the traditional Schrödinger equation have the same physical meaning. The difference is that the Hamilton in time fractional Schrödinger equation is time-dependence.

### 5.7.5    The Path Integral Principle and the Schrödinger Equation

Before introducing space fractional Schrödinger equation, we first introduce the relationship between Feynman path integral theory and its relationship with traditional Schrödinger equation [114]. Then, we will introduce the space fractional Schrödinger equation based on the theory of the path integral.

The path integral theory was proposed by the famous American physicist Freeman [20]. In classical mechanics, the true path is uniquely determined by the Hamilton principle, and the evolution of the physical processes has the only true path. The physicists of quantum mechanics think that the orbits of the particles are random, and there is not a specific particle orbit. We can only obtain the possibility of a particle at a specific position rather than the location of the particle at a certain moment. For a large number of particles, the Schrödinger equation can derive their accurate probability wave [113]. For microscopic particles, Feynman thought that all possible paths affect the amplitude of probability wave. In other words, if you want to know the amplitude of the probability wave at any time, you should consider the effects of all possible paths [114]. According to the paper by Dirac on Lagrangian, Feynman introduced the classical Lagrangian and established the path integral theory, and regarded the conversion function k as a propagator, where the definition of $k(x, x')$ is

$$k(x, x') = \exp\left[i\varepsilon L\left(\frac{x'-x}{\varepsilon}\right), x\right], \tag{5.7.17}$$

then he obtained

$$\psi(x', t + \mathrm{d}t) = \int\limits_{-\infty}^{+\infty} A\exp\left[i\frac{\mathrm{d}t}{\hbar}L\left(\frac{x'-x}{\mathrm{d}t}\right), x\right]\psi(x, t)\mathrm{d}x. \tag{5.7.18}$$

Integral over time of L, the exponential function in the formula can be rewritten as [114]

$$\exp\{iS[x(t)]/\hbar\}, \tag{5.7.19}$$

where S is the Hamiltonian main function in the classical mechanics, also known as the Hamiltonian. Meanwhile, Feynman thought that for the amplitude of final state of probability wave, the complex amplitudes contributed by each path are the same, the only difference is the phase of each path $S/\hbar$. However, only the paths very close to the extreme path, they may affect the amplitude of the final state of probability wave, while other paths' contribution will offset each other [114].

In order to consider the time evolution of the physical process, Feynman introduced Lebesgue integral operator, then we can obtain the propagator and wave function

$$K(b, a) = \int_a^b \exp\{i\,S[b, a]/\hbar\}\mathscr{D}x(t), \qquad (5.7.20)$$

$$\psi(x', t') = \int_{-\infty}^{+\infty} K(x', t'; x, t)\psi(x, t)\mathrm{d}x, \qquad (5.7.21)$$

where $K(x', t'; x, t)$ is the propagator. If $t' = t + \varepsilon$, $\varepsilon$ is small, we can introduce the difference approximation, and (5.7.21) can be rewritten as [114]

$$\psi(x', t') = \int_{-\infty}^{+\infty} \frac{1}{A}\exp\left\{i\varepsilon L\left(\frac{x'+x}{2}, \frac{x'-x}{\varepsilon}, t\right)/\hbar\right\}\psi(x, t)\mathrm{d}x, \qquad (5.7.22)$$

where the integrand in the propagator is the weight contributed by each possible path. From formula (5.7.22), the path integral theory established by Feynman indicates that the distribution of different paths' contribution is a normal distribution.

We have briefly described the path integral theory in the previous section, the following part will introduce the relationship between path integral theory and Schrödinger equation, thus further describes the space fractional Schrödinger equation. Firstly we assume $x' = x - \eta$, then formula (5.7.22) can be rewritten as

$$\psi(x', t') = \int_{-\infty}^{+\infty} \frac{1}{A}\exp\left[\frac{i}{\hbar}\frac{m\eta^2}{2\varepsilon} - \frac{i\varepsilon}{\hbar}V\left(x' + \frac{\eta}{2}, t\right)\right]\psi(x' + \eta, t)\mathrm{d}x, \qquad (5.7.23)$$

then we use Taylor expansion of first order with $\varepsilon$ and second order with $\eta$, ignoring second-order small quantity of $\varepsilon$, we can obtain [114]

$$\psi(x', t) + \varepsilon \frac{\partial \psi}{\partial t} = \int\limits_{-\infty}^{+\infty} \frac{1}{A} \exp\left(\frac{i}{\hbar} \frac{m\eta^2}{2\varepsilon}\right)\left[1 - \frac{i\varepsilon}{\hbar} V(x', t)\right]$$

$$\cdot \left[\psi(x', t) + \eta \frac{\partial \psi}{\partial x} + \frac{\eta^2}{2} \frac{\partial^2 \psi}{\partial x^2}\right] d\eta \qquad (5.7.24)$$

To ensure the establishment of equation, the coefficients of $\psi(x', t)$ must be the same.

$$\frac{1}{A} \int\limits_{-\infty}^{+\infty} \exp\left(\frac{im\eta^2}{2\hbar\varepsilon}\right) d\eta = \frac{1}{A}\left(\frac{2\pi i\hbar\varepsilon}{m}\right)^{1/2} = 1, \text{ i.e. } A = \left(\frac{2\pi i\hbar\varepsilon}{m}\right)^{1/2}; \qquad (5.7.25)$$

We can obtain [114]

$$\frac{1}{A} \int\limits_{-\infty}^{+\infty} \exp\left(\frac{im\eta^2}{2\hbar\varepsilon}\right) \eta d\eta = 0 \text{ 和 } \frac{1}{A} \int\limits_{-\infty}^{+\infty} \exp\left(\frac{im\eta^2}{2\hbar\varepsilon}\right) \eta^2 d\eta = \frac{i\hbar\varepsilon}{m}. \qquad (5.7.26)$$

When getting rid of the second-order small quantities, we can obtain the traditional Schrödinger equation [114]

$$i\hbar \frac{\partial \psi}{\partial t} = -\frac{\hbar^2}{2m} \nabla^2 \psi + V\psi. \qquad (5.7.27)$$

### 5.7.6 Space Fractional Schrödinger Equation

Last section has briefly introduced the path integral and the relationship between path integral and Schrödinger equation; in this section, we will introduce the space fractional Schrödinger equation based on the fundamental idea of the path integral. The path integral theory proposed by Feynman indicates that each path for final state contribution is Gauss distribution. However, in many practical particle fields, such as anomalous diffusion, Gauss distribution cannot describe the distribution well. The particles number distribution exhibits a very sharp top, rather than the exhibits an approximate bell shape. And unlike the sharp attenuation at tails of Gaussian distribution, the particles number distribution shows obvious tailing phenomenon [116, 117]. Some researchers use Lévy distribution to describe the particles number distribution. Lévy distribution actually is a kind of stable distribution which can be obtained by extending the Gaussian distribution, and the Gaussian distribution function is

$$\exp(-a|k|^2). \tag{5.7.28}$$

Lévy thought that if the distribution satisfied

$$\exp(-a|k|^\alpha), \tag{5.7.29}$$

the distribution form is stable [118]. When $\alpha = 2$, it is a Gaussian distribution, which means that the Gaussian distribution is a special case of the Lévy distribution. Based on the stable distribution Lévy, some researchers expanded the original central limit theorem and obtained the generalized central limit theorem; they thought that every distribution satisfying formula (5.7.29) tends to be stable, not only Gaussian distribution [119]. Laskin et al. [116, 117] have proposed the amendments to the Feynman path integral principle; they thought that each path for the final state contribution should satisfy the Lévy distribution. This theory is briefly described below.

According to formula (5.7.22), we have the path integral formula [116, 117] at time $t = t + n\varepsilon$, supposing $x_A, t_A$, $x_B, t_B$ are endpoints

$$\psi(x_B, t_B) = \int_{-\infty}^{+\infty} K(x_B, t_B; x_A, t_A)\psi(x_A, t_A)\mathrm{d}x, \tag{5.7.30}$$

$$K(x_B, t_B; x_A, t_A) = \int \exp\{iS[t_B, t_A]/\hbar\}\mathscr{D}x(t)$$
$$= \lim_{N\to\infty} \frac{1}{A^N} \int \mathrm{d}x_1 \cdots \mathrm{d}x_N$$
$$\exp\left\{\frac{im}{2\hbar\varepsilon} \sum_{j=1}^{N} (x_j - x_{j-1})^2 - \frac{i}{\hbar} \int_{t_A}^{t_B} V[x, \tau]\mathrm{d}\tau\right\},$$

where $\exp\left\{\frac{im}{2\hbar\varepsilon} \sum_{j=1}^{N} (x_j - x_{j-1})^2\right\}$ is considered to be the Gaussian probability density function. Laskin [117, 118] amended Gaussian distribution term as Lévy distribution term; here shows the fractional propagator corresponding to Lévy distribution

$$K_L(x_B, t_B; x_A, t_A) = \lim_{N\to\infty} \int \mathrm{d}x_1 \cdots \mathrm{d}x_N \left(\frac{iD_\alpha\varepsilon}{\hbar}\right)^{-N/\alpha}$$
$$\cdot \prod_{j=1}^{N} L_\alpha\left\{\left(\frac{\hbar}{iD_\alpha\varepsilon}\right)^{1/\alpha} |x_j - x_{j-1}|\right\} \cdot \exp\left\{-\frac{i}{\hbar} \int_{t_A}^{t_B} V[x, \tau]\right\}, \tag{5.7.31}$$

where $\left(\frac{D_\alpha t}{\hbar}\right)^{-1/\alpha} L_\alpha \left\{\left(\frac{\hbar}{D_\alpha \varepsilon}\right)^{1/\alpha}|x|\right\} = \frac{\pi}{\alpha|x|} H_{2,2}^{1,1}\left(\frac{\hbar}{D_\alpha \varepsilon}\right)^{1/\alpha}|x| \left|\begin{array}{l}(1,1/\alpha),(1,1/2)\\(1,1),(1,1/2)\end{array}\right.\right], \alpha$

is the index of the characteristic function in Lévy distribution. If taking $t_B = t_A + \varepsilon$, $x_B = x_A$-$\eta$, fractional propagator can be simplified as [116, 117]

$$K_L(x_B, t_B; x_A, t_A) = \int dx \left(\frac{iD_\alpha \varepsilon}{\hbar}\right)^{-1/\alpha} \cdot L_\alpha\left\{\left(\frac{\hbar}{iD_\alpha \varepsilon}\right)^{1/\alpha}|\eta|\right\} \cdot \exp\left\{-\frac{i\varepsilon}{\hbar}V[x,\tau]\right\}.$$

(5.7.32)

According to the definition of fractional propagator and Fourier integral, the propagator can be written as [116, 117]

$$K_L(x_B, t_B; x_A, t_A) = \int \frac{1}{2\pi\hbar}\int dp \cdot \exp\left[\frac{ip\eta}{\hbar} - \frac{iD_\alpha \varepsilon |p|^\alpha}{\hbar}\right]$$
$$\cdot \exp\left\{-\frac{i\varepsilon}{\hbar}V\left[x' + \frac{\eta}{2}, t\right]\right\} dx,$$

(5.7.33)

and by

$$i\hbar\frac{\partial}{\partial t}\left\{\frac{1}{2\pi\hbar}\int dp \cdot \exp\left[\frac{ip\eta}{\hbar} - \frac{iD_\alpha \varepsilon |p|^\alpha}{\hbar}\right]\right\}$$
$$= -D_\alpha(\hbar\nabla)^\alpha\left\{\frac{1}{2\pi\hbar}\int dp \cdot \exp\left[\frac{ip\eta}{\hbar} - \frac{iD_\alpha \varepsilon |p|^\alpha}{\hbar}\right]\right\},$$

(5.7.34)

where

$$(\hbar\nabla)^\alpha = -\frac{1}{2\pi\hbar}\int dp \cdot \exp[ipx/\hbar]|p|^\alpha.$$

We can obtain the Space fractional Schrödinger equation [116, 117]

$$i\hbar\frac{\partial\psi}{\partial t} = -D_\alpha(\hbar\nabla)^\alpha\psi + V(x)\psi.$$

(5.7.35)

### 5.7.7   *Fractal Time–space Origin of Fractional Schrödinger Equation*

Time and space are the most basic concepts of nature, and also the origin of different mathematical theories and physical quantities. Fractal time–space transformation (2.5.4) introduced in Sect. 2.5.1 is a basic mathematical physical transformation.

This section will introduce the fractal time–space transformation (2.5.4) into quantum mechanics.

According to the fractal invariable hypothesis in the basic physical processes about two 'abnormal' (anomalous) introduced by Wen Chen, energy and frequency, momentum and wave number quantum relations remain unchanged, i.e.

$$E = \hat{h}_\alpha \hat{v}, \tag{5.7.36}$$

$$p = \hat{h}_\beta \hat{k}, \tag{5.7.37}$$

where $E$ stands for energy, $p$ represents the momentum, $\hat{h}_\alpha$ and $\hat{h}_\beta$ is the Planck constant after scale change, $\hat{k}$ and $\hat{v}$ are the wave number and frequency after scale change, respectively. By fractal time–space transformation (2.5.4), we can connect the frequency with the wave number in integer dimensional spatiotemporal and fractal spatiotemporal

$$\hat{v} = v^\alpha \text{ and } \hat{k} = k^\beta. \tag{5.7.38}$$

Thus, the quantum relation between fractional Planck frequency and energy is

$$E = \hat{h}_\alpha v^\alpha, 0 < \alpha \leq 1 \tag{5.7.39}$$

the quantum relation between fractional momentum and wavelength is

$$p = \hat{h}_\beta k^\beta, 0 < \beta \leq 1, \tag{5.7.40}$$

where $\hat{h}_\alpha$ and $\hat{h}_\beta$ are Planck constant in the fractal spatiotemporal. In Sect. 2.5.1, we point out that index $\alpha$ and $\beta$ are fractional Brownian motion (time) and Lévy statistics (space) index, respectively. Thus, Lévy statistics (space) and fractional Brownian motion (time) are essentially related to momentum and energy. The expression of kinetic is $E_k = |p|^2/2m$, where $m$ represents the mass of the particle. However, the existing literature (*Phy. Rev. Lett.* 90, 170,601, 2003; *Phy. Rev. E.* 66, 056,108, 2002) give the wrong formula of energy and momentum: $E_k = D_\beta |p|^{2\beta}$. $D_\beta$ is the fractal spatiotemporal constant, its physical dimension is $\text{erg}^{1-2\beta} \times \text{m}^{2\beta} \times \text{s}^{-2\beta}$. We find that Lévy statistics only has relationship with momentum, while the fractional Brown motion has relationship with energy.

Considering the quantum plane wave $\Psi(x, t) = A e^{i\vec{k}\vec{r} - ivt}$, fractional quantum relation (5.7.39) and (5.7.40) we can obtain

$$(-\Delta)^\beta \Psi = |k|^{2\beta} \Psi = \frac{p^2}{h_\beta^2} \Psi. \tag{5.7.41}$$

Therefore, we have

$$E_k = \frac{p^2}{2m} = (-\Delta)^\beta \frac{\hat{h}_\beta^2}{2m}. \tag{5.7.42}$$

In addition, we obtain

$$\frac{\partial^\alpha \Psi}{\partial t^\alpha} = (-i\nu)^\alpha \Psi = e^{-i\pi\alpha/2} \frac{E}{\hat{h}_\alpha} \Psi \rightarrow e^{i\pi\alpha/2} \hat{h}_\alpha \frac{\partial^\alpha \Psi}{\partial t^\alpha} = E\Psi. \tag{5.7.43}$$

According to the fractal invariable hypothesis [84] introduced in Sect. 2.5.1, the classic Hamilton energy of a particle whose potential energy is $V\left(\vec{r}\right)$ is

$$E = E_k + E_p = \frac{p^2}{2m} + V\left(x^\beta\right). \tag{5.7.44}$$

According to (5.7.41), (5.7.42) and (5.7.43), replacing the energy item in formula (5.7.44) with energy operator, we can obtain the fractional Hamilton quantum mechanics expression, namely fractional Schrödinger equation:

$$e^{i\pi\alpha/2} \hat{h}_\alpha \frac{\partial^\alpha \Psi}{\partial t^\alpha} = \frac{\hat{h}_\beta^2}{2m}(-\Delta)^\beta \Psi + V\left(x^\beta\right)\Psi, \quad 0 < \alpha, \beta \leq 1. \tag{5.7.45}$$

There is also another derivation method: the time–space Fourier transformation of the fractional Schrödinger Eq. (5.7.45) is

$$\hat{h}_\alpha \nu^\alpha \widehat{\Psi} = \left(\frac{\hat{h}_\beta^2 k^{2\beta}}{2m} + V\right)\widehat{\Psi}. \tag{5.7.46}$$

According to the physical interpretation of the Schrödinger equation, (5.7.45) is essentially a quantum Hamilton function $E = E_k + V$, where $E_k = p/2\ m$. Thus, we can directly obtain the fractional quantum relationships (5.7.39) and (5.7.40). The fractional Schrödinger Eq. (5.7.45) can be derived by a reverse derivation.

According to the fractal equivalent hypothesis introduced in Sect. 2.5.1, fractional Schrödinger Eq. (5.7.45) is quantum effect of time–space structure.

European and American scholars based on the Feynman integral of Lévy path, fractional Brown motion or pure mathematics independently proposed fractional Schrodinger equation respectively. Our work [84] unique in that the equation is derived from the basic assumptions of physics.

The physical and mechanics properties of soft matter are determined by the macro-molecules on the mesoscopic scale. These macromolecules are usually composed of a huge number of basic atoms and molecules; the mesoscopic scale fractal structure has a decisive effect on their physical processes. The fractal time–space transfor-mation parameters $\alpha$ and $\beta$ which are introduced by Wen Chen [84] reflect the time–space structure of soft material macromolecules, thus they depict the essence of the soft matter's "anomalous" physics behavior. This section describes that the Fractional Schrödinger equation derivation is an application of these basic physical assumptions. Fractional Schrödinger equation has been used to explain the complex quantum process of polymer materials. In the interpretation and description of various quantum anomalous diffusion (e.g. laser cooling), it also has potential significance.

## 5.8  Other Application Fields

### 5.8.1  Applications of Fractional Calculus in the Fracture Mechanics

Barpi and Valente [120] mentioned the use of the combination of fractional rheological model and micromechanics model of the crack propagation region in fracture mechanics to describe the creep rupture of quasi-static large volume concrete (such as dams, because we can ignore the effect of inertia force in the quasi-static problem) (Fig. 5.32).

The above picture demonstrates the fractional viscoelastic model in the paper [120], its corresponding constitutive equation is stated as.

For elastic component

$$\begin{aligned} \sigma_1 &= E_1(\varepsilon - \varepsilon_1) \\ \sigma_2 &= E_2\varepsilon \end{aligned}; \tag{5.8.1}$$

For viscous component

$$\frac{\partial^\alpha \varepsilon_1}{\partial t^\alpha} = \frac{\sigma_1}{E_1\tau_1^\alpha} = \frac{\varepsilon - \varepsilon_1}{\tau_1^\alpha} \quad (0 < \alpha < 1). \tag{5.8.2}$$

The micromechanics model of crack propagation [120]:

**Fig. 5.32**  Creep model [120]

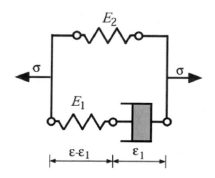

$$\frac{w}{w_c} = \frac{(K_{Ic}^{\mathrm{hom}})^2}{E_c(1-V_f)f_t} \frac{f_t}{\sigma} \left[1 - \left(\frac{\sigma}{f_t}\right)^3\right] = \beta \frac{f_t}{\sigma} \left[1 - \left(\frac{\sigma}{f_t}\right)^3\right], \qquad (5.8.3)$$

where $w$ is the Crack opening displacement, $w_c$ is the critical crack opening displacement (No stress transferred in the cohesive zone), $K_{Ic}^{\mathrm{hom}}$ is the fracture toughness of Isotropic material, $V_f$ is the total volume fraction, $f_t$ is the tensile strength, $\beta$ is the concrete microstructural parameter, $E_c$ is the elastic modulus, $\sigma$ is the cohesive zone stress. In this paper, type I crack model is simulated with finite element method, and is compared with the three-point bending experiment of three points of the different load. The results of failure life and load–displacement are good with real processes.

### 5.8.2 Applications of Fractional Order Calculus in System Control [78, 121]

The controller is a kind of instrument or device which enables the system to reach a predetermined steady state through the feedback and control information. Fractional controller uses fractional differential equation to control the system, and the relevant research and application in control system still belongs to a new field. The traditional integer-order system controllers are mainly used in practice [78, 121–124].

The PID controller was widely used in the industrial system control. Cao et al. [121] mentioned in the application research on pneumatic position servo control: the introduction of fractional derivative operator improves the traditional PID controller, and also significantly enhances the control performance of the system. In their paper, the fractional controller is applied to the pneumatic servo control to overcome the difficulties caused by the pneumatic drive strong nonlinear time-varying system. They also gave the corresponding control equation, analyzed and simulated the result [121].

$$u(t) = K_P e(t) + K_I \frac{\partial^{-\lambda}}{\partial t^{-\lambda}} e(t) + K_D \frac{\partial^\delta}{\partial t^\delta} e(t), \qquad (5.8.4)$$

where $K_P$ is the proportional constant, $K_I$ is the integration time constant, $K_D$ is the differential time constant, $\lambda$ is the integral order, $\delta$ is the differential order. According to the above equation, the control system transfer function is [121]

$$G_c(s) = K_P + K_I s^{-\lambda} + K_D s^\delta, \qquad (5.8.5)$$

when $\lambda = 0$, $\delta = 0$, $G_c(s)$ is the integer-order P controller; when $\lambda = 1$, $\delta = 0$, $G_c(s)$ is the integer-order PI controller; when $\lambda = 0$, $\delta = 1$, $G_c(s)$ is the integer-order PD controller; when $\lambda = 1$, $\delta = 1$, $G_c(s)$ is the integer-order PID controller. Thus, the traditional controller based on integer-order differential operator can be seen as a

**Fig. 5.33** Comparison between integer-order PID controller and fractional-order PID controller [121]

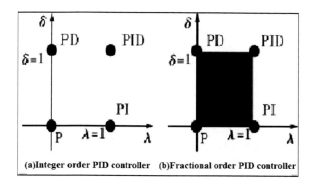

(a)Integer order PID controller   (b)Fractional order PID controller

special case of the fractional controller. The Fractional controller has more flexibility, while the results of control simulation indicate that fractional-order controller has a very good performance [121] (Fig. 5.33).

## *5.8.3   Inverse Problems on Fractional Derivative Models*

The inverse problem usually refers to the establishment of mathematical model and information reconstruction based on indirect, incomplete or noisy known data. Such as medical imaging, exploration geophysics, non-destructive evaluation etc. To solve these problems, we need to speculate the nature of the internal properties and characteristics based on external performance of the object.

The inverse problem of differential equation can be roughly classified into five categories: inverse problem on source control, inverse problem on parameter control, inverse problem on boundary condition control, inverse problem on initial condition and inverse problem on shape control. For example, pollution source control on environment hydraulics is a kind of inverse problem on source control. The inverse problem on parameter control includes inferring Earth structure from the size of the seismic waves in the stratum physical exploration; seeking the thermal conductivity of heterogeneous materials by the distribution of their surface temperature, etc. In recent years, the inverse problems are put forward and studied in more and more disciplines and fields. Due to the ill-posedness and nonlinearity of the inverse problem, it is much more difficult to solve the inverse problem than positive problem does. The inverse problem on explaining the observed data in geophysics has great economic and social benefit, but it is also very difficult. A large number of geologists, physicists and mathematicians are interested in it, thus it becomes one of the most active branches in the inverse problem. The proposed methods for solving the inverse problem in the field, such as direct discrete inversion method, optimal control method, perturbation method [125] promote the development of the inverse problem. But at present, most studies on inverse problem are focused on the integer derivative model; along with the development of the fractional calculus, inverse problems of fractional derivative model have also been proposed.

## 1   Inverse problem on fractional derivative model

In recent years, fractional derivative (integral) has been widely applied in continuous medium mechanics [126], viscoelastic and viscoplastic flow [126] and anomalous diffusion [127, 128] problems and so on. A lot of literature shows that the fractional derivative can be used in the field of physical, financial and hydrology, etc. With the fractional derivative model in practical application becoming more and more extensive, more attention is paid to the study of it. Battaglia et al. used non-integer-order identification system to solve the heat conduction inverse problem, and used fractional-order identification system for turning machine tool cutting metal in the process of heat conduction heat flux function inverse problem as an example to analyze. The results show that compared with integer-order identification system, non-integer-order identification system has superiority to solve this kind of problem [129]. Murio [130] developed a stable numerical solution of a fractional diffusion inverse heat conduction problem and presented an effective method. In addition, Murio [131] analyzed the Caputo time-fractional inverse heat conduction problem (TFIHCP). Sivaprasad et al. [132] investigated the identification of fractional dynamic damping systems via inverse sensitivity analysis. At present, the research about the integer-order differential equation of the linear inverse problem tends to be mature, and the study of nonlinear inverse problem has also made significant achievements [133, 134]. However, the research about the inverse problem of fractional differential equation is still at an early stage, and the literature in this area is not much.

In this section, taking inverse source problem of one-dimensional spatial fractional derivative anomalous diffusion equation, for example, we introduce a numerical method to readers, namely, the best perturbation.

## 2   Inverse source problem of spatial fractional anomalous diffusion equation

Space anomalous diffusion equation used to describe solute anomalous migration and transformation process is mainly a mathematical model. In this study, we only discuss the inverse source problem of one-dimensional spatial fractional derivative anomalous diffusion and the form of variable coefficient spatial fractional anomalous diffusion equation as follows [135]

$$\frac{\partial u}{\partial t} = d(x)\frac{\partial^\alpha u}{\partial x^\alpha} + q(x, t), \quad 1 < \alpha < 2, t > 0, 0 < x < L \tag{5.8.6}$$

where $u = u(x, t)$ is the concentration of the solute of the x point at time $t$, $d(x)$ is the variable coefficient, $q(x)$ is the source term. Here, the definition of the fractional derivative is Grünwald definition, namely

$$\frac{\partial^\alpha u(x, t)}{\partial x^\alpha} = \frac{1}{\Gamma(-\alpha)} \lim_{N \to \infty} \frac{1}{h^2} \sum_{k=1}^{N} \frac{\Gamma(k - \alpha)}{\Gamma(k + 1)} u(x - (k - 1)h, t)$$

where $N$ is a positive integer, $h = \Delta x = (L - x_0)/N$, $\Gamma(*)$ is the Gamma function. If $\alpha = 2$, the equation becomes classical diffusion equation with variable coefficient.

The corresponding initial condition and boundary conditions as

$$\begin{cases} u(x, 0) = f_0(x) \\ u(0, t) = b_1(t) \\ u(L, t) = b_2(t) \end{cases} . \tag{5.8.7}$$

If the source term $q(x, t)$ in formula (5.8.6) is known, formula (5.8.6) and (5.8.7) is a positive problem to solve $u(x, t)$. Using the numerical methods of fractional differential equation, we can obtain the numerical solution of $u(x, t)$ [135–137].

If the source term $q(x, t)$ in formula (5.8.6) is unknown, we need to determine the $u(x, t)$ and $q(x, t)$. In order to estimate the source term from the known data, an additional condition has to be added, the concentration at time $t = T$ is used, namely

$$u(x, T) = u_T(x), \quad 0 \le x \le L. \tag{5.8.8}$$

Thus formulas (5.8.6), (5.8.7) and (5.8.8) form a variable coefficient inverse source problem of anomalous diffusion.

**(1)   Transformation of the inverse source problem**

For Eqs. (5.8.6) and (5.8.7), $d(x, t)$, $f_0(x)$, $b_1(t)$ and $b_2(t)$ are all known, the problem is to determine the source term $q(x, t)$, therefore an additional condition (5.8.8) need to be added. Assume $q^*(x, t)$ is the exact solution of $q(x, t)$. The corresponding exact solution to the direct problem as $c^*(x, t)$. $Q$ is assumed to be a complete linear real function space, where $q^*(x, t) \in Q$, basis functions $\{\phi_1(x, t), \phi_2(x, t), \ldots, \phi_n(x, t), \ldots\}$ belong to $Q$, we have

$$q^*(x, t) = \sum_{i=1}^{\infty} k_i \phi_i(x, t). \tag{5.8.9}$$

In a real-world application, the source term is usually estimated with finite terms, namely

$$q^*(x, t) = \sum_{i=1}^{n} k_i \phi_i(x, t), \tag{5.8.10}$$

where $n$ depends on the approximate precision. Therefore, the inverse source problem is transformed into determining an n-dimensional real vector

$$K^T = (k_1, k_2, \ldots, k_n) \in R^n, \tag{5.8.11}$$

to ensure that

$$q^*(x, t) = \sum_{i=1}^{n} k_i \phi_i(x, t) = \boldsymbol{K}^T \boldsymbol{\Phi}(x, t), \qquad (5.8.12)$$

satisfies the governing equation and the conditions (5.8.8), where

$$\boldsymbol{\Phi}(x, t) = (\phi_1(x, t), \phi_2(x, t), \dots, \phi_n(x, t))^T, \qquad (5.8.13)$$

As mentioned above, for a given source term $q(x, t)$, a solution u(x,t) and an additional condition $u(x, T) = u_T(x)$ can be defined correspondingly. Namely, there exists a nonlinear operator $A$, such that

$$A[q(x, t)] = u_T(x) = \varphi(x). \qquad (5.8.14)$$

The inverse source problem is transformed into an inverse problem of parameter identification. As the operator Eq. (5.8.13) is ill-posed, we can transform it into a nonlinear functional optimization problem using the Tikhonov regularization algorithm [137]. The form of the nonlinear functional optimization problem can be written as

$$\boldsymbol{F}[q(x, t)] = \|A[q(x, t)] - \varphi(x)\|_{[0, L] \times [0, T]}^2 + \gamma \boldsymbol{H}[q(x, t)], \qquad (5.8.15)$$

where $\gamma$ is a regularization parameter and H is a stabilization functional of $q(x)$, can be written as

$$H[q(x, t)] = \|q(x, t)\|^2. \qquad (5.8.16)$$

Here the definition of $\|*\|$ is

$$\|f(x)\| = \left( \int_0^L f^2(x) dx \right)^{1/2}.$$

## (2)  General process of the best perturbation method [138, 139]

According to the previous section, we consider formula (5.8.6), (5.8.7) with boundary condition $u(x, t; q(x, t))$, assume $q_0(x, t) = \sum_{i=1}^{n} k_i^0 \phi_i(x, t) = \boldsymbol{K}_0^T \boldsymbol{\Phi}(x, t)$ is a function close to $q^*(x, t)$, and add a tiny perturbation $\delta q_0(x, t) = \sum_{i=1}^{n} \delta k_i^0 \phi_i(x, t) = \delta \boldsymbol{K}_0^T \boldsymbol{\Phi}(x, t)$ to $q_0(x, t)$. For $q_0(x, t) + \delta q_0(x, t)$, the corresponding solution to formula (5.8.6) and (5.8.7) with condition is $u(x, t; q_0(x, t) + \delta q_0(x, t))$. Therefore, the inverse problem is transformed into determination of the parameter vector $\delta \boldsymbol{K}_0$, which can be confirmed by minimizing locally the following objective function:

$$F[\delta \boldsymbol{K}_0] = \|u(x, T; q_0(x, t) + \delta q_0(x, t)) - \varphi(x)\|^2 + \gamma H[\delta \boldsymbol{K}_0] \qquad (5.8.17)$$

where $\gamma$ is a regularization parameter and $H[\delta K_0]$ is stabilization functional.

Since $\delta q_0(x)$ is very small, we have

$$
u(x, t; q_0(x, t) + \delta q_0(x, t)) = u(x, t; q_0(x, t))
$$
$$
+ \nabla_{\delta K_0}^T u(x, t; q_0(x, t))\delta K_0 + o(\|\delta q_0(x, t)\|) \qquad (5.8.18)
$$

$$
F[\delta K_0] = \left\| u(x, T; q_0(x, t)) - \varphi(x) + \nabla_{\delta K_0}^T u(x, T; q_0(x, t))\delta K_0 \right\|^2 + \gamma H[\delta K_0].
$$
$$
(5.8.19)
$$

Discretizing the domain $\Omega \times [0, T]$ ($\Omega = \{x | 0 < x < L\}$), defining $(x_m = m\Delta x \; m = 0, 1, 2, \ldots, N)$, and letting $H[\delta K_0] = \delta K_0^T \delta K_0$, we can obtain

$$
F[\delta K_0] = \sum_{m=1}^{M} \frac{\left\| u(x_m, T; q_0(x, t)) - \varphi(x_m) + \nabla_{\delta K_0}^T u(x_m, T; q_0(x, t))\delta K_0 \right\|^2}{+\gamma \delta K_0^T \delta K_0}
$$

assuming

$$
A = (a_{m,i})_{M \times n}, \quad a_{m,i} = \frac{\partial}{\partial k_i} u(x_m, T; q_0(x, t)); \quad U = \begin{bmatrix} u(x_1, T; q_0(x, t)) \\ u(x_2, T; q_0(x, t)) \\ \vdots \\ u(x_M, T; q_0(x, t)) \end{bmatrix};
$$

$$
U^* = \begin{bmatrix} \varphi(x_1) \\ \varphi(x_2) \\ \vdots \\ \varphi(x_M) \end{bmatrix};
$$

This relation can be simplified to

$$
F[\delta K_0] = \delta K_0^T A^T A \delta K_0 + 2\delta K_0^T A^T (U - U^*)
$$
$$
+ (U - U^*)^T (U - U^*) + \gamma \delta K_0^T \delta K_0 \qquad (5.8.20)
$$

We can find that $F[\delta K_0]$ gets the local minimum, the vector $\delta K_0$ satisfies the following linear equation:

$$
(A^T A + \gamma I)\delta K_0 = A^T (U^* - U) \qquad (5.8.21)
$$

To solve Eqs. (5.8.21), we can get $\delta K_0$, then we substitute $\delta q_0(x, t) = \sum_{i=1}^{n} \delta k_i^0 \phi_i(x, t) = \delta K_0^T \Phi(x, t)$, the perturbation $\delta q_0(x)$ can be obtained.

In this section, we use implicit finite difference method to solve the positive problem.

**Numerical example:** With a one-dimensional anomalous diffusion problem considered,

$$\frac{\partial u}{\partial t} = d(x)\frac{\partial^{\alpha} u}{\partial x^{\alpha}} + q(x, t), t > 0, 0 < x < 1, \tag{5.8.22}$$

the corresponding initial and boundary conditions

$$\begin{cases} u(x, 0) = x^3 \\ u(0, t) = 0 \\ u(1, t) = e^{-t} \end{cases}, \tag{5.8.23}$$

where $\alpha = 1.2$, $d(x) = 0.1x^{2.4}$, $q(x, t) = \sin x$ is independent of time. Figure 5.31 shows the numerical solution to the governing equation with the conditions at the time $T = 1$. For the inverse source problem, the corresponding additional condition is $\tilde{u}(x_m, 1)$ $(m = 1, \ldots, M)$, where $M = 1/\Delta x$, $\Delta x = \Delta t = 0.05$, i.e. $\varphi(x_m) = \tilde{u}(x_m, 1)$, $m = 1, \ldots, M$, and the set of basis functions $\Phi(x) = \{1, x, x^2, x^3, x^4\}$ is selected. Error criterion is

$$\begin{aligned} s^2 &= \frac{\sum\limits_{m=1}^{N} (u(x_m, T; q^{(k)}(x, t)) - \varphi(x_m))^2}{N + 1} < \text{Eps and Rel} \\ &= \frac{\|q^{(k)}(x, t) - q^*(x, t)\|_2}{\|q^*(x, t)\|_2} < \text{Eps}, \end{aligned}$$

where $q^{(k)}(x, t)$ is the approximate solution of the source term after iterating $k$ times, Eps is error requirement.

Firstly, we select $q_0(x) = 1 + x + x^2$ as the initial guess function, then we compare the convergent rate of the numerical method under different regularization parameters. Taking the regularization parameters as $\gamma = 10^{-2}$ and $\gamma = 10^{-4}$, respectively. Figure 5.34 shows the comparison of the computational solution to the exact solution of the source term. It is obvious that the computational solution matches well with the exact solution in this case. Through comparing the data in Tables 5.2 and 5.3, we know the parameter vector is stable after iterating more than four thousand times with $\gamma = 10^{-2}$, but the convergent rate is faster when the regularization parameter is $10^{-4}$. Note that the relative error is less than $10^{-4}$ after iterating several times, and then it reaches the level of $3.7671 \times 10^{-5}$. Figure 5.35 shows the relative errors of different regularization parameters at each iterative step.

Choosing another initial guess function as $q_0(x) = 100 + 100x - 100x^2 - 100x^3 + 100x^4$ and the regularization parameter as $\gamma = 10^{-4}$, and comparing the data in Tables 5.3 and 5.4, we can see that the convergent rate in this case is faster than that of the first one. Figure 5.36 shows the relative errors of different initial guess functions at each iterative step. The results show that the relative error gets the level

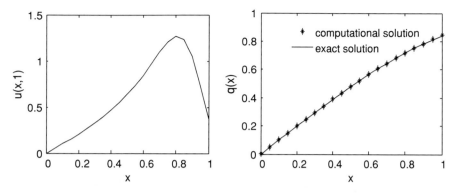

**Fig. 5.34** Comparison result of numerical solution and exact solution (Left: Numerical solution; Right: comparison result)

**Table 5.2** $q_0(x) = 1 + x + x^2$ and $\gamma = 10^{-2}$

| Iterative time | $K_{j+1} = \{k_1, k_2, k_3, k_4, k_5\}^{j+1}$ | Rel | s |
|---|---|---|---|
| 1000 | [0.0003,0.9952,0.0230,−0.2097,0.0331] | 1.9617e-004 | 0.0028 |
| 4000 | [0.0000,0.9991,0.0057,−0.1824,0.0189] | 3.7540e-005 | 9.4454e-004 |
| 5000 | [0.0000,0.9991,0.0057,−0.1824,0.0189] | 3.7657e-005 | 9.4452e-004 |
| 10,000 | [0.0000,0.9992,0.0057,−0.1824,0.0189] | 3.7671e-005 | 9.4452e-004 |
| 20,000 | [0.0000,0.9992,0.0057,−0.1824,0.0189] | 3.7671e-005 | 9.4452e-004 |

**Table 5.3** $q_0(x) = 1 + x + x^2$ and $\gamma = 10^{-4}$

| Iterative time | $K_{j+1} = \{k_1, k_2, k_3, k_4, k_5\}^{j+1}$ | Rel | s |
|---|---|---|---|
| 10 | [0.0003,0.9943,0.0272,−0.2165,0.0366] | 2.4851e-004 | 0.0035 |
| 50 | [0.0000,0.9992,0.0057,−0.1824,0.0189] | 3.7629e-005 | 9.4452e-004 |
| 100 | [0.0000,0.9992,0.0057,−0.1824,0.0189] | 3.7671e-005 | 9.4452e-004 |
| 500 | [0.0000,0.9992,0.0057,−0.1824,0.0189] | 3.7671e-005 | 9.4452e-004 |

of $3.7671 \times 10^{-5}$ and the parameter vector $K$ is [0.0000, 0.9992, 0.0057, −0.1824, 0.0189].

## 5.9   Variable-Order, Distributed-Order and Random-Order Fractional Derivative Models with Its Applications

This subsection will give a brief introduction on recent developments of variable-order fractional derivative, distributed-order fractional derivative and random-order fractional derivative in modeling and applications.

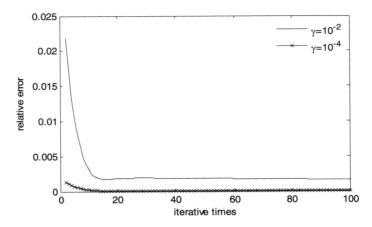

**Fig. 5.35** The relative errors of different regularization parameters at each iterative step

**Table 5.4** $q_0(x) = 100 + 100x\text{-}100x^2\text{-}100x^3 + 100x^4$ and $\gamma = 10^{-4}$

| Iterative time | $K_{j+1} = \{k_1, k_2, k_3, k_4, k_5\}^{j+1}$ | Rel | s |
|---|---|---|---|
| 50 | [0.0001,0.9989,0.0068,−0.1841,0.0197] | 3.2514e-005 | 0.0010 |
| 100 | [0.0000,0.9992,0.0057,−0.1824,0.0189] | 3.7670e-005 | 9.4452e-004 |
| 500 | [0.0000,0.9992,0.0057,−0.1824,0.0189] | 3.7671e-005 | 9.4452e-004 |
| 1000 | [0.0000,0.9992,0.0057,−0.1824,0.0189] | 3.7671e-005 | 9.4452e-004 |

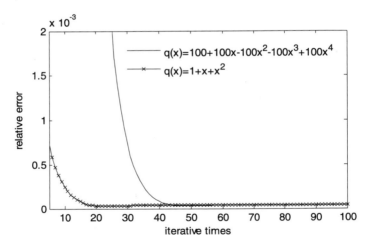

**Fig. 5.36** The relative errors of different initial guess functions at each iterative step

## 5.9.1  Variable-Order Fractional Derivative Modeling and Applications

Samko et al. first proposed the concept of variable-order (VO) operator and investigated the mathematical properties of VO integration and differentiation operators of Riemann–Liouville type [140–142]. Lorenzo and Hartley generalized different types of VO fractional operator definitions and made some theoretical studies via the iterative Laplace transform [143]. Coimbra et al. investigated the dynamics and control of nonlinear viscoelasticity oscillator via VO operator [144–147]. Ingman et al. employed the time-dependent VO operator to model the viscoelastic deformation process [148, 149]. Pedro et al. studied the motion of particles suspended in a viscous fluid with drag force is determined using the VO calculus [150]. Chechkin et al. introduced the space-dependent VO derivative into the differential equation of diffusion process in inhomogeneous media with the assumption that the waiting-time probability density function (PDF) is space dependent in the continuous time random walk (CTRW) scheme [151]. Nowadays, variable-order fractional derivative modeling and application have become a research hotspot in fractional calculus research.

**Viscoelasticity**

Variable-order fractional Voigt model

$$\sigma(t) = E_1 \tau_1^{q_1(t)} D^{q_1(t)} \varepsilon(t) + E_2 \tau_2^{q_2(t)} D^{q_2(t)} \varepsilon(t), \tag{5.9.1}$$

where $\sigma(t)$ represents the stress, $\varepsilon(t)$ denotes the strain, $E$ is the modulus of elasticity, $\tau_1, \tau_2$ are the relaxation time, $q_1, q_2$ are the orders of variable-order fractional derivative, the definition of the above equation is (2.5.26). If $q_1 = q_2$, then the above model changes into

$$\sigma(t) = E D^{q(t)} \varepsilon(t). \tag{5.9.2}$$

It can be further changed into more generalized form

$$\alpha(\varepsilon, \dot{\varepsilon}, t) D^{q_1(\varepsilon, \dot{\varepsilon}, t)} \sigma(t) = \beta(\varepsilon, \dot{\varepsilon}, t) D^{q_2(\varepsilon, \dot{\varepsilon}, t)} \varepsilon(t). \tag{5.9.3}$$

In Ref. [145], the authors have analyzed the previous experimental results which have been presented in Fig. 5.37. The interested readers can refer to the cited references.

Variable-order fractional viscous-viscoelasticity oscillator model

$$m D^2 x(t) + c_q D^{q(x(t))} x(t) + k x(t) = F(t), \tag{5.9.4}$$

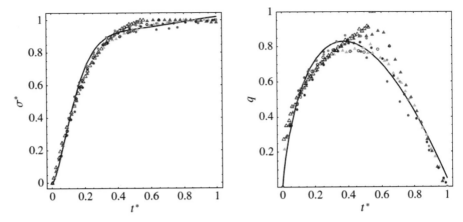

**Fig. 5.37** left part is the relationship between normalized stress and time. The points are all the data sets and the solid line is the order function. The right part is the variable-order evolution curve, based on the comparison result of variable-order model and experimental data (from [145])

where $c_q$ and $k$ are the visco-elastic coefficient and the spring coefficient, $m$ is the mass, $F()$ denotes the damping force. The analysis and numerical results of the above model are drawn in Fig. 5.38 [146].

**Anomalous Diffusion Modeling**

The comprehensive investigation of the variable-order operators in the anomalous diffusion modeling is still not received enough attention in the papers. Considering different situations of diffusion process, we classify the variable-order fractional derivative models into four different types: time dependent, space dependent, concentration dependent and system parameter dependent models [152]. The variable-order fractional derivative model can serve as an effective mathematical framework for

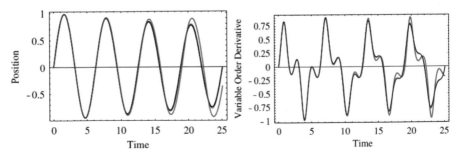

**Fig. 5.38** The comparison result of numerical and analytical results (5.9.4) when the damping force and related parameters are given. The solid line denotes the analytical result and the dashed line is the numerical result [146]

the description of various real-world anomalous diffusion processes in transitional regimes or other particular situations.

Four different types of variable-order models in one dimension are stated as follows:

(1)   Time-dependent variable-order derivative model

$$D_0^{\alpha(t)}c(x, t) = K\frac{\partial^2 c(x, t)}{\partial x^2}, \ 0 < \alpha(t) < 1, \tag{5.9.5}$$

where $\alpha(t)$ the variable-order, $c(x, t)$ denotes the density, $K$ is the diffusion coefficient. In the above equation, if $\alpha(t) = c$, $(0 < c < 1)$, then it changes into constant-order fractional diffusion equation. This model can better describe the diffusion processes whenever they get more anomalous or more Fickian in the course of time (Fig. 5.39).

(2)   Space-dependent variable-order derivative model

$$D_0^{\alpha(x)}c(x, t) = K\frac{\partial^2 c(x, t)}{\partial x^2}, 0 < \alpha(x) < 1. \tag{5.9.6}$$

This model is suitable for describing the diffusion process of the particles in non-homogeneous media, which means the diffusion capacity is different in different spatial locations. It corresponds to the random walk model, in which the waiting-time probability density function of particle jumps is dependent on the spatial position. This assumption is also consistent with the features of real-world diffusion processes and related issues.

**Fig. 5.39** The diffusion curve of variable-order time fractional diffusion model at fixed point. In the numerical simulation, the initial condition $c(x,0) = sin(x\pi/L)$, the boundary condition $c(0,t) = c(L,t) = 0$. *the expression of variable-order is* $\alpha(t) = \alpha_0 + pt/C$, $\alpha_0 = 0.6$, $p = 0.2$, $C = 10$

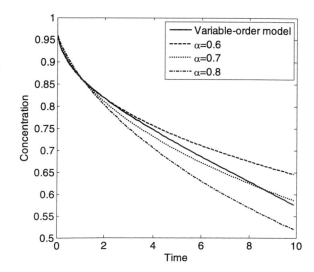

(3)    Concentration-dependent variable-order derivative model

$$D_0^{\alpha[c(x,t)]}c(x,t) = K\frac{\partial^2 c(x,t)}{\partial x^2}, 0 < \alpha[c(x,t)] < 1, \qquad (5.9.7)$$

this type of variable-order model may be useful for the anomalous diffusion modeling in chemistry or biology fields. The concentration-dependent variable-order means the diffusion behavior is related with concentration. Please refer to the reference for more details about this model [152].

(4)    System parameter-dependent variable-order derivative model

$$D_0^{\alpha[f(x,t)]}c(x,t) = K\frac{\partial^2 c(x,t)}{\partial x^2}, 0 < \alpha[f(x,t)] < 1. \qquad (5.9.8)$$

For example, when we consider the anomalous diffusion in the turbulence, because it is a dissipative system, the Reynolds number $Re$ determines the diffusion pattern especially in the laminar turbulent transition. The similar situation happens in the transport of passive tracers carried by fluid flow in porous medium or in the transmission medium with fractal structure. The fractal dimension D or the Hurst number $H$ changes with time or space in its transport process. The behavior of these diffusion or transport processes in response to system parameter changes can be better described using VO elements rather than time or space varying coefficients. Some applications of fractional operator also imply that the derivative order perhaps is not a constant, but a function of system parameters.

A class of spatial variable-order fractional diffusion equation model can also be established as follows:

$$\frac{\partial c(x,t)}{\partial t} = K\frac{\partial^{2\beta(x,t)}c(x,t)}{\partial|x|^{2\beta(x,t)}}, 0 < \beta(x,t) < 1. \qquad (5.9.9)$$

The spatial derivative order $\beta(x,t)$ can also be determined by the experimental or field measurement results, and further describes a class of real-world diffusion processes.

If you need to use the variable-order modeling method in the study of solute transport in porous media, the following variable-order derivative model can be established

$$\frac{\partial c(x,t)}{\partial t} = k(x,t)R_{\alpha(x,t)}c(x,t) - \upsilon(x,t)\frac{\partial c(x,t)}{\partial x} + f(c,x,t), \qquad (5.9.10)$$

in which, $k$, $\upsilon$ are the diffusivity coefficient and velocity respectively, $R_{\alpha(x,t)}$ denotes the Riesz spatial derivative, the corresponding definition is [153, 154]

$$-(-\Delta)^{\alpha(x,t)/2}f(x) = -\frac{1}{2\cos\frac{\pi\alpha(x,t)}{2}}\left[{}_aD_x^{\alpha(x,t)}f(x) + {}_xD_b^{\alpha(x,t)}f(x)\right],$$

$$m - 1 < \alpha(x, t) < m,$$

$$_{a+}D_x^{\alpha(x,t)} f(x) = \sum_{j=0}^{m-1} \frac{f^{(j)}(a)(x-a)^{j-\alpha(x,t)}}{\Gamma(-\alpha(x,t)+j+1)}$$

$$+ \frac{1}{\Gamma(m-\alpha(x,t))} \int_a^x \frac{f^{(m)}(\eta)}{(x-\eta)^{\alpha(x,t)-m+1}} d\eta,$$

$$_xD_{b-}^{\alpha(x,t)} f(x) = \sum_{j=0}^{m-1} \frac{(-1)^{m-j} f^{(j)}(b)(b-x)^{j-\alpha(x,t)}}{\Gamma(-\alpha(x,t)+j+1)}$$

$$+ \frac{1}{\Gamma(m-\alpha(x,t))} \int_x^b \frac{f^{(m)}(\eta)}{(\eta-x)^{\alpha(x,t)-m+1}} d\eta.$$

$$(5.9.11)$$

Here also gives more references about variable-order fractional derivative modeling for interesting readers.

### 5.9.2 Distributed-Order Fractional Derivative Modeling and Applications

The definition of the distributed-order fractional derivative is proposed by Caputo in 1960s, but the related research on its application in anomalous diffusion just received attentions in recent years. The distributed-order model is a natural extension of constant-order model, the constant-order fractional term is replaced by a series of fractional orders with different weights [155].

There are different forms of constitutive equation models in describing viscoelasticity of materials. The distributed-order model should be employed when the single constant-order fractional model cannot well describe experimental data. This section will present several kinds of distributed-order derivative viscoelastic constitutive equations. A simple form of distributed -order derivative model can be expressed as [156]

$$\sigma(t) = \int_0^1 \phi_\varepsilon(\gamma) \varepsilon^{(\gamma)}(t) d\gamma, \quad t > 0, \tag{5.9.12}$$

in which $\sigma$, $\varepsilon$ are the stress and strain, respectively, $\phi_\varepsilon$ is the weight function, $\gamma$ denotes the fractional derivative order.

Distributed-order fractional viscoelasticity model [157]

$$\int_0^1 \phi_\sigma(\gamma)\sigma^{(\gamma)}(x,t)d\gamma = \int_0^1 \phi_\varepsilon(\gamma)\varepsilon^{(\gamma)}(x,t)d\gamma, \qquad (5.9.13)$$

in which $\sigma$, $\varepsilon$ are the stress and strain, respectively, $\phi_\sigma$, $\phi_\varepsilon$ are the weight functions, $\gamma$ denotes the fractional derivative order. Another form of viscoelasticity model is [158]

$$\int_0^1 \phi_\sigma(\gamma)\sigma^{(\gamma)}(x,t)d\gamma = \int_0^1 \phi_\varepsilon(\gamma)y^{(\gamma)}(x,t)d\gamma, \qquad (5.9.14)$$

where $y$ is displacement. Please read more presented references about the distributed-order fractional derivative viscoelasticity model.

Distributed-order fractional diffusion model can also be used to describe different kinds of decelerating and accelerating diffusion processes [159–162]. Various kinds of diffusion processes can be described by changing the weight functions, especially for the trans-scale diffusion process in multi-fractal media. Here just present a distributed-order model for one-dimension diffusion

$$\int_0^1 p(\alpha)D_0^\alpha u(x,t)d\alpha = K\frac{\partial^2 u(x,t)}{\partial x^2}, \quad 0 < \alpha < 1, \qquad (5.9.15)$$

in which $p(\alpha)$ is the weight function, $\alpha$ is the time derivative order, $K$ is the diffusion coefficient, $D_t^\alpha(\cdot)$ is the Caputo fractional derivative.

The corresponding discretization form can be expressed as

$$\sum_{i=1}^m p(\alpha_i)D_0^{\alpha_i}u(x,t) = K\frac{\partial^2 u(x,t)}{\partial x^2}, 0 < \alpha < 1. \qquad (5.9.16)$$

If there are only two terms in the left side, then the expression is written as

$$p(\alpha_1)D_0^{\alpha_1}u(x,t)+p(\alpha_2)D_0^{\alpha_2}u(x,t) = K\frac{\partial^2 u(x,t)}{\partial x^2}, 0 < \alpha < 1. \qquad (5.9.17)$$

By employing the Fourier transform and Laplace transform on the above equation, the mean squared displacement expression can be written as ($0 < \alpha_1 < \alpha_2 < 1$)

$$\begin{cases} <x^2(t)> \propto t^{\alpha_2}, t \to 0, \\ <x^2(t)> \propto t^{\alpha_1}, t \to \infty. \end{cases} \qquad (5.9.18)$$

The generalized form of distributed-order model which includes time and space fractional derivatives can be stated as

$$\begin{cases} \int\limits_0^1 p(\alpha)D_0^\alpha u(x,t)\mathrm{d}\alpha = \int\limits_0^1 q(\beta)K(\beta)\frac{\partial^{2\beta}u(x,t)}{\partial|x|^{2\beta}}, \\ \int\limits_0^1 p(\alpha)\mathrm{d}\alpha = 1,\ \int\limits_0^1 p(\beta)\mathrm{d}\beta = 1,\ \alpha,\beta \in (0,1], \end{cases} \qquad (5.9.19)$$

in which $p(\cdot)$, $q(\cdot)$ are the weight functions.

### 5.9.3  Random-Order Fractional Derivative Modeling and Applications

Here introduces a new type of fractional derivative, named random-order fractional derivative, to describe the relaxation, oscillation and diffusion phenomena. In the random-order fractional derivative model, the fractional derivative order includes a constant and a random variable. The random term is mainly caused by the random fluctuations of external field or system parameters. The model can accurately analyze the system behavior under random fluctuations. Possible application fields include environmental pollute, project risk estimation, system stability analysis [163].

One-dimensional Relaxation equation

$$_tD_*^{\alpha_0+\varepsilon_t}u(t) = -\lambda u(t),\ p(\varepsilon_t|0 < \alpha_0 + \varepsilon_t < 1) = 1, \qquad (5.9.20)$$

where $_tD_*^{\alpha_0+\varepsilon_t}$ is the Caputo random-order fractional derivative, $p()$ denotes the probability density function, $\varepsilon_t$ is the random noise term ($\varepsilon_t$ is a small quantity, compared with $\alpha_0$). In the above model, it assumes the noise term is random and not dependent on time and space (Fig. 5.40).

The random-order fractional diffusion equation in one dimension can be written as

$$\frac{\partial^{\alpha_0+\varepsilon_{x,t}}u(x,t)}{\partial t^{\alpha_0+\varepsilon_{x,t}}} = K\frac{\partial^2 u(x,t)}{\partial x^2},\ p(\varepsilon_{x,t}|0 < \alpha_0 + \varepsilon_{x,t} < 1) = 1, \qquad (5.9.21)$$

in which $\alpha_0$ is the constant-order term, $\varepsilon_{x,t}$ is the random noise term. In the real-world diffusion processes, it may be disturbed by the external field or system parameters, the corresponding influence is characterized by the random noise term. The model can estimate and quantify the system fluctuation.

More details and information about random-order derivative modeling are presented in the Refs. [163, 164].

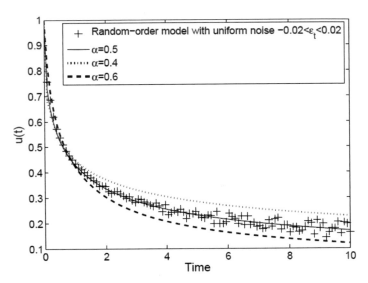

**Fig. 5.40** The relaxation curve of random-order fractional relaxation model. The initial condition is $u(0) = 1.0$ in the model, time step is $\Delta t = 0.1$

## 5.10 Some Applications of Fractional Calculus in Biomechanics

By applying the mechanics principles and methods, the biomechanics studied the relationships between the mechanics characteristics and the functions of biologic tissues and organs. It is important for understanding the mechanism of human body and providing the basis for designing of artificial organs and tissues. The viscoelastic and the constitutive equations of organism is one of the most important content in biomechanics. In the fourth section of this chapter, the fractional derivative models viscoelastic materials have been introduced. As the continuation and expansion, some applications of the equations for viscoelastic materials in biological modeling are introduced in this section.

Vestibule system is the receptor of the state of motion and spatial location in the inner ear. It contains semicircular canal and otolith system, where the semicircular canal is the receptor of rotary motion and the otolith system is the receptor of variable aligning motion. In this section, we will firstly introduce the fractional models for otolith organs and semicircular canal, and then introduce the fractional St. Venant model for human cranial bone.

### 5.10.1  Generalized Fractional Viscoelastic Dynamical Model of Otolith Organs

Otolith organs are the receptors of motion in the inner ear. It consists of the utricle and saccule. The classical elastic model of otolith organs is presented by Grant et al. [165]. They considered the endolymph, otolith layer and gel layer as Newtonian fluid, rigid slab and elastic body, respectively. In 1932, Grant [166] modified the above model and considered the gel layer as Klein-Voight viscoelastic body. However, the modified model can't fit the real data. Using fractional calculus and introducing the strain energy function, Su and Xu [52] generalized the models of Grant [165, 166] and obtained the analytical solutions of the model.

As for viscoelastic body, we assume that it satisfies the following non-dimension equation

$$_0D_t^\alpha W = \frac{d^\alpha W}{dt^\alpha} = \sigma_{ij}\varepsilon_{ij} + p\varepsilon_{ij}\delta_{ij}, \tag{5.10.1}$$

where $0 \le \alpha \le 1$, $_0D_t^\alpha$ is the Riemann—Liouville operator, $\sigma_{ij}$ and $\varepsilon_{ij}$ are the stress tensor and the strain tensor. $p$ is the static pressure. $\delta_{ij}$ is the Kronecker tensor. According to the theory of the continuum mechanics, the viscoelastic body includes viscoelastic solid and viscoelastic fluid. In the case of small deformation and homogeneous, we have the following constitutive equations,

$$\text{(for viscoelastic solid)} \quad \sigma_{ij} = -p\delta_{ij} + \lambda_e T^\alpha \frac{d^\alpha \varepsilon_{kk}}{dt^\alpha}\delta_{ij} + 2GT^\alpha \frac{d^\alpha \varepsilon_{ij}}{dt^\alpha} \tag{5.10.2}$$

$$\text{(for viscoelastic fluid)} \quad \sigma_{ij} = -p\delta_{ij} + \lambda_j T^{\beta-1} \frac{d^\beta \varepsilon_{kk}}{dt^\beta}\delta_{ij} + 2\mu T^\alpha \frac{d^\beta \varepsilon_{ij}}{dt^\beta} \tag{5.10.3}$$

where $\lambda_e$ and $G$ are Lame coefficient of the viscoelastic solid. $\lambda_j$ and $\mu$ are the viscosity coefficient of the viscoelastic fluid. $\alpha$ and $\beta$ are the viscoelastic parameters of the viscoelastic solid and viscoelastic fluid. $T$ is the non-dimensional time.

If we consider the otolith layer as a rigid slab of thickness $b$ and density $\rho_0$ and consider the strata gelatinosum and the Endolymph as viscoelastic solid and viscoelastic fluid, respectively, the following model of the otolith organs can be obtained as

$$\rho_f \frac{\partial u}{\partial t} = \mu T^{\beta-1} \frac{d^\beta}{dt^\beta} \int_0^t \frac{\partial^2 u}{\partial t^2} dt, \tag{5.10.4}$$

$$\rho_0 b \frac{\partial v}{\partial t} + (\rho_0 - \rho_f) b \left( \frac{\partial V_s}{\partial t} - g_x \right) =$$

$$\mu T^{\beta-1} \frac{d^\beta}{dt^\beta} \int_0^t \left. \frac{\partial u}{\partial y} \right|_0 dt - \frac{1}{2} \frac{E}{1+v} T^\alpha \frac{d^\alpha}{dt^\alpha} \int_0^t \left. \frac{\partial w}{\partial y} \right|_0 dt, \qquad (5.10.5)$$

$$\rho_f \frac{\partial w}{\partial t} = \frac{1}{2} \frac{E}{1+v} T^\alpha \frac{d^\alpha}{dt^\alpha} \int_0^t \frac{\partial^2 w}{\partial y^2} dt. \qquad (5.10.6)$$

The initial and boundary conditions are

$$u(0, t) = v(t), u(\infty, t) = 0, w(c, t) = v(t), w(0, t) = 0, \qquad (5.10.7)$$

$$u(y, 0) = v(0) = w(y, 0) = 0, \qquad (5.10.8)$$

where $u$, $v$ and $w$ are the velocities of the endolymph, the otolith layer and the gel layer with respect to the basement membrane. $\rho$ and $\mu$ are the density and viscosity coefficient of the endolymph. $E$, $v$ and $c$ are Young's modulus, Poisson's ratio and the depth of the gel layer. $V_s$ is the velocity of the basement membrane. $g_x$ is the x-component gravity acceleration.

By denoting $V$ the characteristic velocity and introducing the non-dimensional variables $y' = \frac{y}{b}$, $t' = \frac{\mu}{\rho_0 b^2} t$, $u' = \frac{u}{V}$, $v' = \frac{v}{V}$, $w' = \frac{w}{V}$, $V_f' = \frac{V_f}{V}$, we can obtain the non-dimensional form of the model. When $\varepsilon$ is less than 1, the approximate expression of the non-dimensional velocity of the otolith layer can be obtained by the Laplace transform method. It is

$$\frac{v(t)}{1-R} \approx \sum_{n=0}^\infty \frac{(-1)^n \varepsilon^n a^{n+1}}{n! R^{\frac{n+1}{2}}} t^{n(2-\alpha)} E_{1-\frac{\beta}{2}, n(1-\alpha+\frac{\beta}{2})+1}^{(n)} \left( -at^{1-\frac{\beta}{2}} \right). \qquad (5.10.9)$$

Here, $R = \frac{\rho_f}{\rho_0}$ and $a = \frac{\sqrt{R}}{1+R/3} > 0$. In order to analyze the characters of the system, using the expansion of the Mittag–Leffler function, we have

$$\frac{v(t)}{1-R} \approx \sum_{n=0}^\infty \sum_{n=0}^\infty \frac{(-1)^{n+m} \varepsilon^n a^{n+m+1} (m+n)!}{n! m! R^{\frac{n+1}{2}} \Gamma \left( 1 + n(2-\alpha) + m \left( 1 - \frac{\beta}{2} \right) \right)} t^{n(2-\alpha)+m(1-\frac{\beta}{2})}.$$

$$(5.10.10)$$

Correspondingly, the displacement $\delta(t) = \int_0^t v(\tau) d\tau$ becomes

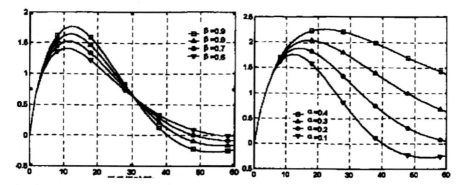

**Fig. 5.41** Curves of the non-dimensional deformations of otolith layer respect to different $R$ and $\varepsilon$, where $\alpha = 0.1$, $\beta = 0.9$, (Left) $R = 0.37$, (Right) $\varepsilon = 0.028$ (from Ref. [52]), the horizontal coordinate is time and the longitudinal coordinate is displacement

$$\frac{\delta(t)}{1-R} \approx \sum_{n=0}^{\infty} \sum_{n=0}^{\infty} \frac{(-1)^{n+m} \varepsilon^n a^{n+m+1} (m+n)!}{n!m!R^{\frac{n+1}{2}} \Gamma\left(2 + n(2-\alpha) + m\left(1 - \frac{\beta}{2}\right)\right)} t^{n(2-\alpha)+m\left(1-\frac{\beta}{2}\right)+1}.$$

$$(5.10.11)$$

To show the efficiencies of different materials of otolith organs, by setting $\alpha = 0.1$ and $\beta = 0.9$, Fig. 5.41a, b can be obtained as follows:

From the above figure, we can see that the maximum deformation of otolith layer $\delta_m$, the total time $T_m$ and the decay rates $V_d$ change significantly if $\varepsilon$ changes. When $\varepsilon$ becomes bigger, $\delta_m$ and $T_m$ become small and $V_d$ becomes bigger. When $\varepsilon$ keeps unchanged, $\delta_m$ and $V_d$ become bigger if $R$ becomes bigger. Therefore, we can draw the following conclusions. The material characters of the cuticle layer of otolith organs control the deformation of otolith layer. The endolymph plays an assistant regulation role which has little influence on the deformation of otolith layer. For more information, the readers can see [52] for reference.

### 5.10.2   Generalized Fractional Dynamic Model of Semicircular Canal

Gaede and Schmaltz firstly presented the model of semicircular canal, but the model is too simple to apply. Buskirk et al. [167] presented the widely used Buskirk model which is an integral–differential model. Xu and Tan [168] presented an analytical solution of high accuracy and three modes of dynamic response of the Buskirk model for the fluid in a single semicircular canal.

Assuming that the wall of the semicircular canal is cyclic annular rigid body and the endolymph is Newtonian fluid, the non-dimensional governing equation of the Buskirk model in case of small curvature is

$$\frac{\partial u}{\partial t} + (1 + \eta)\frac{\alpha(t)}{\Omega} = -\varepsilon \int_0^t \int_0^1 ur\,dr\,dt + \frac{1}{r}\frac{\partial}{\partial r}\left(r\frac{\partial u}{\partial r}\right), \qquad (5.10.12)$$

where $u$ is the axial velocity in the cylindrical coordinate. $r$ and $R$ are radial coordinate and the curvature radius. $\Omega$ and $\alpha(t)$ are characteristic angular velocity and the component of the angular acceleration perpendicular to the semicircular canal. $\eta = \gamma/\theta$ is the ratio of the field angel of sacculo-utricular and crista ampullaris. $\varepsilon = 2K\pi a^6/(\rho\theta R\nu^2)$ is the parameter of the model where $K$, $\rho$ and $\nu$ is the stiffness coefficient of the crista ampullaris, the density of the endolymph and the kinematic viscosity coefficient. $a$ is the radius of the semicircular canal. Su et al. [169] considered the crista ampullaris and the endolymph as viscoelastic solid and non-Newtonian fluid whose viscoelastic parameters are $\alpha$ and $\beta$ ($0 \le \alpha, \beta \le 1$), respectively. The following fractional equation for the motion of the endolymph is obtained

$$\frac{\partial u}{\partial t} + (1 + \eta)\frac{\alpha(t)}{\Omega} = -\varepsilon_0 D_t^\alpha \int_0^t \int_0^1 u(r, t)\,dr\,dt + \frac{1}{r}{}_0D_t^\alpha \int_0^t \frac{\partial}{\partial r}\left(r\frac{\partial u}{\partial r}\right)dt.$$

$$(5.10.13)$$

The initial and boundary conditions are

$$u(a, t) = 0, \quad \frac{\partial u(0, t)}{\partial r} = 0, u(r, 0) = 0. \qquad (5.10.14)$$

Assuming that when $t = 0$, $\alpha(t) = \Omega\delta(t)$ and using the Laplace transform method, the approximate expression of $u(r, t)$ can be obtained.

$$u(r, t) \approx (1 + \eta)(r^2 - 1) \sum_{n=0}^\infty \frac{1}{n!}\left(-\frac{\varepsilon}{4}t^{2-\alpha}\right)^n E_{2-\beta, n(\beta-\alpha)+1}^{(n)}(-4t^{2-\beta}). \qquad (5.10.15)$$

Using the asymptotic expansion of the Mittag–Leffler function, the velocity profile of semicircular canal can be obtained as

$$u(r, t) \approx \frac{(1 + \eta)(r^2 - 1)}{4}t^{\beta-2}E_{\beta-\alpha, \beta-1}\left(-\frac{\varepsilon}{16}t^{\beta-\alpha}\right) \approx \frac{4(1 + \eta)(r^2 - 1)}{\varepsilon\Gamma(\alpha - 1)}t^{\alpha-2}.$$

$$(5.10.16)$$

Consequently, the final state of motion of the endolymph is governed by the viscous properties of the crista ampullaris. Considering that the accurate expression of the velocity is difficult to obtain, we will study the frequency response properties of the system. Considering the average angular displacement of the endolymph $\theta(t)$ and the angular speed $\omega(t)$ as the input and output of the system and using the transfer function

$$G(p) = \frac{-(1+\eta)pI_2(\sqrt{p^2-\beta})}{2p^2 I_0(\sqrt{p^2-\beta}) + \varepsilon p^2 I_2(\sqrt{p^2-\beta})}, \tag{5.10.17}$$

the following corresponding frequency response can be obtained

$$G(j\omega) = \frac{-(1+\eta)j\omega I_2((j\omega)^{1-\beta/2})}{2(j\omega)^2 I_0((j\omega)^{1-\beta/2}) + \varepsilon(j\omega)^2 I_2((j\omega)^{1-\beta/2})} = M(\omega)\exp(j\varphi(\omega)), \tag{5.10.18}$$

Here $M(\omega)$ and $\varphi(\omega)$ are the argument and phase angle. In Fig. 5.42, The Bode diagrams for different values of $\alpha$ and $\beta$ are given. The data used here is presented by Buskirk. From the figure we can see that the frequency almost independent of $\alpha$ and $\beta$ at high frequency section. While in the low-frequency section, different $\alpha$ and $\beta$ will lead to different characteristics of the system. At the low-frequency section, as the increase of $\alpha$, the argument will increase and the phase angular will decrease. When the input frequency increasing, the argument will increase and the phase angular will decrease too. The changes of $\beta$ mainly affect the argument properties of medium frequency and the phase angular properties of low and medium frequency. Combining the physical meanings of the parameters, we can see from the figure that the viscoelastic effect of the endolymph mainly governs the output of

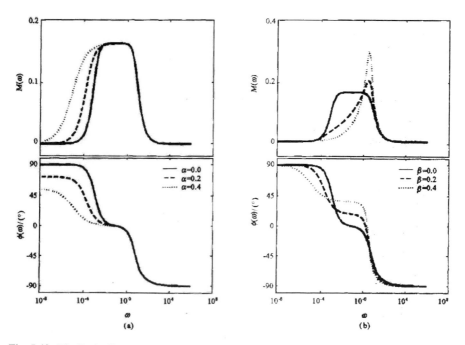

**Fig. 5.42** The Bode diagram of the frequency character in the logarithmic coordinates (from Ref. [169])

the system, while the steady state response is mainly governed by the viscoelastic characters of the crista ampullaris.

### 5.10.3 Fractional Model of Human Cranial Bone

A large number of experimental data and theoretical results show that bone is anisotropic viscoelastic body. Zhu et al. [170] did the compressive and tensile relaxation and creep experiment of small specimen of human cranial bone. Using the standard St. Venant model and linear viscoelastic theory, they obtained the quasi-steady state constitutive equations and draw the profiles of relaxation and creep of the bone. By denoting $E_1$ and $E_2$ the elastic coefficients of two spring, $\eta$ the viscous coefficient of a dashpot, the standard St. Venant model is illustrated in Fig. 5.43.

The constitutive equation is

$$\sigma(t) + \tau\dot{\sigma}(t) = E\varepsilon(t) + E_1\tau\dot{\varepsilon}(t), \tag{5.10.19}$$

where $\sigma(t)$ and $\varepsilon(t)$ are the stress and strain. $E = \frac{E_1 E_2}{E_1+E_2}$, $\tau_r = \frac{\eta}{E_1+E_2}$, $\tau_d = \frac{\eta}{E_2}$. If the loading processes in the stress relaxation and creep are

$$\varepsilon(t) = \begin{cases} At & 0 \le t \le t_1 \\ \varepsilon_1 & t > t_1 \end{cases} \tag{5.10.20}$$

and

$$\sigma(t) = \begin{cases} Bt & 0 \le t \le t_1 \\ \sigma_1 & t > t_1 \end{cases} \tag{5.10.21}$$

and using the Boltzmann superposition principle, the following stress relaxation and creep functions can be obtained

$$\sigma(t) = \begin{cases} AEt + A\tau_r(E_1 - E)(1 - \exp(-t/\tau_r)) & 0 \le t \le t_1 \\ AEt + A\tau_r(E_1 - E)(\exp(t/\tau_r) - 1)\exp(-t/\tau_r) & t > t_1 \end{cases} \tag{5.10.22}$$

**Fig. 5.43** Illustration of standard St. Venant model

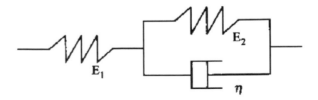

$$\varepsilon(t) = \begin{cases} \frac{Bt}{E} - \frac{B\tau_d}{E_2}(1 - \exp(-t/\tau_d)) \ 0 \le t \le t_1 \\ \frac{Bt}{E} - \frac{B\tau_d}{E_2}(\exp(t/\tau_d) - 1)\exp(-t/\tau_d) \ t > t_1 \end{cases}. \tag{5.10.23}$$

Liu and Xu [171] generalized the standard St. Venant model using the Riemann–Liouville fractional operator and obtained the fractional St. Venant model

$$\tau_r^{-q}{}_0D_t^{-q}\sigma(t) + \sigma(t) - \sigma_0 = E\tau_r^{-\mu}{}_0D_t^{-\mu}\varepsilon(t) + E_1(\varepsilon(t) - \varepsilon_0),$$

and obtained the stress relaxation and creep function under the same loadings

$$\sigma(t) = \begin{cases} \frac{AE_1 t}{q} H_{1,2}^{1,1}\left[\frac{t}{\tau_r} \Big| \begin{matrix} \left(0, \frac{1}{q}\right) \\ \left(0, \frac{1}{q}\right);(-1,1) \end{matrix}\right] + \frac{AEt}{q}\left(\frac{t}{\tau_r}\right)^{\mu} H_{1,2}^{1,1}\left[\frac{t}{\tau_r} \Big| \begin{matrix} \left(0, \frac{1}{q}\right) \\ \left(0, \frac{1}{q}\right);(-1-\mu,1) \end{matrix}\right] & 0 \le t \le t_1 \\ \\ \frac{AE_1(t-t_1)}{q} H_{1,2}^{1,1}\left[\frac{t}{\tau_r} \Big| \begin{matrix} \left(0, \frac{1}{q}\right) \\ \left(0, \frac{1}{q}\right);(-1,1) \end{matrix}\right] + \frac{AEt}{q}\left(\frac{t}{\tau_r}\right)^{\mu} H_{1,2}^{1,1}\left[\frac{t}{\tau_r} \Big| \begin{matrix} \left(0, \frac{1}{q}\right) \\ \left(0, \frac{1}{q}\right);(-1-\mu,1) \end{matrix}\right] - & t > t_1 \\ \quad \frac{AE(t-t_1)}{q}\left(\frac{t-t_1}{\tau_r}\right)^{\mu} H_{1,2}^{1,1}\left[\frac{t-t_1}{\tau_r} \Big| \begin{matrix} \left(0, \frac{1}{q}\right) \\ \left(0, \frac{1}{q}\right);(-1-\mu,1) \end{matrix}\right] \end{cases}$$

$$\tag{5.10.24}$$

$$\varepsilon(t) = \begin{cases} \frac{Bt}{E_1\mu} H_{1,2}^{1,1}\left[\frac{Wt}{\tau_r} \Big| \begin{matrix} \left(0, \frac{1}{\mu}\right) \\ \left(0, \frac{1}{\mu}\right);(-1,1) \end{matrix}\right] + \frac{Bt}{E_1\mu}\left(\frac{t}{\tau_r}\right)^{\mu} H_{1,2}^{1,1}\left[\frac{t}{\tau_r} \Big| \begin{matrix} \left(0, \frac{1}{\mu}\right) \\ \left(0, \frac{1}{\mu}\right);(-1-q,1) \end{matrix}\right] & 0 \le t \le t_1 \\ \\ \frac{Bt}{E_1\mu} H_{1,2}^{1,1}\left[\frac{Wt}{\tau_r} \Big| \begin{matrix} \left(0, \frac{1}{\mu}\right) \\ \left(0, \frac{1}{\mu}\right);(-1,1) \end{matrix}\right] + \frac{Bt}{E_1\mu}\left(\frac{t}{\tau_r}\right)^{\mu} H_{1,2}^{1,1}\left[\frac{t}{\tau_r} \Big| \begin{matrix} \left(0, \frac{1}{\mu}\right) \\ \left(0, \frac{1}{\mu}\right);(-1-q,1) \end{matrix}\right] - \\ \frac{B(t-t_1)}{E_1\mu} H_{1,2}^{1,1}\left[\frac{W(t-t_1)}{\tau_r} \Big| \begin{matrix} \left(0, \frac{1}{\mu}\right) \\ \left(0, \frac{1}{\mu}\right);(-1,1) \end{matrix}\right] - & t > t_1 \\ \frac{B(t-t_1)}{E_1\mu}\left(\frac{t-t_1}{\tau_r}\right)^{\mu} H_{1,2}^{1,1}\left[\frac{(t-t_1)}{\tau_r} \Big| \begin{matrix} \left(0, \frac{1}{\mu}\right) \\ \left(0, \frac{1}{\mu}\right);(-1-q,1) \end{matrix}\right] \end{cases}$$

$$\tag{5.10.25}$$

Figures 5.44 and 5.45 are the data fitting of the above two functions with the experiment data in ref. [170].

It is obvious that the fractional St. Venant model can describe the viscoelastic character of human cranial bone more accurately. We also can see that $q > \mu$ is always holding for both the stress relaxation function and creep function. It coincides with the requirements of thermodynamic stability of viscoelastic materials for fractional differential equations.

(Note: parts of Sects. 5.1–5.3 are from references [52, 169] and [171]).

**Fig. 5.44** stress relaxation
(from ref. [171])

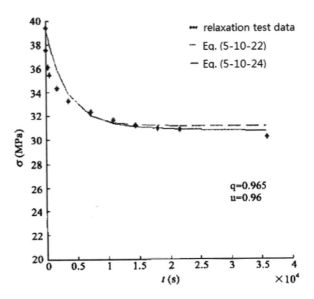

**Fig. 5.45** creep (from ref.
[171])

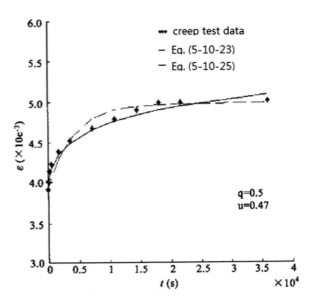

## 5.11 Some Applications of Fractional Calculus in the Modeling of Drug Release Process

Mathematical modeling and simulation are quantitative research methods in pharmacokinetics. They play important roles in the research of mechanism of action and the development of new drugs. The pharmacokinetics models can predict the drug concentration–time curves by relating the dosage or the frequency of dosage with the drug concentration. There are many methods to classify the pharmacokinetics models. According to the properties of the system, we can classify them as static or dynamic models, linear or nonlinear models, continuous or discrete models, and so on. We can also classify them as empirical or mechanism models. The empirical models are difficult to be generalized because they need small quantity of mechanism assumption and the data generating process is not clear. The mechanism models are based on the physical and biological principles and can describe kinds of characters of the system. They usually use the form of differential equations. In this section, we will introduce a fractional generalization of an empirical model and a mechanism model of drug release from a slab matrix.

### 5.11.1 Empirical Models

The empirical models usually ignore various physical and chemical processes and describe the results of drug release, i.e. the accumulative release of drug [172] (or fractional release). The classical empirical models for zero-order and one-order release are

$$\frac{\mathrm{d}X}{\mathrm{d}t} = K_0 \text{ and } \frac{\mathrm{d}X}{\mathrm{d}t} = -K_1 X.$$

The initial conditions are $X(0) = 0$ (0-order release) and $X(0) = X_0$ (1-order release). The corresponding solutions are $X = K_0 t$ and $X = X_0 \exp(-K_1 t)$ respectively. By introducing the fractional derivatives, the fractional models of 0-order and 1-order release are.

${}_0^C D_t^\alpha X = K_{f0}$ and ${}_0^C D_t^\alpha X = -K_{f1} X$, where ${}_0^C D_t^\alpha$ is the Caputo type fractional derivative operator. The corresponding results of drug release are $X = K_{f0} t^\alpha / \Gamma(\alpha + 1)$ and $X = X_0 E_\alpha(-K_{f1} t^\alpha)$, where $E_\alpha(t)$ is the Mittag–Leffler function. Dokoumetzidis and Macheras [173] compared the theoretical results with the experimental data in different references and draw the result that no matter for long time or short time release, the fractional model fits the data well.

### 5.11.2 Mechanism Models

Controlled release formulations can be used to reduce the amount of drug necessary to cause the same therapeutic effect in patients. Higuchi [174] firstly developed a remarkable simple model to simulate the drug release process from ointment in a planar system and obtained the formula for computing the amount of drug release based on the quasi-steady state assumption. Koizumi et al. [175] studied the drug release from the spherical matrix. Using the numerical methods, many models of drug release have been presented and calculated. In controlled drug delivery system,

**Fig. 5.46** Illustration of drug release from a slab matrix

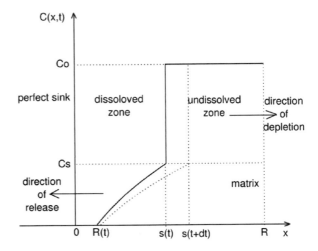

diffusion is the basic mechanism. Here, we will introduce the applications of fractional diffusion equations in the modeling of drug release from planar matrix. Liu and Xu [176] were the first who introduced the time-fractional diffusion equation to the drug release process and obtained the analytical solutions of the model. Li et al. [177, 178] discussed the time–space fractional drug release models in which the fractional derivative operators are in different definitions and obtained the scale invariant solutions. Yin and Xu [179] studied the release of drug from the slow degradable matrix and presented a model with two moving boundaries (Fig. 5.46).

The release mechanism of sustained and controlled release systems is usually divided into the diffusion type, swelling type, erosion type, penetration type and others [172]. Here we will consider the diffusion type only. The concentration profile at time t is shown in Fig. 5.41, $R$ is the scale of the polymer matrix. $S(t)$ is the diffusion front. $R(t)$ is the boundary of the matrix, for non-degradable matrix, we have $R(t) = 0$. $C_0$ is the initial concentration of drug distributed in the matrix. $C_s$ is the solubility of the solute in the solvent. In the following, we will assume that $C_0 > C_s$. $K$ is the diffusion coefficient of the drug. When the solvent penetrates into the matrix in which the initial concentration is higher than the solubility, only part of the drug dissolves and releases from the matrix. So, two zones, the dissolved zone and un-dissolved zone exist in the matrix. The boundary of these two zones, i.e. the diffusion front, moves inward as time progresses until all the drugs are dissolved. Considering the anomalous properties of the drug diffusion and using the fractional diffusion equation, the governing equation in the dissolved zone is

$$
{}_0^C D_t^\alpha C = -K \nabla_x^\beta C \tag{5.11.1}
$$

where $\nabla_x^\beta$ is the space fractional derivative. It can be the Caputo type, the Riemann–Liouville type, the Riesz type or others. At the boundary of the matrix, we use the perfect sink condition $C(0, t) = 0$. At the moving boundary, we use the condition

$C(S(t), t) = C_s$. In order to describe the progress of the moving boundary, considering the principle of conservation of mass and the generalized Fick law, we can introduce the following Stefan condition $\frac{dS(t)}{dt} = K_0^C D_t^{1-\alpha} \nabla_x^\beta C$ at $x = S(t)$. The existed fractional models can be summarized as

$$_0^C D_t^\alpha C = -K \nabla_x^\beta C, \quad (0 < \alpha \leq 1 < \beta \leq 2, R(t) \leq x \leq S(t) \tag{5.11.2}$$

$$C(x, t) = 0, \quad (x = R(t)) \tag{5.11.3}$$

$$C(x, t) = C_s, \quad (x = S(t)) \tag{5.11.4}$$

$$(C_0 - C_s)\frac{dS(t)}{dt} = K_0 D_t^{1-\alpha} \nabla_x^\beta C, \quad (t > 0, x = s(t)) \tag{5.11.5}$$

$$(C_0 - C_s)_0 D_t^\alpha S(t) = K \nabla_x^\beta C, \quad (t > 0, x = s(t)) \tag{5.11.6}$$

$$S(0) = 0, \tag{5.11.7}$$

Model 1[176]: $R(t) = 0, 0 < \alpha \leq 1, \beta = 2, \nabla_x^\beta = \frac{\partial^2}{\partial x^2}$. The boundary condition is (5.11.6).

Model 2[1]: $R(t) = 0, 0 < \alpha \leq 1, 1 < \beta \leq 2$. $\nabla_x^\beta$ is the Riesz type fractional derivative. The boundary condition is (5.11.5).

Model 3[178]: $R(t) = 0, 0 < \alpha \leq 1, 1 < \beta \leq 2$. $\nabla_x^\beta$ is the Caputo type (Case I) or the Riemann–Liouville type (Case II) fractional derivative. The boundary condition is (5.11.5).

Model 4[179]: $R(t) = \zeta t, 0 < \alpha \leq 1, \beta = 2, \nabla_x^\beta = \frac{\partial^2}{\partial x^2}$. The boundary condition is (5.11.6).

It is worth to note that when $\alpha = 1, \beta = 2$, the above models are the model of drug release from planner matrix of integer order.

From the mathematical viewpoint, these models are moving boundary problem [180] whose analytical solution is difficult to obtain. As for the integer-order model, under the quasi-steady state assumption, the well-known Higuchi formula is

$$Q = \sqrt{2(C_0 - C_s)C_s K t}, \tag{5.11.8}$$

where $Q$ is the amount of drug absorbed at time $t$ per unit area. Deleting the quasi-steady state condition, Paul [181, 182] presented the accurate solution of the integer-order model. It is

$$C = C_s \frac{erf(\delta)}{erf(\delta*)}, \tag{5.11.9}$$

where $erf()$ is the error function. $\delta = \frac{x}{2\sqrt{Kt}}$, $\delta = \frac{S(t)}{2\sqrt{Kt}}$, $\sqrt{\pi}\delta * \exp(\delta*^2)erf(\delta*) = \frac{C_s}{C_0 - C_s}$.

As for non-degradable matrix (models 1–3) and only considering the first stage that the drug is not fully dissolved (i.e. $S(t) < R$), Liu and Xu [176] obtained the analytical solutions in the form of Wright function using the Laplace and Fourier transform method. Li et al. [177] obtained the analytical in the form of $H$ function using the same method. Under certain conditions, the results of Liu and Xu [176] are the special case of Li et al. [177]'s results. The concentration is [1]

$$C(x, t) = \frac{2q}{\beta} C_0 H_{4,4}^{2,2} \left[ \frac{x}{D^{1/\beta} t^{\alpha/\beta}} \middle| \begin{array}{l} (1, 1/\beta)(1, 1)(1, \alpha/\beta)(1, 1/2) \\ (1, 1/\beta)(1, 1)(0, 1)(1, 1/2) \end{array} \right]. \quad (5.11.10)$$

The expression of the moving boundary is $S(t) = pt^{\alpha/\beta}$, where $p$ and $q$ are constant to be determined. The above solutions are concise, but the computations must use the computer software since the $H$ function is complex. If the space fractional derivative is the Caputo type or the Riemann–Liouville type, the scale invariant solutions can be obtained [178]. The scale invariant variables and the moving boundary are $z = xt^{-\alpha/\beta}$ and $S(t) = pt^{\alpha/\beta}$, respectively. The concentration is

$$C(x, t) = C_1 z W_{(-\alpha, 1-\alpha/\beta)(\beta, 2)}(z^\beta) \quad (5.11.11)$$

where $W_{(\mu,a)(v,b)}(z) = \sum_{k=0}^{\infty} \frac{z^k}{\Gamma(a+\mu k)\Gamma(b+vk)}$ is the generalized Wright function. $p$ and $C_1$ are constants to be determined by the boundary conditions. It is obvious that the solution (5.11.10) by the integral transform method has the nature of scale invariance. As been pointed in Ref. [183], most of the analytical solutions of moving boundary problems that can be obtained are scale invariant. This view is right for the fractional cases.

The non-dimensional diffusion front position $S(\tau)$ for model 2 is shown in Fig. 5.47. Curves 1–3 correspond to space fractional diffusion models, curve 4 corresponds to the ordinary diffusion model. Curves 5 and 6 correspond to time-fractional diffusion models. It is obvious that the time-fractional diffusion model and the space fractional one describe sub-diffusion and super-diffusion, respectively. It is consistent with the results of Metzler and Klafter [184].

Using $M_t$ and $M_\infty$ to denote the amount of the drug release at time $t$ and the total amount of drug and computing the fractional release of models 1-3, we have

$$\text{Model } 1 \quad : \quad M_t = \frac{k\lambda t^{1-\alpha/2}}{\Gamma(2 - \alpha/2)} \quad (5.11.12)$$

where $\lambda^2 = K$. $k$ is a constant which can be obtained by solving the equations.

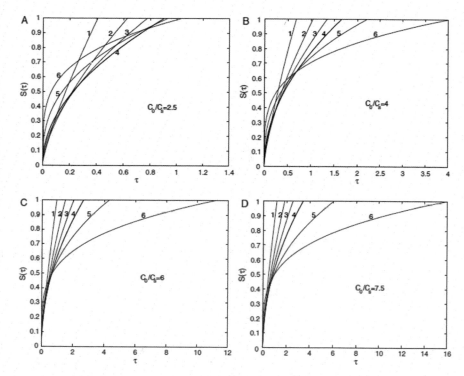

**Fig. 5.47** Non-dimensional diffusion front position $S(\tau) = \frac{S}{R}$ versus non-dimensional time. $\tau = \left(K/R^{\beta}\right)^{1/\alpha} t$ at various solute loading levels. Curves 1–6 correspond to the cases that $(\alpha, \beta)$ equals to (1,1.25), (1,1.5), (1,1.75), (1,2), (0.75,2) and (0.5,2)

$$\text{Model 2} \quad : \quad \frac{M_t}{M_{\infty}} = \left[ p - \frac{C_s}{C_0} \frac{2pq}{\beta} H_{4,4}^{2,2}[p] \right] t^{\alpha/\beta} \tag{5.11.13}$$

where $p, q$ are constants to be determined. $H_{4,4}^{2,2}[p]$ is the H function.

$$\text{Model 3} \quad : \quad (\text{case 1}) \quad \frac{M_t}{M_{\infty}} = \left[ p - \frac{C_s}{C_0} C_1 p^2 W_{(-\alpha,1-\alpha/\beta)(\beta,3)}(p^{\beta}) \right] t^{\alpha/\beta}$$
$$\tag{5.11.14}$$

$$\text{Model 3} \quad : \quad (\text{case 2}) \quad \frac{M_t}{M_{\infty}} = \left[ p - \frac{C_s}{C_0} C_2 p^{\beta} W_{(-\alpha,1-\alpha/\beta-\alpha)(\beta,\beta+1)}(p^{\beta}) \right] t^{\alpha/\beta}$$
$$\tag{5.11.15}$$

where $p, C_1$ and $C_2$ are constants to be determined. $W_{(a,b)(c,d)}()$ is the generalized Wright function.

The dimensionless fractional releases for the two cases of model 3 are shown in Fig. 5.48. Apparently, for every set of parameters, case 1 needs less time for the

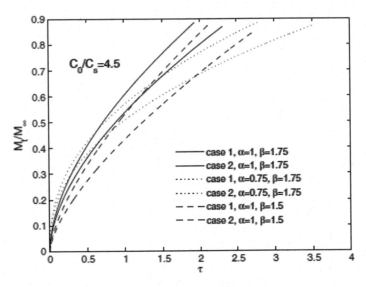

**Fig. 5.48** The fractional release of Model 3 respects to different time

diffusion interface to reach R. To show the effects of the initial drug loading, the fractional releases of case 1 of model 3 in the case $\alpha = 0.75$, $\beta = 1.75$ are shown in Fig. 5.49. It is obvious that the maximum fractional release increases as the increase of the initial drug loading.

As for the case of degradable matrix (model 4), there's no analytical solution presented the cause of the existence of two moving boundaries. If the speed of

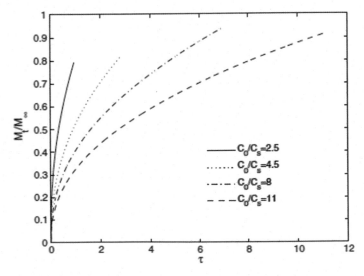

**Fig. 5.49** The curves of the fractional release of model 3 (case 1) when $\alpha = 0.75$, $\beta = 1.75$

the matrix dissolution is very slow and assuming that $R(t) = \zeta t$ and $\zeta \, o(1)$, the approximate solution can be obtained by the two-parameter regular perturbation method.

It is obvious that the results (5.11.12)–(5.11.15) coincide with the semi-empirical formula $\frac{M_t}{M_\infty} = kt^n$ presented by Ritger and Peppas [185]. Moreover, the results here can describe the anomalous diffusion processes of the index $n \in [0.5, 1)$. So the fractional models explain the semi-empirical formula using the diffusion mechanism. It is another proof that the fractional equations are powerful tools to describe anomalous diffusion.

Until now, only some simple cases of fractional diffusion equation in drug release process are investigated. Mathematical models considering more factors like matrix of other geometric shapes or swelling matrix will be focused on in the future. Therefore, the clinical applications of the fractional derivative models still have many difficulties to be overcome.

(Part of this section comes from [186]).

# References

1. H.B. Hu, L. Wang, A brief history of power law distributions [J]. Physics **34**(12), 889–896 (2005)
2. L.A. Adamic, Zipf, Power-laws, and Pareto-a ranking tutorial [J]. http://blogimages.bloggen. be/stijfvreter/attach/12750.pdf
3. C. Kuhnert, Scaling laws in urban supply networks [J]. Phys. A **363**, 96–103 (2006)
4. J.Z. Zhang, *Fractal [M]* (Tsinghua University Press, Beijing, 1995)
5. X.W. Zhang, *The Constitution Theory [M]* (China Science and Technology University Press, Hefei, 2003)
6. R. Albert, A.L. Barabasi, Statistical mechanics of complex networks [J]. Rev. Mod. Phys. **74**, 47–97 (2002)
7. A.L. Barabasi, R. Albert, Emergence of scaling in random networks [J]. Science **286**(5439), 509–512 (1999)
8. P. Bak, *How Nature Works [M]* (Central China Normal University Press, Wuhan, 2001)
9. S. Kauffman, *Investigations[M]* (Hunan Science and Technology Press, Changsha, 2004)
10. J.M. Carlson, J. Doyle, Highly optimized tolerance: robustness and design in complex systems [J]. Phys. Rev. Lett. **84**(11), 2529–2532 (2000)
11. T.Q. Yang, *Theory of Viscoelasticity [M]* (Huazhong University of Science and Technology Press, Wuhan, 1990)
12. R.L. Bagley, Power law and fractional calculus model of viscoelasticity [J]. AIAA J. **27**(10), 1412–1417 (1989)
13. K. Adolfsson, M. Enelund, P. Olsson, On the fractional order model of viscoelasticity [J]. Mech. Time-Depend. Mater. **9**, 15–34 (2005)
14. Q.Y. Wu, J. Wu. Polymer Rheology[M] (Higher Education Press, Beijing, 2002)
15. C.F. Chan, *Non-newtonian Fluid Mechanics [M]* (Science Press, Beijing, 1984)
16. H.M. Wang, Z.X. Zhao, *Engineering Fluid Mechanics [M]* (Hohai University Press, Nanjing, 2005)
17. T.L. Szabo, Time domain wave equations for lossy media obeying a frequency power law [J]. J. Acoust. Soc. Am. **96**(1), 491–500 (1994)
18. W. Chen, S. Holm, Modified Szabo's wave equation models for lossy media obeying frequency power law [J]. J. Acoust. Soc. Am. **114**(5), 2570–2574 (2003)

19. Z.H. Lin, *Thermodynamics and Statistical Physics [M]* (Peking University Press, Beijing, 2007)
20. Y.L. Guo, H.J. Shen, *History of Physics [M]* (Tsinghua University Press, Beijing, 2005)
21. Z.X. Wang, *Introduction to Statistical Physics [M]* (High Education Press, Beijing, 1988)
22. G. Hu, *Stochastic Forces and Nonlinear Systems [M]* (Shanghai Science and Technology Education Press, Shanghai, 1994)
23. D. Lianke, *Fractal Dynamics [M]* (Liaoning Science and Technology Press, Shenyang, 1994)
24. E. Lutz, Fractional Langevin equation [J]. Phys. Rev. E **64**, 051106 (2001)
25. S. Burov, E. Barkai, The critical exponent of the fractional langevin equation is $a_c \approx 0.402$ [J]. arXiv:0712.3407v1 (2007)
26. G.B. Suffritti, A. Taloni, P. Demontis, Some considerations about the modelling of single file diffusion [J]. Diffus. Fundam. **7**, 1–2 (2007)
27. F. Mainardi, F. Tampieri, Diffusion regimes in Brownian motion indued by the basset history force. Fracalmo Pre-Print 0102 Technical Report\ No 1, ISA0-TR-1/99 (1999)
28. V. Kobelev, E. Romanov, Fractional Langevin equation to describe anomalous diffusion [J]. Prog. Theor. Phys. Suppl. **139**, 470–479 (2000)
29. K.M. Kolwankar, Brownian motion of fractal particles: Lévy flights from white noise [J] (2005). arXiv: cond-mat/0511307v1
30. S.C. Lim, Fractional derivative quantum fields at positive temperature [J]. Phys. A **363**, 269–281 (2006)
31. B.J. West, Thoughts on modeling complexity [J]. Complexity **11**(3), 34–43 (2006)
32. X.J. Li, Nonlinear site seismic response analysis methods [D]. PhD thesis (Engineering Mechanics Institute, China Earthquake Administration, Harbin, 1993)
33. F. Mainardi, Fractional relaxation-oscillation and fractional diffusion-wave phenomena [J]. Chaos Solitons Fractals **7**(9), 1461–1477 (1996)
34. A. Tofighi, The intrinsic damping of the fractional oscillator [J]. Phys. A **329**, 29–34 (2003)
35. Y.E. Ryabov, A. Puzenko, Damped oscillation in view of the fractional oscillator equation [J]. Phys. Rev. B **66**, 184–201 (2002)
36. X.Z. Zhang, Y. Cheng, L. Xie, A new explicit solution of dynamic response analysis [J]. Earthq. Eng. Eng. Vib. **22**(3), 1–8 (2002)
37. Z.D. Yuan, J. Fei, D. Liu et al., *Numerical Solution of Rigid Ordinary Differential Problems [M]* (Science Press, Beijing, 1987)
38. S.J. Shen, F.W. Liu, A computat ionally effective numerical method for the fractional order Bagley-Torvik equation [J]. J. Xiamen Univ. (Nat. Sci.) **43**(3), 306–311 (2004)
39. P.H. Zhuang, F.W. Liu, Explicit approximation for space-time fractional diffusion equation [J]. Comput. Math. Chin. Univ. **27**, 223–228 (2005)
40. C.H. Yang, F.W. Liu, A fractional predictor-corrector method of the fractional relaxation-oscillation equation [J]. J. Xiamen Univ. (Nat. Sci.) **44**(6), 761–765 (2005)
41. L.C. Liu, Q.F. Yan, W. Zhang, X.Y. Huang, The dynamics equation and FE numerical solution of high speed CD-ROM viscoelastic damping frame structure [J]. Noise Vib. Control **10**, 14–17 (2006)
42. E. Cai, *Foundation of Viscoelastic Material [M]* (Beijing University of Aeronautics and Astronautics, Beijing, 1989)
43. L. Gaul, The influence of damping on waves and vibrations [J]. Mech. Syst. Signal Process. **13**(1), 1–30 (1999)
44. S. Kempfle, I. Sch, H. Beyer, Fractional calculus via functional calculus: theory and applications [J]. Nonlinear Dyn. **29**(1), 99–127 (2002)
45. D.S. Yin, J.J. Ren, C.L. He, W. Chen, A new rheological model element for geomaterials [J]. Chin. J. Rock Mech. Eng. **26**(9), 1899–1903 (2007)
46. J.H. Qian, Z.Z. Yin, *Geotechnical Principles and Calculation [M]* (China Water Conservancy and Hydropower Press, Beijing, 1996)
47. Z.H. Liu, W.M. Zhang, The viscoelastic solid model with fractional order derivative and its applications [J]. J. Zhuzhou Inst. Technol. **16**(4), 23–25 (2002)

48. W.M. Zhang, A new rheological model theory with fractional order derivatives [J]. Nat. Sci. J. Xiangtan Univ. **23**(1), 30–36 (2001)
49. H.T. Qi, M.Y. Xu, Creep compliance of fractional viscoelastic models: generalized Zener and Poynting-Thomson model [J]. J. Shandong Univ. **39**(3), 42–48 (2004)
50. M.Y. Xu, W.C. Tan, Representation of the constitutive equation of viscoelastic materials by the generalized fractional element networks and its generalized solutions [J]. Sci. China Ser. G: Phys. Mech. Astron. **46**(2), 145–157
51. W.X. Pan, W.C. Tan, An unsteady flow of a viscoelastic fluid with the fractional maxwell model between two parallel plates [J]. Mech. Eng. **25**(1), 19–22 (2003)
52. H.J. Su, M.Y. Xu, Generalized visco-elastic model of otolith organs with fractional order [J]. Chin. J. Biomed. Eng. **20**(2), 46–52 (2001)
53. C.Y. Zhang, W.M. Zhang, P. Zhang, Viscoelastic fractional derivative model of concrete with aging [J]. Chinese J. Appl. Mech. **21**(1), 1–4 (2004)
54. K.X. Hu, K.Q. Zhu, Mechanical analogies of fractional elements [J]. Chin. Phys. Lett. **26**(10), 1083011–1083013 (2009)
55. M.J. Zhang, L.P. Zhang, S.B. Zhang et al., Study on rheological constitutive relations for structural soft soils [J]. J. Jilin Univ. (Earth Sci. Ed.) **1**(34), 242–246 (2004)
56. K.X. Hu, K.Q. Zhu, The exact solution of Stokes' second problem including start-up progress with fractional element [J]. Acta. Mech. Sin. **25**, 577–582 (2009)
57. J.L. Davis, *Mathematics of Wave Propagation [M]* (Princeton University Press, New Jersey, 2000)
58. H.L. Zhang, *The Theoretical Acoustic* [M] (Higher Education Press, Beijing, 2007)
59. S.P. Timoshenko, J.N. Goodier, *Theory of Elasticity [M]* (Tsinghua University Press, Beijing, 2004)
60. B.Y. Xu, X.P. Shen, Z.S. Cui, *Solid Mechanics* [M] (China Environmental Science Press, Beijing, 2003)
61. N.E. Tatar, A blow up result for a fractionally damped wave equation [J]. Nonlinear Differ. Equ. Appl. **12**, 216–226 (2005)
62. A.M.A. El-Sayed, Fractional-order diffusion-wave equation [J]. Int. J. Theor. Phys. **35**(2), 311–322 (1996)
63. W. Chen, Physical interpretation of fractional diffusion-wave equation via lossy media obeying frequency power law [J]. Simula Research Laboratory, P. O. Box. 134, 1325 Lysaker, Norway, 2003.
64. R. Rusovici, Modeling of shock wave propagation and attenuation in viscoelastic structures [D]. Dissertation for Doctor Degree. (Virginia Polytechnic Institute and State University, Blacksburg, 1999)
65. Y.Z. Feng, *Introduction to Continuum Mechanics* [M]( Science Press, Beijing, 1984)
66. T.L. Szabo, J. Wu, A model for longitudinal and shear wave propagation in viscoelastic media [J]. J. Acoust. Soc. Am. **107**(5), 2437 (2000)
67. J. Lighthill, *Waves in Fluids [M]* (Cambridge UP, Cambridge, 1980)
68. Y.A. Rossikhin, M.V. Shitikova, Applications of fractional calculus to dynamic problems of linear and nonlinear hereditary mechanics of solids [J]. Appl. Mech. Rev. **50**, 15–67 (1997)
69. S. Ginter, Numerical simulation of ultrasound-thermotherapy combining nonlinear wave propagation with broadband soft-tissue absorption [J]. Ultrasonics **27**, 693–696 (2000)
70. A.I. Nachman, J. Smith, R.C. Waag, An equation for acoustic propagation in inhomogeneous media with relaxation losses [J]. J. Acoust. Soc. Am. **88**(3), 1584–1595 (1990)
71. G. Wojcik, J. Mould, Jr. F. Lizzi, N. Abboud, M. Ostromogilsky, D. Vaughan, Nonlinear modelling of therapeutic ultrasound, in *1995 IEEE Ultrasonics Symposium Proceedings* (1995), pp 1617–1622
72. W. Chen, S. Holm, A. Bounaim, Å. Ødegård, A. Tveito, A frequency decomposition time domain model of broadband frequency-dependent absorption [J]. Preprint (Universitetet i Oslo, Institutt for informatikk) (2002). http://urn.nb. no/URN: NBN: no-35455
73. M. Caputo, F. Mainardi, A new dissipation model based on memory mechanism [J]. Pure Appl. Geophys. **91**, 134–147 (1971)

74. M.Y. Xu, W.C. Tan, Middle process, critical phenomena-fractional operator theory, methods, progress and its application in modern mechanics [J]. Sci. China (Ser. G) **36**(3), 225–238 (2006)
75. F.X. Chang, J. Chen, W. Huang, Anomalous diffusion and fractional advection-diffusion equation [J]. ACTA Phys. Sin. **54**(03), 1113–1117 (2005)
76. R. Gorenflo, F. Mainardi, D. Moretti, G. Pagnini, P. Paradisi, Discrete random walk models for space-time fractional diffusion [J]. Chem. Phys. **284**, 521–541 (2002)
77. R.L. Magin, Anomalous diffusion expressed through fractional order differential operators in the Bloch-Torrey equation [J]. J. Magn. Reson. **190**, 255–270 (2008)
78. I. Podlubny, *Fractional Differential Equations [M]* (Academic Press, San Diego and London, 1999)
79. Y. Lin, C. Xu, Finite difference/spectral approximations for the time-fractional diffusion equation [J]. J. Comput. Phys. **225**, 1533–1552 (2007)
80. A. Bounaïm et al., Detectability of breast lesions with CARI ultrasonography using a bioacoustic computational approach [J]. Comput. Math. Appl. **54**, 96–106 (2007)
81. A. Bounaïm et al., Quantification of the CARI breast imaging sensitivity by 2D/3D numerical time-domain ultrasound wave propagation [J]. Math. Comput. Simul. **65**, 521–534 (2004)
82. W. Weiwad et al., Direct measurement of sound velocity in various specimens of breast tissue [J]. Invest. Radiol. **35**(12), 721–726 (2000)
83. F.T. D'astrous, F.S. Foster, Frequency dependence of ultrasound attenuation and backscatter in breast tissue [J]. Ultrasound Med. Biol. **12**(10), 795–808 (1986)
84. W. Chen, Lévy stable distribution and [0, 2] power law dependence of acoustic absorption on frequency in various lossy media [J]. Chin. Phys. Lett. **22**(10), 2601–2603 (2005)
85. B.B. Mandelbrot, *The Fractal Geometry of Nature* (W. H. Freeman, San Francisco, 1982)
86. K.I. Sato, *Lévy Processes and Infinitely Divisible Distributions* [M] (Cambridge University Press, 1999)
87. B.I. Henry, S.L. Wearne, Fractional reaction–diffusion [J]. Phys. A **276**, 448–455 (2000)
88. M. Ochmann, S. Makarov, Representation of the absorption of nonlinear waves by fractional derivative [J]. J. Acoust. Soc. Am. **94**(6), 3392–3399 (1993)
89. S.G. Samko, A.A. Kilbas, O. I. Marichev, *Fractional Integrals and Derivatives: Theory and Applications* [M] (Gordon and Breach Science Publishers, 1993)
90. W. Chen, S. Holm, Fractional Laplacian time-space models for linear and nonlinear lossy media exhibiting arbitrary frequency dependency [J]. J. Acoust. Soc. Am. **115**(4), 1424–1430 (2004)
91. Q.Y. Guan, Z.M. Ma, Reflected symmetric $\alpha$-stable processes and regional fractional Laplacian [J]. Probab. Theory Relat. Fields **134**, 649–694 (2006)
92. D. Del-Castillo-Negrete, B.A. Carreras, V.E. Lynch, Front dynamics in reaction-diffusion systems with Lévy flights: a fractional diffusion approach [J]. Phys. Rev. Lett. **91**(1), 018301–018304 (2003)
93. H.M. Zhang, F.W. Liu, Numerical solution for the Lévy-Feller diffusion equation [J]. Comput. Math. Chin. Univ. **27**, 238–241 (2005)
94. M. Meerschaert, C. Tadjeran, Finite difference approximations for two-sided space-fractional partial differential equations [J]. Appl. Numer. Math. **56**(1), 80–90 (2006)
95. W. Chen, X.D. Zhang, The comparison analysis of Szabo model and space fractional derivative acoustic wave dissipation equation (complete) (2010)
96. T.L. Szabo, Time domain nonlinear wave equations for lossy media, in *Advances in Nonlinear Acoustics: Proc. of 13th ISNA*, ed. H. Hobaek, (World Scientific, Singapore, 1993), pp. 89–94
97. D.T. Blackstock, Generalized burgers equation for plane waves [J]. J. Acoust. Soc. Am. **77**(6), 2050–2053 (1985)
98. N. Sugimoto, Burgers equation with a fractional derivative; hereditary effects on nonlinear acoustic waves [J]. J. Fluid Mech. **225**, 631–653 (1991)
99. P. Biler, G. Karch, W.A. Woyczynski, Asymptotics and high dimensional approximations for nonlinear pseudodifferential equations involving Lévy generators [J]. Demonstratio Math. **34**(2), 403–413 (2001)

100. J.M. Carcione, F. Cavallini, F. Mainardi, A. Hanyga, Time-domain modeling of constant-q seismic waves using fractional derivatives [J]. Pure Appl. Geophys. **159**, 1719–1736 (2002)
101. E. Kjartansson, Constant Q-wave propagation and attenuation [J]. J. Geophys. Res.: Solid Earth **84**, 4737–4748 (1979)
102. Y.B. Yin, K.Q. Zhu, Oscillating flow of a viscoelastic fluid in a pipe with the fractional Maxwell model [J]. Appl. Math. Comput. **173**, 231–242 (2006)
103. J.S. Zhuo, *Generalized Variational Principles in Elasticity and Plasticity [M]* (China Water Power Press, Beijing, 2002)
104. QX. Wu, F. Qi, *Theoretical Mechanics* [M] (Higher Education Press, Beijing, 2003)
105. D. Baleanu, O.P. Agrawal, Fractional Hamilton formalism within Caputo's derivative [J]. Czech J. Phys. **56**(10/11), 1087–1092 (2006)
106. O.P. Agrawal, Formulation of Euler-Lagrange equations for fractional variational problems [J]. J. Math. Anal. Appl. **272**, 368–379 (2002)
107. E.M. Rabei et al., Hamilton-Jacobi formulation of systems within Caputo's fractional derivative [J]. Phys. Scr. **77**(1), 015101 (2008)
108. E.M. Rabei, Fractional Hamilton-Jacobi equation and WKB approximation [D]. Department of Physics, Mutah University, Al-Karak, Jordan
109. E.M. Rabei, B.S. Ababneh, Hamilton-Jacobi fractional sequential mechanics [J]. Arxiv preprint (2007), arXiv:0704.0519
110. A.-R. El-Nabulsi, A fractional approach to nonconservative Lagrangian dynamical systems [J]. FIZIKA A **14**(4), 289–298 (2005)
111. I. Podlubny, Y.Q. Chen, Adjoint fractional differential expressions and operators, in *Proceedings in ASME 2007 International Design Engineering Technical Conferences and Computers and Information in Engineering Conference (IDETC/CIE2007)* (2007), pp. 1385–1390
112. A. Mejia, *Quantum Mechanics* [M] (Science Press, Beijing, 1986)
113. T.S. Cheng, *Modern Quantum Mechanics Tutorial* [M] (Peking University Press, Beijing, 2006)
114. Feynman, *Quantum Mechanics and Path Integral* [M] (Science Press, Beijing, 1986)
115. M. Naber, Time fractional Schrödinger equation [J]. J. Math. Phys. **45**, 3339 (2004)
116. N. Laskin, Fractional quantum mechanics and Lévy path integrals [J]. Phys. Lett. A **268**(4–6), 298–305 (2000)
117. N. Laskin, Fractional Schrödinger equation [J]. Phys. Rev. E **66**, 056108 (2002)
118. H. Nakao, Multi-scaling properties of truncated Lévy flights [J]. Phys. Lett. A **266**(4–6), 282–289 (2000)
119. B. Mandelbrot, *Fractal Object: Shape, Opportunities and Dimension [M]* (World Publishing Company, Beijing, 1999)
120. F. Barpi, S. Valente, Creep and fracture in concrete: a fractional order rate approach [J]. Eng. Fract. Mech. **70**, 611–623 (2003)
121. J.Y. Cao, B.G. Cao, Digital realization and characteristics of fractional order controllers [J]. Control Theory Appl. **23**(5), 791–799 (2006)
122. Y.Q. Chen, T. Bhaskaran, D.Y. Xue, Practical tuning rule development for fractional order proportional and integral controllers [J]. ASME J. Comput. Nonlinear Dyn. **3**(2), 020201.1–021404.7 (2008)
123. Y.Q. Chen, H.-S. Ahn, D.Y. Xue, Robust controllability of interval fractional order linear time invariant systems [J]. Signal Process. **86**, 2794–2802 (2006)
124. Y. Li, Y.Q. Chen, I. Podlubny, Mittag-Leffler stability of fractional order nonlinear dynamic systems [J]. Automatica **45**(8), 1965–1969 (2009)
125. W.C. Qian, *Singular Perturbation Theory and Its Applications in Mechanics [M]* (National Defense Industry Press, Beijing, 1981)
126. A. Carpinteri, F. Mainardi, *Fractals and Fractional Calculus in Continuum Mechanics* [M] (Springer, Berlin, 1997), pp. 291–348
127. F. Liu, V. Anh, I. Turner, Numerical solution of the space fractional Fokker-Planck equation [J]. J. Comput. Appl. Math. **166**, 209–219 (2004)

128. M.M. Meerschaert, C. Tadjeran, Finite difference approximations for fractional advection–dispersion flow equations [J]. J. Comput. Appl. Math. **172**, 65–77 (2004)

129. J.L. Battaglia, O. Cois, L. Puigsegur, A. Oustaloup, Solving an inverse heat conduction problem using a non-integer identified model [J]. Int. J. Heat Mass Transf. **44**, 2671–2680 (2001)

130. D.A. Murio, Stable numerical solution of a fractional-diffusion inverse heat conduction problem [J]. Comput. Math. Appl. **53**, 1492–1501 (2007)

131. D.A. Murio, Time fractional IHCP with Caputo fractional derivatives [J]. Comput. Math. Appl. **56**, 2371–2381 (2008)

132. R. Sivsprasad, S. Venkatesha, C.S. Manohar, Identification of dynamical systems with fractional derivative damping models using inverse sensitivity analysis [J]. Comput. Mater. Contin. **298**(1), 1–29 (2009)

133. Y.F. Wang et al., *Computational Ethods for Inverse Problems and Their Applications [M]* (Higher Education Press, Beijing, 2007), p. 1

134. T.Y. Xiao, S.G. Yu, Y.F. Wang et al., *The Numerical Solution of the Inverse Problem [M]* (Science Press, Beijing, 2003)

135. C. Tadjeran, M.M. Meerschaert, H.-P. Scheffler, A second-order accurate numerical approximation for the fractional diffusion equation [J]. J. Comput. Phys. **213**, 205–213 (2006)

136. J.H. Chen, F.W. Liu, Analysis of stability and convergence of numerical approximation for the Riesz fractional reaction dispersion equation [J]. J. Xiamen Univ. (Nat. Sci.) **45**(4), 466–469 (2006)

137. K. Diethelm, N.J. Ford, A.D. Freed, Yu. Luchko, Algorithms for the fractional calculus: a selection of numerical methods [J]. Comput. Methods Appl. Mech. Eng. **194**, 743–773 (2005)

138. Y.M. Peng, Research on numerical motheds of the inverse problem for partial differential equations [D]. Xi'an University of Technology, Master Thesis 2005, 3

139. G.S. Li, Y.J. Tan, X.Q. Wang, Inverse problem method on determining magnitude of groundwater pollution sources [J]. Math. Appl. **18**(1), 92–98 (2005)

140. S.G. Samko, B. Ross, Integration and differentiation to a variable fractional order [J]. Integral Transform. Spec. Funct. **1**(4), 277–300 (1993)

141. S.G. Samko, Fractional integration and differentiation of variable order [J]. Anal. Math. **21**, 213–336 (1995)

142. B. Ross, S.K. Samko, Fractional integration operator of variable order in the Holder spaces $H^x$ [J]. J. Math. Math. Sci. **18**(4), 777–788 (1995)

143. C.F. Lorenzo, T.T. Hartley, Variable order and distributed order fractional operators [J]. Nonlinear Dyn. **29**, 57–98 (2002)

144. C.F.M. Coimbra, Mechanics with variable-order differential operators [J]. Ann. Phys. (Leipz.) **12**(11–12), 692–703 (2003)

145. L.E.S. Ramirez, C.F.M. Coimbra, A variable order constitutive relation for viscoelasticity [J]. Ann. Phys. (Leipz.) **16**(7–8), 543–552 (2007)

146. C.M. Soon, C.F.M. Coimbra, M.H. Kobayashi, The variable viscoelasticity oscillator [J]. Ann. Phys. (Leipz.) **14**(6), 378–389 (2005)

147. G. Diaz, C.F.M. Coimbra, Nonlinear dynamics and control of a variable order oscillator with application to the Van der Pol equation [J]. Nonlinear Dyn. **56**(1–2), 145–157 (2009)

148. D. Ingman, J. Suzdalnitsky, M. Zeifman, Constitutive dynamic-order model for nonlinear contact phenomena [J]. J. Appl. Mech. **67**, 383–390 (2000)

149. D. Ingman, J. Suzdalnitsky, Application of differential operator with servo-order function in model of viscoelastic deformation process [J]. J. Eng. Mech. **131**(7), 763–767 (2005)

150. H.T.C. Pedro, M.H. Kobayashi, J.M.C. Pereira, C.F.M. Coimbra, Variable order modeling of diffusive-convective effects on the oscillatory flow past a sphere [J]. J. Vib. Control **14**, 1659–1672 (2008)

151. A.V. Chechkin, R. Gorenflo, I.M. Sokolov, Fractional diffusion in inhomogeneous media [J]. J. Phys. A: Math. Gen. **38**, L679–L684 (2005)

152. H.G. Sun, W. Chen, Y.Q. Chen, Variable-order fractional differential operators in anomalous diffusion modeling [J]. Phys. A **388**, 4586–4592 (2009)

153. P. Zhuang, F. Liu, V. Anh, I. Turner, Numerical methods for the variable-order fractional advection-diffusion equation with a nonlinear source term [J]. SIAM J. Numer. Anal. **47**(3), 1760–1781 (2009)

154. Y.L. Kobelev, L.Y. Kobelev, Yu.L. Klimontovich, Anomalous diffusion with time-and coordinate-dependent memory [J]. Dokl. Phys. **48**(6), 264–268 (2003)

155. M. Caputo, Linear models of dissipation whose Q is almost frequency independent. Part 2 [J]. Geophys. J. Int. **13**, 529–539 (1967)

156. T.M. Atanackovic, L. Oparnica, S. Pilipovi, On a nonlinear distributed order fractional differential equation [J]. J. Math. Anal. Appl. **328**, 590–608 (2007)

157. T.M. Atanackovic, A generalized model for the uniaxial isothermal deformation of a viscoelastic body [J]. Acta Mech. **159**, 77–86 (2002)

158. T.M. Atanackovic, M. Budincevic, S. Pilipovic, On a fractional distributed-order oscillator [J]. J. Phys. A: Math. Gen. **38**, 6703–6713 (2005)

159. F. Mainardi, G. Pagnini, R. Gorenflo, Some aspects of fractional diffusion equations of single and distributed order [J]. Appl. Math. Comput. **187**, 295–305 (2007)

160. A.V. Chechkin, V.Yu. Gonchar, R. Gorenflo, N. Korabel, I.M. Sokolov, Retarding subdiffusion and accelerating superdiffusion governed by distributed-order fractional diffusion equations [J]. Phys. Rev. E **78**, 021111 (2008)

161. I.M. Sokolov, A.V. Chechkin, J. Klafter, Distributed-order fractional kinetics [J] (2004), arXiv preprint cond-mat/0401146

162. A.N. Kochubei, Distributed order calculus and equations of ultraslow diffusion [J]. J. Math. Anal. Appl. **340**, 252–281 (2008)

163. H.G. Sun, Y.Q. Chen, W. Chen, Random-order fractional differential equation models [J]. Signal Process. **91**, 525–530 (2011)

164. H.G. Sun, W. Chen, H. Sheng, Y.Q. Chen, On mean square displacement behaviors of anomalous diffusions with variable and random orders [J]. Phys. Lett. A **374**, 906–910 (2010)

165. J.W. Grant, Mechanics of the otolith organs[J], in *Handbook of Bioengineering*, ed. By R. Skalak, C. Shu (McGraw-Hill Book Co., New York, 1987), pp. 31:1–31.

166. J.W. Grant, J.R. Conton, A model for otolith dynamic response with a viscoelastic ge layer [J]. J. Vestib. Res. **1**, 139 (1991)

167. W.C. Buskirk, R.G. Watts, Y.K. Liu, The fluid mechanics of the semicircular canal [J]. J. Fluid Mech. **78**(1), 87–98 (1976)

168. M.Y. Xu, W.C. Tan, The fluid dynamic problems in semicircular canal [J]. Sci. China (A) **30**(3), 272–280 (2000). ((in Chinese))

169. H.J. Su, C.Y. Yang, Z.G. Yang, M.Y. Xu, Generalized fractional dynamic model of semicircular canal [J]. J. Shandong Univ. **40**(1), 37–41 (2005). ((in Chinese))

170. X.H. Zhu, X. Dong, L. Liu, W.C. Xiu, Study on viscoelasticity of human cranial bone [J]. Chin. J. Biomed. Eng. **12**(1), 35–42 (1993). ((in Chinese))

171. J.G. Liu, M.Y. Xu, Study on a fractional model of viscoelasticity of human cranial bone [J]. Chin. J. Biomed. Eng. **24**(1), 12–16 (2005). ((in Chinese))

172. J.W. Zhang, J.K. Gu, *Pharmacokinetics of Controlled/Sustaine Release Dosage Form [M]* (Science Press, Beijing, 2009)

173. A. Dokoumetzidis, P. Macheras, Fractional kinetics in drug absorption and disposition processes [J]. J. Pharmacokinet Pharmacodyn. **36**, 165–178 (2009)

174. T. Higuchi, Rate of release of medicaments from ointment bases containing drugs in suspension [J]. J. Pharm. Sci. **50**, 874–875 (1961)

175. T. Koizumi, P.A. Suwannee, Release of medicaments from spherical matrices containing drug in suspension [J]. Int. J. Pharm. **116**, 45–49 (1995)

176. J. Liu, M. Xu, An exact solution to the moving boundary problem with fractional anomalous diffusion in drug release devices [J]. ZAMM-J. Appl. Math. Mech./Z. Für Angew. Math. Mech.: Appl. Math. Mech. **84**, 22–28 (2004)

177. X. Li, M. Xu, S. Wang, Analytical solutions to the moving boundary problems with space–time-fractional derivatives in drug release devices [J]. J. Phys. A: Math. Theor. **40**, 12131 (2007)

178. X. Li, M. Xu, S. Wang, Scale-invariant solutions to partial differential equations of fractional order with a moving boundary condition [J]. J. Phys. A: Math. Theor. **41**, 155202 (2008)

179. C. Yin, M. Xu, An asymptotic analytical solution to the problem of two moving boundaries with fractional diffusion in one-dimensional drug release devices [J]. J. Phys. A: Math. Theor. **42**, 115210 (2009)

180. J. Crank, *Free and Moving Boundary Problems [M]* (Clarendon Press, Oxford, 1987)

181. D. Paul, Modeling of solute release from laminated matrices [J]. J. Membr. Sci. **23**, 221–235 (1985)

182. D. Paul, S. McSpadden, Diffusional release of a solute from a polymer matrix [J]. J. Membr. Sci. **14**, 33–48 (1976)

183. M.J. Abdekhodaie, Y.L. Cheng, Diffusional release of a dispersed solute from planar and spherical matrices into finite external volume [J]. J. Control. Release **43**, 175–182 (1997)

184. R. Metzler, J. Klafter, The random walk's guide to anomalous diffusion: a fractional dynamics approach [J]. Phys. Rep. **339**, 1–77 (2000)

185. P.L. Ritger, N.A. Peppas, A simple equation for description of solute release. I. Fickian and non-Fickian release from non-swellable devices in the form of slabs, spheres, cylinders or discs [J]. J. Control. Release **5**, 23–36 (1987)

186. C. Yin, X. Li, Anomalous diffusion of drug release from a slab matrix: Fractional diffusion models [J]. Int. J. Pharm. **418**, 78–87 (2011)

# Chapter 6
# Numerical Methods for Fractional Differential Equations

This chapter presents some typical numerical methods for time and space fractional differential equations. Discretization schemes for the Grünwald–Letnikov, Riemann–Liouville, Caputo, fractal and positive time-fractional derivatives are separately discusse, and validated by easy-to-follow numerical examples. Despite being different from "fractional derivative", fractal derivatives are still included in this chapter. For the convenience of discussions, we call the equations having fractional derivatives with respect to time/space variable the time/space fractional differential equations (TFDEs/SFDEs for short), respectively. If both time and space fractional derivatives are involved, we call the equations time–space fractional differential equations (TSFDEs).

Nowadays, different solution techniques have been applied to fractional differential equations (FDEs), such as finite difference, finite element and mesh-free methods. Herein we only elaborate on finite difference methods and integral equation methods. For solving fractional ordinary differential equations, only the most typical method, i.e. Predictor–Corrector method, is introduced. On the other hand, the diversity of the solution techniques for fractional partial differential equations makes us simply emphasize the discretization schemes for the fractional differential operators mentioned in the preceding paragraph. For more details of the numerical methods for fractional partial differential equations, we refer the readers to the end-of-chapter references.

## 6.1 Time-Fractional Differential Equations (TFDEs)

Many complex phenomena such as anomalous diffusion feature relatively strong memory attributes and manifest themselves in terms of history dependence. In a statistical sense, the physical or mechanical systems of interest are generally non-Markov processes and should be described by fractional Brown motions. The derivative of integer order with local definition fails to describe the history dependency,

© Science Press 2022
W. Chen et al., *Fractional Derivative Modeling in Mechanics and Engineering*,
https://doi.org/10.1007/978-981-16-8802-7_6

while the fractional derivative whose definition involves an integral term allows the history dependency in the evolution of system functions. Hence, fractional calculus is usually considered an effective tool for describing the physical or mechanics process having memory attributes. On the other hand, the discretization schemes of fractional integrals and derivatives have drawn a fast-growing attention recently. Podlubny discussed the numerical approximation approaches for several typical time-fractional derivatives. He pointed out that unlike handling derivatives of integer order, one can no longer arbitrarily truncate the Taylor expansion of system function using the step-by-step method and that the computation cost increases rapidly with the increasing length of time interval considered. To lower the computational complexity, he put forward the "short memory" principle and only considered the behavior of system function in the "nearest past". This is something partly like the conventional multistep methods in the solution of ordinary differential equations. However, Ford has argued that the "short memory" principle simply works well for some specific problems. To our knowledge, the "short memory" principle is somewhat simple and intuitive either in physics/mechanics or in mathematics and is thereby not a reliable method. In fact, the fractional derivative is a convolution of the operand and the weakly singular kernel. It is commonly recognized that a direct convolution computation is usually time-consuming. Furthermore, since the approximant of the time-fractional derivative includes many addends, the iteration matrix is accordingly dense. This imposes an adverse effect on the computation efficiency, which is an open issue to be solved by matrix sparsity and preconditioning techniques.

### 6.1.1 Finite Difference and Integral Equation Methods

In this subsection, easy-to-follow examples are given to show the validity of the methods.

#### 1. Finite difference methods (FDMs)

The finite difference approximations for the time-fractional derivatives in sense of Grünwald–Letnikov, Riemann–Liouville and Caputo, denoted by $_a^{GL}D_t^\alpha$, $_a^{RL}D_t^\alpha$ and $_a^C D_t^\alpha$, are given. Due to the less applicability, the approximation for Weyl [1, 2] fractional derivative is omitted.

#### (1) Grünwald–Letnikov derivative

The derivative

$$D_t^\alpha f(t) = \lim_{N\to\infty} \frac{1}{(\Delta t)^\alpha} \sum_{k=0}^{N} \frac{\Gamma(k-\alpha)}{\Gamma(-\alpha)\Gamma(k+1)} f(t - k\Delta t) \text{ for } t - a = N\Delta t$$

(6.1.1)

with time step $\Delta t$, differential order $\alpha$ and Gamma function $\Gamma(\cdot)$, is a natural extension of integer-order derivative in terms of the differential order and is widely considered in numerical computation. The finite difference truncation of (6.1.1) is given by

$$a^{GL} D_t^\alpha f(t_i) \approx \frac{(\Delta t)^{-\alpha}}{\Gamma(-\alpha)} \sum k = 0i \frac{\Gamma(k-\alpha)}{\Gamma(k+1)} f_{i-k} \text{ for } i = 1, 2, \ldots \quad (6.1.2)$$

where $f_{i-k}$ is an approximant at the $(i-k)$th grid point and $t_i = i\Delta t$. To improve the stability, a right-shifted version of (6.1.2) is proposed as [*]

$$_a D_t^\alpha f(t_i) \approx \frac{(\Delta t)^{-\alpha}}{\Gamma(-\alpha)} \sum_{k=0}^{i} \frac{\Gamma(k-\alpha)}{\Gamma(k+1)} f_{i-k+1} \text{ for } i = 1, 2, \ldots. \quad (6.1.3)$$

(6.1.2) is proven conditionally convergent and low accurate, i.e. $(O(\Delta t))$ [3]. Adding some modified terms to (6.1.2) yields the high-order accurate fractional linear multistep method which is applied to solve the nonlinear TFDEs [4].

## (2)  Riemann–Liouville derivative

This type of derivative is a typical fractional-order derivative since it is a direct extension of integer-order derivative and can readily reduce to its integer-order counterpart when the differential order is taken as an integer. The derivative compounds the conventional differentiation and integration as

$$_a^{RL} D_t^\alpha f(t) = \frac{1}{\Gamma(n-\alpha)} \left(\frac{d}{dt}\right)^n \int_a^t (t-\tau)^{n-\alpha-1} f(\tau) d\tau, \ (n-1 \le \alpha < n). \quad (6.1.4)$$

Diverse finite difference schemes for (6.1.4) can be seen in [5, 6]. For $\alpha \in (0,1) \cup (1,2)$, the commonly used difference schemes are shown as follows.

## (1)  $0 < \alpha < 1$

Integration (6.1.4) by part leads to

$$_a^{RL} D_t^\alpha f(t) = \frac{f(0)}{\Gamma(1-\alpha)} t^{-\alpha} + \frac{1}{\Gamma(1-\alpha)} \int_0^t \frac{f'(\tau)}{(t-\tau)^\alpha} d\tau. \quad (6.1.5)$$

Dividing the integral in the right-hand side of (6.1.5) into sub-integrals and using backward and central differences to approximate $f'$ in each sub-integral, one can obtain the approximation formulas with one- and two-order accuracy, respectively.

One-order formula:

$$_a^{RL} D_t^\alpha f(t_i) \approx \frac{f(0)}{\Gamma(1-\alpha)} t_i^{-\alpha}$$

$$+ \frac{1}{\Gamma(2-\alpha)\Delta t^2} \sum_{j=0}^{i-1} [(j+1)^{1-\alpha} - j^{1-\alpha}](f_{i-j+1} - f_{i-j}), \quad (6.1.6)$$

where $t_i = i\Delta t$.

Two-order formula:

$$
{}_{a}^{RL}D_t^\alpha f(t_i) \approx \frac{f(0)}{\Gamma(1-\alpha)}t_i^{-\alpha}
$$

$$
+ \frac{1}{\Gamma(2-\alpha)\Delta t^2}\sum_{j=0}^{i-1}[(j+1)^{1-\alpha} - j^{1-\alpha}]\frac{f_{i-j+1} - f_{i-j-1}}{2}. \quad (6.1.7)
$$

Formula (6.1.6) has the larger truncated error yet the better stability than (6.1.7).

(2)    $1 < \alpha < 2$

Integrating (6.1.4) by part gives

$$
{}_{a}^{RL}D_t^\alpha f(t) = \frac{f(0)}{\Gamma(1-\alpha)}t^{-\alpha} + \frac{f'(0)}{\Gamma(2-\alpha)}t^{1-\alpha} + \frac{1}{\Gamma(2-\alpha)}\int_0^t \frac{f''(\tau)d\tau}{(t-\tau)^{\alpha-1}}. \quad (6.1.8)
$$

A central difference approximation for $f''$ leads to

$$
{}_{a}^{RL}D_t^\alpha f(t_i) \approx \frac{f(0)}{\Gamma(1-\alpha)}t_i^{-\alpha} + \frac{f'(0)}{\Gamma(2-\alpha)}t_i^{1-\alpha}
$$

$$
+ \frac{1}{\Gamma(3-\alpha)\Delta t^2}\sum_{j=0}^{i-1}[(j+1)^{2-\alpha} - j^{2-\alpha}](f_{i-j+1} + f_{i-j-1} - 2f_{i-j}).
$$

$$
(6.1.9)
$$

Formula (6.1.9) is unconditionally stable as well as a truncated error varying with $\alpha$. For instance, the truncated error is $O(\Delta t)$ for $\alpha = 1$ but $O(\Delta t^2)$ for $\alpha = 2$. Moreover, due to the requirement of the value of $f'(0)$, (6.1.9) is more suitable to solve problems with sufficiently smooth initial conditions.

(3)    **Caputo derivative**

Unlike the Riemann–Liouville definition, the Caputo fractional derivative is defined by

$$
{}_{a}^{C}D_t^\alpha f(t) = \frac{1}{\Gamma(n-\alpha)}\int_a^t (t-\tau)^{n-\alpha-1}f^{(n)}(\tau)d\tau, \quad (n-1 < \alpha < n), \quad (6.1.10)
$$

which includes the integer-order derivative in the integrand, thereby in association with the initial conditions that are easy to measure. Among the various discretization schemes [7], two typical schemes are listed below.

(1)    Direct discretization

$$
{}_a^C D_t^\alpha f(t_i) \approx \frac{1}{\Gamma(n+1-\alpha)} \sum_{j=0}^{i-1} \frac{f(t_{i+1-j}) - f(t_{i-j})}{(\Delta t)^\alpha} [(j+1)^{1-\alpha} - j^{1-\alpha}],
$$

(6.1.11)

with $t_i = a + i\Delta t$ $(i = 1,2,...)$ and $\alpha \in (n-1,n)$ is commonly used.

(2)    Improved discretization

$$
{}_a^C D_t^\alpha f(t_i) \approx \frac{1}{\Gamma(n+1-\alpha)} \sum_{j=0}^{i-1} f^{(n-1)}(t_{i+1-j}) [(j+1)^{1-\alpha} - j^{1-\alpha}]
$$

(6.1.12)

often enjoys higher accuracy and better stability than (6.1.11). The Houbolt method and the linear acceleration methods based on the Newmark and Wilson-$\theta$ methods [8–12] can be used to approximate $f^{(n-1)}(t_{i-j+1})$.

It should be noted that in solving TFDEs, the finite difference temporal discretization combined with different space discretization schemes will lead to different solution techniques. Lin and Xu [13] obtained an accurate and stable method by applying the Legendre spectral method to space discretization. Special attention has also been paid to combining the finite element method or the mesh-free method with the finite difference methods [14, 15].

Among the three types of fractional derivatives aforementioned, the Caputo fractional derivative is usually considered in the TFDEs in that the corresponding finite difference approximation has the highest accuracy and is most stable [16]. Besides, the inherent finite difference form enables the Grünwald–Letnikov fractional derivative to be an alternative for solving TFDEs. Despite the FDMs for solving TFDEs are relatively mature, three open issues still need further consideration: (1) large errors of the numerical solutions in the initial time range; (2) large computation complexity for long time range evaluation due to the history dependency; (3) proofs of stability and convergence of the methods. These issues shall be revisited in Sect. 6.3.

(4)    Numerical examples

**Example 1.** Attenuation or relaxation equation

$$
\begin{cases} {}_a^C D_t^\alpha u(t) = Bu(t), \ \alpha \in (0, 1], \\ u(0) = 10 \end{cases}
$$

(6.1.13)

can describe the anomalous attenuation or relaxation behaviors in mechanism, semiconductor, electromagnetics and optics. Using (6.1.11) and the initial condition, one can obtain the numerical solutions for different $\alpha$ as shown in Fig. 6.1.

**Fig. 6.1** Numerical solution
for Eq. (6.1.12) ($B = 1$)

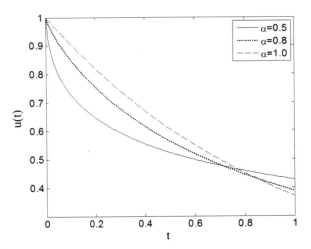

**Example 2** The Bagley–Torvik equation (Damped vibration equation)

$$m\ddot{u}(t) + c_a^{GL} D_t^\alpha u(t) + ku(t) = f(t)\,(1 < \alpha < 2) \qquad (6.1.14)$$

is able to describe the vibration in complex media, where $m\ddot{u}(t)$ is the inertia force,
$c_a^{GL} D_t^\alpha u(t)$ the damping force, $ku(t)$ the elastic force and $f(t)$ the external force [17].
Use (6.1.1) and the standard difference to discretize (6.1.14) as

$$m(\Delta t)^{-2}(u_n - 2u_{n-1} + u_{n-2}) + c(\Delta t)^{-\alpha} \sum_{r=0}^{n} \omega_r^{(\alpha)} u_{n-r} + ku_{n-1} = f_{n-1},$$
$$(6.1.15)$$

which leads to

$$u_n = \frac{f_{n-1} - ku_{n-1} - c(\Delta t)^{-\alpha} \sum_{r=1}^{n} \omega_r^{(\alpha)} u_{n-r} + m(\Delta t)^{-2}(2u_{n-1} - u_{n-2})}{m(\Delta t)^{-2} + c(\Delta t)^{-\alpha}}.$$
$$(6.1.16)$$

The coefficients $\omega_r^{(\alpha)}$ are determined by [17]

$$\omega_0^{(\alpha)} = 1;\; \omega_j^{(\alpha)} = \left(1 - \frac{\alpha + 1}{j}\right)\omega_{j-1}^{(\alpha)},\; j = 1, 2, \ldots.$$

As a comparison, we also consider the Caputo fractional derivative, replacing
$_a^{GL} D_t^\alpha u(t)$ with $_a^C D_t^\alpha u(t)$. Given $\alpha = 1.5, f(t) = 8$ for $t \in [0,1]$ (0, otherwise), $m = 1$
and $c = 0.5$, Figs. 6.2 and 6.3 show the numerical solutions of the TFDEs having the
Grünwald–Letnikov (G-L) and Caputo operators under different initial conditions.

**Fig. 6.2** Numerical solution for the Bagley–Torvik equation of the order 1.5. ($\Delta t = 0.1, u(0) = 0, \dot{u}(0) = 0$)

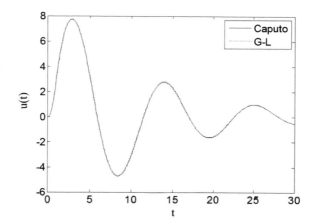

**Fig. 6.3** Numerical solution for the Bagley–Torvik equation of the order1.5. ($\Delta t = 0.1, u(0) = 1, \dot{u}(0) = 0.1$)

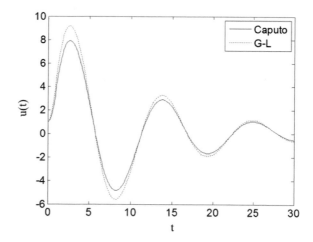

## 2. Integral equation methods (IEMs)

It is another common experience to solve a TFDE by transforming it into an equivalent integral equation. Using the integral transform can reduce the singularity in numerical manipulation and moreover simplify the solution process. The Abel integral equation and the Volterra integral equation of the second kind play a crucial role in IEMs.

Fractional integrate the TFDE

$$\begin{matrix} C \\ 0 \end{matrix} D_t^\alpha u(t) = f(t) \text{ for } \alpha \in (n-1, n) \text{ and } t > 0 \qquad (6.1.17)$$

on both sides and use the relation in Sect. 2.2.3 to obtain the equivalent Abel integral equation [18, 19]

$$u(t) - \sum_{j=0}^{n-1} \frac{t^j}{j!} u^{(j)}(a) = \frac{1}{\Gamma(\alpha)} \int_a^t (t-\tau)^{\alpha-1} f(\tau) ds. \tag{6.1.18}$$

It follows that one just needs to evaluate the integral on the right-hand side of (6.1.18) where the expression of $f(t)$ is known. Adding a linear term to (6.1.17) yields

$$_0^C D_t^\alpha u(t) + Bu(t) = f(t) \text{ for } \alpha \in (n-1, n) \text{ and } t > 0. \tag{6.1.19}$$

It becomes the fractional relaxation equation when $\alpha \in (0, 1)$ which characterizes slow dissipation and becomes the fractional oscillation equation when $\alpha \in (1, 2)$ which governs a damped vibration. In a similar manner to handling (6.1.17), one derives the Volterra integral equation of the second kind:

$$u(t) - \sum_{j=0}^{n-1} \frac{t^j}{j!} u^{(j)}(a) = \frac{1}{\Gamma(\alpha)} \int_a^t (t-\tau)^{\alpha-1} (f(\tau) - Bu(\tau)) d\tau. \tag{6.1.20}$$

To obtain $u(t)$, the quadrature formula

$$\frac{1}{\Gamma(\alpha)} \int_{t_j}^{t_{j+1}} (t-\tau)^{\alpha-1} (f(\tau) - Bu(\tau)) d\tau \approx (\Delta t)^{-\alpha} \sum_{i=n-1}^{n} \omega_{j,i} (f(t_j) - Bu(t_j)) \tag{6.1.21}$$

is commonly utilized to evaluate the right-hand side of (6.1.20). It should be noticed that the weight factor $\omega_{j,i}$ has included the information of the weakly singular kernel $(t-\tau)^{\alpha-1}*$.

## Predictor–Corrector (PC) Method for Fractional Relaxation–Oscillation Equation

PC method, being widely used at present, is an extension of the Adams–Bashforth method. The predicted and the corrected values are, respectively, determined by the explicit and implicit Adams–Bashforth methods, followed by an extrapolation that accelerates the convergence. PC method enjoys relatively high accuracy but requires much computational effort [20–22]. Deng [23] has improved the solution accuracy of the time-fractional Fokker–Planck equation using the modified Adams PC method; Adolfsson [24] further modified the PC method using the discontinuous Galerkin approach and put forward a variable-step scheme to handle the long-duration evaluation. Below are the basic procedures of the PC method as well as the error analyses.

Consider the fractional relaxation–oscillation Eq. (6.1.20) and approximate the integral in its right-hand side by the rectangle rule [20]

$$\int_0^{t_k} (t_k - \tau)^{\alpha-1}[f(\tau) - Bu(\tau)]d\tau \approx \sum_{j=0}^{k-1} b_{j,k}[f(t_j) - Bu(t_j)], \qquad (6.1.22)$$

with $b_{j,k} = h^\alpha/\alpha[(k-j)^\alpha-(k-j-1)^\alpha]$, thus giving the predictor $u^P(t_k)$:

$$u^P(t_k) = \sum_{i=0}^{n-1} \frac{t_k^i}{i!}u^{(i)}(0) + \frac{1}{\Gamma(\alpha)}\sum_{j=0}^{k-1} b_{j,k}[f(t_j) - Bu(t_j)]. \qquad (6.1.23)$$

The corrector $u^c(t_k)$ is obtained by replacing the rectangle rule (6.1.22) with the trapezoidal rule [20]

$$\int_0^{t_k} (t_k - \tau)^{\alpha-1}[f(\tau) - Bu(\tau)]d\tau \approx \sum_{j=0}^{k} a_{j,k}[f(t_j) - Bu(t_j)], \qquad (6.1.24)$$

where $a_{j,k}$

$$= \frac{h^\alpha}{\alpha(\alpha+1)}\begin{cases} (k-1)^{\alpha+1} - (k-\alpha-1)k^\alpha & j=0 \\ (k-j+1)^{\alpha+1} - 2(k-j)^{\alpha+1} + (k-j-1)^{\alpha+1} & 1 \le j < k. \\ 1 & j=k \end{cases}$$

This yields the corrector:

$$u^c(t_k) = \sum_{i=0}^{n-1} \frac{t_k^i}{i!}u^{(i)}(0)$$

$$+ \frac{1}{\Gamma(\alpha)}\left(\sum_{j=0}^{k-1} a_{j,k}[f(t_j) - Bu(t_j)] + a_{k,k}[f(t_k) - Bu^P(t_k)]\right). \qquad (6.1.25)$$

Totally $2(n+1)^2 + n$ memory storage and $2(n+1)^2 + n(n+3)$ operations are demanded to compute values of $n$ grid points. Figures 6.4 and 6.5, respectively, show the numerical solutions for fractional relaxation and fractional oscillation equations for different $\alpha$, $B = 1$, u(0) = 1, u'(0) = 0 and $f(t) = 0$.

The absolute and relative errors for fractional relaxation equation with $\alpha = 0.5$, $\Delta t = 0.01, 0.1$ and 1 are given in Figs. 6.6, and 6.7, respectively. The specific data of analytical/numerical solutions and the numerical errors can be seen in Tables 6.1, 6.2 and 6.3.

**Fig. 6.4** Numerical solution
for fractional relaxation
equation

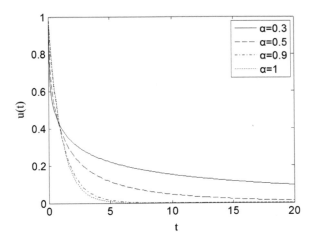

**Fig. 6.5** Numerical solution
for fractional oscillation
equation

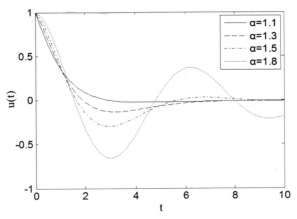

**Fig. 6.6** Absolute error plot
for fractional relaxation
equation of the order 0.5
with different time steps

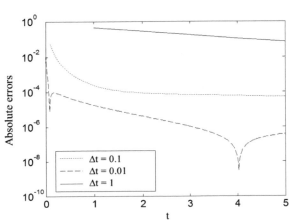

**Fig. 6.7**  Relative error plot for fractional relaxation equation of the order 0.5 with different time steps

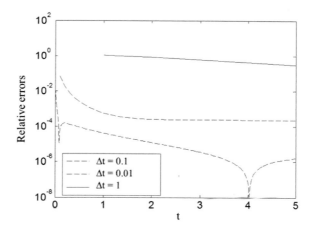

**Table 6.1** Analytical solutions, numerical solutions, absolute errors and relative errors for fractional relaxation equation with $\Delta t = 0.1$

| time(s) | Exact solution | Numerical solution | Absolute errors | Relative errors |
|---|---|---|---|---|
| 0.2 | 6.43788E-1 | 6.55275E-1 | 1.14862E-2 | 1.78416E-2 |
| 0.4 | 5.53606E-1 | 5.55904E-1 | 2.29818E-3 | 4.15129E-3 |
| 0.6 | 4.98025E-1 | 4.98874E-1 | 8.49727E-4 | 1.70620E-3 |
| 0.8 | 4.58246E-1 | 4.58661E-1 | 4.14584E-4 | 9.04720E-4 |
| 1.0 | 4.27584E-1 | 4.27829E-1 | 2.45444E-4 | 5.74026E-4 |

**Table 6.2** Analytical solutions, numerical solutions, absolute errors and relative errors for fractional relaxation equation with $\Delta t = 0.01$

| Time(s) | Exact solutions | Numerical solutions | Absolute errors | Relative errors |
|---|---|---|---|---|
| 0.2 | 6.43788E-1 | 6.43693E-1 | 9.51748E-5 | 1.47836E-4 |
| 0.4 | 5.53606E-1 | 5.53546E-1 | 6.03097E-5 | 1.0894E-4 |
| 0.6 | 4.98025E-1 | 4.97987E-1 | 3.80455E-5 | 7.63929E-5 |
| 0.8 | 4.58246E-1 | 4.58221E-1 | 2.54952E-5 | 5.56364E-5 |
| 1.0 | 4.27582E-1 | 4.27566E-1 | 1.78739E-5 | 4.18022E-5 |

**Table 6.3** Analytical solutions, numerical solutions, absolute errors and relative errors for fractional relaxation equation with $\Delta t = 1$

| Time(s) | Exact solutions | Numerical solutions | Absolute errors | Relative errors |
|---|---|---|---|---|
| 2 | 3.36204E-1 | 6.23776E-1 | 2.87572E-1 | 8.5535E-1 |
| 4 | 2.55396E-1 | 3.64208E-1 | 1.08812E-1 | 4.26053E-1 |
| 6 | 2.14626E-1 | 2.60051E-1 | 4.54246E-2 | 2.11645E-1 |
| 8 | 1.88821E-1 | 2.10842E-1 | 2.20212E-2 | 1.16624E-1 |
| 10 | 1.70578E-1 | 1.82982E-1 | 1.24043E-2 | 7.27196E-2 |

Denoting by $\varepsilon_a$ and $\varepsilon_r$ the maximum absolute and relative errors, we have.

(a)  in  Fig.  6.6,  $\varepsilon_a$  $=$  $\begin{cases} 0.4487, \Delta t = 1 \\ 0.0538, \Delta t = 0.1 \\ 0.0148, \Delta t = 0.01 \end{cases}$  (b)  in  Fig.  6.7,  $\varepsilon_r$  $=$

$\begin{cases} 1.049, \Delta t = 1 \\ 0.0743, \Delta t = 0.1 \\ 0.0165, \Delta t = 0.01 \end{cases}$ .

It can be seen from Figs. 6.6 and 6.7 that as the time step decreases, the numerical solutions converge to the analytical ones despite the fact that the errors of the approximants evaluated near $t = 0$ are rather obvious. Since being an explicit method, the PC method is conditionally stable. The stable condition is related to the differential order$\alpha$.

From the small errors for the cases of $\Delta t = 0.01$ and $\Delta t = 0.1$, we learn that the PC method is sufficiently accurate. The sensitivity of the accuracy to the value of time step accounts for the fact that the error for the case $\Delta t = 0.01$is much smaller than that for the case $\Delta t = 0.1$.

The absolute error estimation of the PC method for the following TFDE

$$ {}_0^C D_t^\alpha u(t) = f(t, u(t)), u^{(k)}(0) = u_0^{(k)}, k = 0, 1, \ldots, \lceil \alpha \rceil - 1, \qquad (6.1.26) $$

has drawn the attention of many researchers such as Li Changpin and Deng Weihua. Below are some well-known results [25, 26].

(1)   If $\alpha > 0$ and ${}_a^C D_t^\alpha u(t) \in C^2[0, T]$, then there exists some $T$, such that

$$ \underset{0 \le i \le N}{Max} |u(t_i) - u_i| = \begin{cases} O(\Delta t^2), \alpha \ge 1, \\ O(\Delta t^{1+\alpha}), 0 < \alpha < 1. \end{cases} $$

(2)   If $\alpha > 1$ and $u(t) \in C^{1+\lceil \alpha \rceil}[0, T]$, then there exists some $T$, such that

$$ \underset{0 \le i \le N}{Max} |u(t_i) - u_i| = O(\Delta t^{1+\lceil \alpha \rceil - \alpha}). $$

(3)   If $0 < \alpha < 1$ and $u(t) \in C^2[0, T]$, then there exists some $T$, such that

$$ \underset{0 \le i \le N}{Max} |u(t_i) - u_i| = \begin{cases} O(\Delta t^{2\alpha}), 0 < \alpha < 0.5, \\ O(\Delta t), 0.5 < \alpha < 1. \end{cases} $$

It further holds that for $\varepsilon \in (0,T)$

$$ \underset{t_i \in [\varepsilon, T]}{Max} |u(t_i) - u_i| = \begin{cases} O(\Delta t^{1+\alpha}), 0 < \alpha < 0.5, \\ O(\Delta t^{2-\alpha}), 0.5 < \alpha < 1. \end{cases} $$

(4)    If $\alpha > 1$ and $f \in C^3(G)$, then

$$\underset{0 \leq i \leq N}{Max}|u(t_i) - u_i| = O(\Delta t^2).$$

(5)    If $0 < \alpha < 1$ and $f \in C^3(G)$, then

$$\underset{0 \leq i \leq N}{Max}|u(t_i) - u_i| = \begin{cases} O(\Delta t^{2\alpha}), & 0 < \alpha < 0.5, \\ O(\Delta t), & 0.5 < \alpha < 1. \end{cases}$$

It further holds that if $\varepsilon \in (0,T)$

$$\underset{t_i \in [\varepsilon, T]}{Max}|u(t_i) - u_i| = \begin{cases} O(\Delta t^{1+\alpha}), & 0 < \alpha < 0.5, \\ O(\Delta t^{2-\alpha}), & 0.5 < \alpha < 1. \end{cases}$$

(6)    If $0 < \alpha < 1$ and $f \in C^2(G)$, then

$$\underset{0 \leq i \leq N}{Max}|u(t_i) - u_i| = O(1).$$

It further holds that if $\varepsilon \in (0,T)$

$$\underset{t_i \in [\varepsilon, T]}{Max}|u(t_i) - u_i| = O(\Delta t^{1-\alpha}).$$

**Modified PC Method**

Note that only the rectangle rule is employed to obtain the predictor in the PC method. To improve the accuracy of the PC method, we can take the trapezoidal rule as a replacement. Apply the trapezoidal rule to the integral in the equivalent integral equation of (6.1.26), i.e.[23]

$$\int_0^{t_{n+1}} (t_{n+1} - \tau)^{\alpha-1} f(u(\tau), \tau) d\tau \approx \frac{(\Delta t)^\alpha}{\alpha(\alpha+1)} \sum_{j=0}^{n+1} a_{j,n+1} f(u(t_j), t_j), \qquad (6.1.27)$$

where $a_{j,n+1} = \begin{cases} n^{\alpha+1} - (n-\alpha)(n+1)^\alpha & j = 0 \\ (n-j+2)^{\alpha+1} + (n-j)^{\alpha+1} - 2(n-j+1)^{\alpha+1} & 1 \leq j \leq n \\ 1 & j = n+1 \end{cases}$.

Next, substitute the above formula into the corrector

$$u^c(t_{n+1}) = \begin{cases} g_0 + \frac{(\Delta t)^\alpha}{\Gamma(\alpha+2)}(f(u^p(t_1), t_1) + \alpha f(u(t_0), t_0)), & n = 0 \\ g_0 + \frac{(\Delta t)^\alpha}{\Gamma(\alpha+2)}(f(u^p(t_{n+1}), t_{n+1}) + (2^{\alpha+1} - 2)f(u(t_n), t_n)), & n \geq 1 \\ \quad + \frac{\Delta t^\alpha}{\Gamma(\alpha+2)} \sum_{j=0}^{n-1} a_{j,n+1} f(u(t_j), t_j), \end{cases}$$

$$(6.1.28)$$

where

$$g_0 = \sum_{i=0}^{n-1} u^{(k)}(0) \frac{t^i}{i!}$$

and then the predictor is derived as

$$u^p(t_{n+1}) = \begin{cases} g_0 + \frac{(\Delta t)^\alpha}{\Gamma(\alpha+2)} f(u(t_0), t_0), & \text{for } n = 0 \\ g_0 + \frac{(\Delta t)^\alpha}{\Gamma(\alpha+2)} (2^{\alpha+1} - 1) f(u(t_0), t_0), & \text{for } n \geq 1 \\ + \frac{(\Delta t)^\alpha}{\Gamma(\alpha+2)} \sum_{j=0}^{n-1} a_{j,n+1} f(u(t_j), t_j). \end{cases} \qquad (6.1.29)$$

It is sensible to take the above corrector as a new predictor and to obtain the final corrector by iteration [23]. The stability and accuracy can be further improved in this fashion. Nevertheless, due to the non-locality of the fractional operator, one has to take into account $u(t_k)$ ($k = 1,2,\dots n\text{-}1$) when calculating $u(t_n)$. This will take much computation effort in successive iterations despite an increase in accuracy. Moreover, the numerical errors accumulated in the iteration process will impose an adverse effect on the accuracy of the approach and even make the convergence curve elevate for a larger iteration number.

Next, we investigate the influence of the iteration number of the modified PC method on the error variations. Consider the fractional relaxation equation $_a^C D_t^{0.5} u(t) + u(t) = 0$ with $u(0) = 1$ and $u'(0) = 0$. The absolute and relative error plots for the cases of iteration and non-iteration are given in Figs. 6.8 and 6.9, respectively.

Note that the curve labeled "No iteration" corresponds to the conventional PC method while the rest curves are associated with the modified PC method. It can be observed from Figs. 6.8 and 6.9 that after several iterations, there exists a positive $T$

**Fig. 6.8** Absolute error plots for fractional relaxation equation of the order 0.5 under different iteration numbers

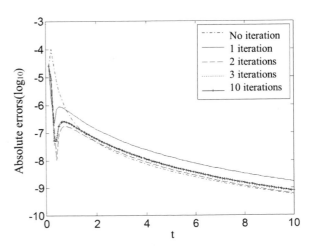

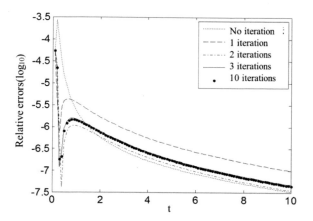

**Fig. 6.9** Relative error plots for fractional relaxation equation of the order 0.5 under different iteration numbers

such that: for $t \in (0,T)$, the error of the modified PC method is obviously smaller than that of the conventional PC method; while for $t > T$, the result turns to the opposite. Besides, we also see that there is no direct connection between the iteration number and the solution accuracy. In fact, among the cases for different iteration numbers, two iterations will lead to the highest accuracy either in $(0,T)$ or in $t > T$. As iteration number exceeds 2, the accuracy in $t > T$ drops, and as the iteration number increases to 10, the accuracy in this time interval almost keeps invariable. It thus seems easy to conclude that a medium number of iterations in the modified PC method will yield acceptable accuracy in the whole evaluation process.

Modified PC method can also be employed to solve multi-term TFDEs [23, 27]. This type of equation takes the form

$$\substack{C \\ a} D_t^{\alpha_n} u(t) = f(t, u(t), \substack{C \\ a} D_t^{\alpha_1} u(t), \cdots, \substack{C \\ a} D_t^{\alpha_{n-1}} u(t)), \tag{6.1.30}$$

with $u^{(k)}(0) = u_0^{(k)}$ for $k = 0,1,\dots,m-1$. Some physics–mechanics governing equations pertain to this type. For instance, the Bagley–Torvik equation which describes the oscillation of an oscillator in viscoelastic media:

$$m\frac{d^2 u(t)}{dt^2} + c \substack{C \\ a} D_t^{3/2} u(t) + ku(t) = f(t); \tag{6.1.31}$$

the Babenko equation which controls the motion of an immersed rigid plate in a Newtonian fluid:

$$\frac{du(t)}{dt} - F(t) \substack{C \\ a} D_t^{1/2} u(t) - G(t)u(t) = H(t); \tag{6.1.32}$$

and the Koeller equation which governs the gas dissolution in the liquid:

$$\substack{C \\ a} D_t^{2\alpha} u(t) - p_1 \substack{C \\ a} D_t^{\alpha} u(t) - p_0 u(t) = f(t). \tag{6.1.33}$$

To use the modified PC method, one needs first to transform the multi-term TFDE into a system of TFDEs. Denoting $\gamma$ by the greatest common divisor of $\{1, a_1, a_2, \ldots, a_n\}$ and letting $n = a_n/\gamma$, one gets from (6.1.30) that

$$
\begin{cases}
{}^C_a D^\gamma_t u_0(t) = u_1(t) \\
{}^C_a D^\gamma_t u_1(t) = u_2(t) \\
\cdots\cdots \\
{}^C_a D^\gamma_t u_{n-2}(t) = u_{n-1}(t) \\
{}^C_a D^\gamma_t u_{n-1}(t) = f\big(t, u_0(t), u_{a_1/\gamma}(t), \ldots, u_{a_{n-1}/\gamma}(t)\big)
\end{cases}
\tag{6.1.34}
$$

with initial conditions

$$
u^{(j)}(0) = u_0^{(j\gamma)} \text{ for } j \leq n.
\tag{6.1.35}
$$

Letting $U = \{u_0, u_1, \ldots, u_{n-1}\}^T$ leads to a compact form of (6.1.34), namely

$$
{}^C_a D^\gamma_t U(t) = F(t, U(t)),
\tag{6.1.36}
$$

with

$$
U(0) = \{u_0(0), u_1\cdot(0), \ldots, u_{n-1}(0)\}^T,
\tag{6.1.37}
$$

and

$$
F(t, U(t)) = \big\{u_1(t), u_2(t), \cdots, u_{n-1}(t), f(t, u_0(t), u_{a_1/\gamma}(t), \ldots, u_{a_{n-1}/\gamma}(t))\big\}^T.
\tag{6.1.38}
$$

Noticing that the compact form (6.1.36) is the same as (6.1.26), one can readily use PC or modified PC method to treat it.

## Other Integral Equation Methods

Different discretization schemes applied to the integral in the equivalent integral equations such as (6.1.18) will lead to different IMEs. Aside from the two quadrature rules used in the PC and modified PC methods, other approximation rules are also permitted. Liu and Lü [28] used the mid-point rectangle quadrature rule and trapezoidal rule to approximate the integral in the second kind Volterra integral equation which is a generalized form of (6.1.18). Leszczynski and Ciesielsk [19] have applied the $R_2$-algorithm proposed in Oldham and Spanier's monograph [] to evaluate the integral in the equivalent integral equation of TFDEs. In this subsection, we briefly introduce the approximation approach presented by Kuar and Agrawal [27]. Consider the equivalent integral equation of (6.1.26), i.e.

$$
u(t) = g(t) + \frac{1}{\Gamma(\alpha)} \int_0^t (t - \tau)^{\alpha-1} f(\tau, u(\tau)) d\tau,
\tag{6.1.39}
$$

with $g(t) = \sum_{i=0}^{n-1} u_0^{(k)} \frac{t^i}{i!}$. To obtain the approximant of $u(T)$, supposing the interval $[0,T]$ is divided into $N$ equal parts, letting $h = T/N$ and $t_j = jh$ for $j = 0,1,\ldots,N$, one can obtain the following approximation formulas of (6.1.39):

$$u(t_{2i+2}) = g(t_{2i+2}) + \frac{1}{\Gamma(\alpha)} \int_0^{t_{2i}} (t_{2i+2} - \tau)^{\alpha-1} f(\tau, u(\tau)) d\tau + \sum_{k=2i}^{2i+2} c_{i,k} f_k, \quad (6.1.40)$$

$$u(t_{2i+1}) = g(t_{2i+1}) + \frac{1}{\Gamma(\alpha)} \int_0^{t_{2i}} (t_{2i+1} - \tau)^{\alpha-1} f(\tau, u(\tau)) d\tau$$

$$+ \sum_{k=2i, 2i+1/2, 2i+1} d_{i,k} f_k \text{ for } i = 0, 1, \ldots, T/2 - 1, \quad (6.1.41)$$

given that the $u(t)$ is known at $t_k$ for $k = 0,1,\ldots,2i$. Since the integrals in the above equations are generally non-singular, they can be numerically evaluated easily using the Newton–Cotes quadrature rules or other rules. The interpolating coefficients $c$ and $d$ above are determined by a weighted integration of the three-point Lagrange interpolating basis. To see this clearly, letting $i = 0$ in the above equations yields the estimations for $u(t_1)$ and $u(t_2)$:

$$u(t_2) = g(t_2) + c_{0,0} f_0 + c_{0,1} f_1 + c_{0,2} f_2, \quad (6.1.42)$$

$$u(t_1) = g(t_1) + d_{0,0} f_0 + d_{0,1/2} f_{1/2} + d_{0,2} f_1, \quad (6.1.43)$$

where

$$\sum_k c_{0,k} f_k = \frac{1}{\Gamma(\alpha)} \int_0^{t_2} (t_2 - \tau)^{\alpha-1} \sum_k \phi_{0,k}(\tau) f_k d\tau$$

$$\approx \frac{1}{\Gamma(\alpha)} \int_0^{t_2} (t_2 - \tau)^{\alpha-1} f(\tau, u(\tau)) d\tau, \quad (6.1.44)$$

$$\sum_k d_{0,k} f_k = \frac{1}{\Gamma(\alpha)} \int_0^{t_1} (t_1 - \tau)^{\alpha-1} \sum_k \psi_{0,k}(\tau) f_k d\tau$$

$$\approx \frac{1}{\Gamma(\alpha)} \int_0^{t_1} (t_1 - \tau)^{\alpha-1} f(\tau, u(\tau)) d\tau. \quad (6.1.45)$$

Note that $\phi_{j,k}, \psi_{j,k}$ (for evaluating $u(t_{2j+1})$) * above are the Lagrange interpolating bases in association with the $k$th grid point at the $(j + 1)$th step of evaluation and

$f_k \approx f(t_k, u(t_k))$ with $t_{k+1/2} = (k + 1/2)h$ *. The unknown $f_{1/2}$ can be determined by the Lagrange interpolant $\sum_k \phi_{0,k}(h/2) f_k$, namely

$$f_{1/2} = \frac{3}{8} f_0 + \frac{3}{4} f_1 - \frac{1}{8} f_2. \tag{6.1.46}$$

Substitute (6.1.46) into (6.1.41) and advance from the first step to the $(T/2-1)$th step to obtain the final approximation formulas for $u(T)$:

$$\begin{cases} u(T) = g(T) + \frac{1}{\Gamma(\alpha)} \int_0^{T-2} (T - \tau)^{\alpha-1} f(\tau, u(\tau)) d\tau + \sum_{k=T-2}^{T} c_{T/2-1,k} f_k, \\ u(T-1) = g(T-1) + \frac{1}{\Gamma(\alpha)} \int_0^{T-2} (T - 1 - \tau)^{\alpha-1} f(\tau, u(\tau)) d\tau + d_{T/2-1,T-2} f_{T-2} \\ \quad + d_{T/2-1,T-2} f_{T-3/2} \left( \frac{3}{8} f_{T-2} + \frac{3}{4} f_{T-1} - \frac{1}{8} f_T \right) + d_{T/2-1,T-1} f_{T-1}. \end{cases} \tag{6.1.47}$$

After $T/2-1$ steps of evaluation from (6.1.40) and (6.1.41), there are only two unknowns $U(T)$ and $U(T-1)$ in (6.1.47). The nonlinear equations can be solved by the Newton–Raphson method, fixed-point iteration or other nonlinear solvers.

Although it requires some mathematics deduction, transforming the differential equations into their equivalent integral equations can lower the computational overhead and effectively handle the singularity. IMEs have been successfully applied to diverse TFDEs, which can be found in [16, 28].

### 3.  FDM-RBF method for solving time-fractional diffusion equations

Unlike the numerical solution of ordinary differential equations which merely requires the discretization of the derivative with respect to one single variable (time or space), the solution of partial differential equations often needs discretization techniques to handle time and space derivatives simultaneously. This section concerns an implicit FDM approximation in the time domain and a radial basis function (RBF)-interpolating discretization in the space domain for solving time-fractional diffusion equation. The validity of this FDE-RBF method has been shown by testing one- and two- dimensional diffusion equations.

Consider the following time-fractional diffusion equation:

$$_a^C D_t^\alpha u(\mathbf{x}, t) = \gamma \Delta u(\mathbf{x}, t), \mathbf{x} \in \mathbb{R}^d, t > 0, \tag{6.1.48}$$

with the homogeneous Dirichlet boundary condition and initial condition $u(\mathbf{x},t) = f(\mathbf{x})$, where $\alpha \in (0,1)$, $_a^C D_t^\alpha$ is the Caputo fractional differential operator, $\Delta$ the Laplacian and $d$ the dimensionality. The unknown quantity $u$ can be explained in terms of concentration or temperature, and the constant factor $\gamma$ is accordingly called the diffusivity or the conductivity. The differential order $\alpha$ characterizes the strength of system memory: stronger memory is related to the smaller $\alpha$. For the convenience of discussion, assume $u$ to be the concentration of particles in the process of anomalous diffusion.

## (1) Temporal discretization by FDM

Using the direct approximation formula (6.1.11) or according to Refs. [29–33], one can evaluate the fractional derivative at $t = t_{n+1}$ as *

$$
\begin{aligned}
{}_a^C D_t^\alpha u(\mathbf{x}, t_{n+1}) &\approx \frac{1}{\Gamma(1-\alpha)} \sum_{k=0}^{n} \frac{u(x, t_{k+1}) - u(x, t_k)}{\Delta t} \int_{k\Delta t}^{(k+1)\Delta t} \frac{d\tau}{(t_{k+1} - \tau)^\alpha} \\
&= \begin{cases} a_0(u^{n+1} - u^n) + a_0 \sum_{k=1}^{n} b_k(u^{n+1-k} - u^{n-k}), & n \geq 1 \\ a_0(u^1 - u^0), & n = 0 \end{cases}
\end{aligned}
$$

(6.1.49)

where * $a_0 = (\Delta t)^{-\alpha}/\Gamma(2-\alpha)$ and $b_k = (k+1)^{1-\alpha} - k^{1-\alpha}$ * for $k = 1,2,...,n$. Substitute (6.1.49) into (6.1.48) to obtain the implicit finite difference scheme:

$$
a_0 u^{n+1} - \gamma \Delta u^{n+1} = \begin{cases} a_0 u^0, & n = 0 \\ a_0[u^n - \sum_{k=1}^{n} b_k(u^{n+1-k} - u^{n-k})], & n \geq 1 \end{cases}
$$

(6.1.50)

where $u^0 = u(\mathbf{x}, t = 0)$. Error estimations for (6.1.50) have been given by researchers but without reaching a complete consensus. Lin et al. [13] claimed that the error term takes the form $O((\Delta t)^{2-\alpha})$ * based on their mathematical deduction and numerical verification, whereas Murio [30] has derived that the error term is of the order $O(\Delta t)$ from both theoretical and numerical analyses. From the numerical results in this section, we are more inclined to accept the latter one.

## (2) Spatial discretization by the Kansa method [34–37]

Consider the domain $\mathbf{x} \in \Omega \subseteq \mathbb{R}^2$ at $t = t_{n+1}$ * and collocate in $\Omega \partial \Omega \subseteq M$ discrete points denoted by $\mathbf{x}_j$ ($j = 1,2,...,M$), in which $\mathbf{x}_1$ and $\mathbf{x}_M$ are on $\partial \Omega$ while the rest are internal points. The functional value at $\mathbf{x}_i$ can be evaluated by a RBF interpolant:

$$
u(\mathbf{x}_i, t_{n+1}) = \sum_{j=1}^{M} \lambda_j^{n+1} \phi(\|\mathbf{x}_i - \mathbf{x}_j\|_2) + \lambda_{M+1}^{n+1} x_i + \lambda_{M+2}^{n+1} y_i + \lambda_{M+3}^{n+1},
$$

(6.1.51)

where $\{\mathbf{x}_i\}$ and $\{\mathbf{x}_j\}$ are the same set of points, $\mathbf{x}_i = (x_i, y_i)$ for $i = 1,2,...,M$, $\lambda$ are unknown coefficients, $\phi(r)$ is the RBF and the monomial addends are regularized terms which guarantee the unique solvability for $\lambda$. Taking all the $\{\mathbf{x}_i\}$ in (6.1.51) and adding constraints

$$
\sum_{j=1}^{M} \lambda_j^{n+1} = \sum_{j=1}^{M} \lambda_j^{n+1} x_j = \sum_{j=1}^{M} \lambda_j^{n+1} y_j = 0
$$

(6.1.52)

lead to the RBF interpolating formula for $u(\mathbf{x}, t_{n+1})$ at $\{\mathbf{x}_i\}$ * whose matrix–vector form is

$$
\begin{Bmatrix}
u_1^{n+1} \\
\vdots \\
u_M^{n+1} \\
x_1 \\
y_1 \\
1
\end{Bmatrix}
=
\begin{bmatrix}
\phi_{11} & \cdots & \phi_{1,M} & x_1 & y_1 & 1 \\
\vdots & \ddots & \vdots & \vdots & \vdots & \vdots \\
\phi_{M,1} & \cdots & \phi_{M,M} & x_M & y_M & 1 \\
x_1 & \cdots & x_M & 0 & \cdots & 0 \\
y_1 & \cdots & y_M & \vdots & \ddots & \vdots \\
1 & \cdots & 1 & 0 & \cdots & 0
\end{bmatrix}
\begin{Bmatrix}
\lambda_1^{n+1} \\
\vdots \\
\lambda_M^{n+1} \\
\lambda_{M+1}^{n+1} \\
\lambda_{M+2}^{n+1} \\
\lambda_{M+3}^{n+1}
\end{Bmatrix},
\tag{6.1.53}
$$

where $u_i^{n+1} = u(\mathbf{x}_i, t_{n+1})$ and $\phi_{ij} = \phi(\|\mathbf{x}_i\text{-}\mathbf{x}_j\|_2)$. It can be easily seen that the interpolating matrix in (6.1.53) is symmetric. Assuming $u$ is known at $\{\mathbf{x}_i\}$ and solving the linear system (6.1.53), one can obtain the interpolating coefficients $\lambda$, thereby deriving functional values at any set of points in $\boldsymbol{\Omega} \subseteq \partial\boldsymbol{\Omega}$ from (6.1.51). Now for the solution of the equation with unknown $u$, one ought to let (6.1.53) satisfy the equation as well as the boundary condition. Noticing this and using (6.1.50), one has the following linear system similar to (6.1.53):

$$
\begin{Bmatrix}
0 \\
F_2^n \\
\vdots \\
F_{M-1}^n \\
0 \\
0 \\
0
\end{Bmatrix}
=
\begin{bmatrix}
L(\phi)_{11} & \cdots & L(\phi)_{1,M} & L(x)_1 & L(y)_1 & L(1) \\
\vdots & \ddots & \vdots & \vdots & \vdots & \vdots \\
L(\phi)_{M,1} & \cdots & L(\phi)_{M,M} & L(x)_M & L(y)_M & L(1) \\
x_1 & \cdots & x_M & 0 & \cdots & 0 \\
y_1 & \cdots & y_M & \vdots & \ddots & \vdots \\
1 & \cdots & 1 & 0 & \cdots & 0
\end{bmatrix}
\begin{Bmatrix}
\lambda_1^{n+1} \\
\vdots \\
\lambda_M^{n+1} \\
\lambda_{M+1}^{n+1} \\
\lambda_{M+2}^{n+1} \\
\lambda_{M+3}^{n+1}
\end{Bmatrix},
\tag{6.1.54}
$$

where

$$
L(\cdot) = \begin{cases} (a_0 - \gamma\Delta)(\cdot), & 1 < i < M \\ (\cdot), & i = 1 \text{ and } i = M \end{cases}
\tag{6.1.55}
$$

and

$$
F_i^n = \begin{cases} \begin{cases} a_0\left(u_i^n - \sum_{k=1}^n b_k(u^{n+1-k} - u^{n-k})\right), & n \geq 1 \\ a_0 u_i^0, & n = 0 \end{cases} & i \in (1, M) \\ 0, & i \in \{1, M\} \end{cases}.
\tag{6.1.56}
$$

The coefficient matrix in (6.1.54) is known once we preset the $\{\mathbf{x}_i\}$ or $\{\mathbf{x}_j\}$. The vector on the left-hand side of (6.1.54) can be derived by recursive relations (6.1.56). Ultimately, we obtain the interpolating coefficient $\lambda$ by solving (6.1.54).

(3)    A one-dimensional example

Consider time-fractional diffusion equation in one dimension:

$$\substack{C\\0}D_t^{0.5}u(x,t) = \frac{\partial^2}{\partial x^2}u(x,t), x \in (0,2), t \in (0,1],$$    (6.1.57)

with boundary condition $u(0,t) = u(2,t) = 0$ and initial condition $u(x,0) = f(x)$. Let

$$f(x) = \begin{cases} 2x, x \in [0, 1/2] \\ \frac{4-2x}{3}, x \in (1/2, 2] \end{cases}$$

and the corresponding analytical solution is [38]

$$u(x,t) = \sum_{n=1}^{\infty} E_{0.5}\left(-\frac{n^2\pi^2}{4}t^{0.5}\right)\sin\left(\frac{n\pi x}{2}\right)\int_0^2 f(x)\sin\left(\frac{n\pi x}{2}\right)dx,$$    (6.1.58)

where the Mittag–Leffler function reads

$$E_\alpha(z) = \sum_{k=0}^{\infty}\frac{z^k}{\Gamma(\alpha k + 1)}.$$    (6.1.59)

To make spatial discretization, use the multiquadratics (MQ) function $\phi(r) = (r^2 + c^2)^{1/2}$ ($c$ is a preset parameter) as the RBF. Let time step $\Delta t$ be 0.01 and space step $h$ 0.1. The parameter $c$ may be sensitive to specific problems. Our numerical results show that for the solution of (6.1.57), higher accurate solutions are usually obtained when $c$ is placed in (0,0.6), in particular when $c = 0.5$. Figure 6.10 plots

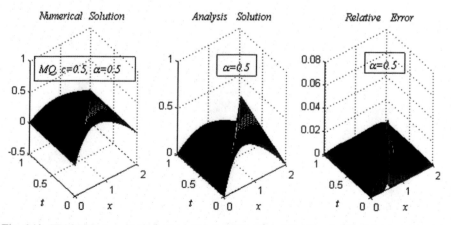

**Fig. 6.10** FDM-MQ ($c = 0.5$) method for one-dimensional fractional diffusion equation: (a) Numerical solution, (b) analytical solution and (c) relative errors

the numerical/analytical solutions and the relative errors. It can be seen from 6.10 (c) that apart from those in the neighborhood of $t = 0$, errors are kept at a low level.

We still take the MQ function ($c = 0.5$) for numerical solution and contrive to derive the error estimation according to the numerical approach given in [39]. Let the space step $h$ be 0.1 and the length of time interval 2. For different time steps, derive the order of the error term by comparing the ratio of successive time steps and the ratio of corresponding maximum absolute errors. Table 6.4 tabulates the maximum absolute errors corresponding to $t = 0.6$ and $t = 1.6$ for time steps 0.1, 0.05, 0.01 and 0.005. It can be seen that the maximum absolute errors drop for decreasing time step, which implies the convergence of the method. It follows from [39] that the order of the maximum absolute error is close to $O(\Delta t)$.

Figure 6.11 depicts the variations of concentrations at different space points with time for $\Delta t = 0.01$ and $h = 0.1$, and Fig. 6.12 plots the variations of concentrations at $x = 0.1$ with time for different $\alpha$.

(4)    A two-dimensional example

Consider the following time-fractional diffusion equation in two space dimensions:

$$
\begin{aligned}
{}_0^C D_t^{1.4} u(x, y, t) = {} & \frac{2t^{1.6}}{\pi^2 \Gamma(0.6)} \frac{\partial^2}{\partial x^2} u(x, y, t) \\
& + \frac{t^{1.6}}{12\pi^2 \Gamma(0.6)} \frac{\partial^2}{\partial y^2} u(x, y, t) + f(x, y, t),
\end{aligned}
\tag{6.1.60}
$$

where $(x,y) \in [0,1]^2$, $t \in (0,1)$, Dirichlet boundary condition is taken, the initial condition is $u(x,y,0) = \sin\pi x \sin\pi y$, the source term

$$
f(x, y, t) = \frac{25t^{1.6}}{12\Gamma(0.6)} (t^2 + 2) \sin \pi x \sin \pi y
$$

and the analytical solution [40] $u(x,y,t) = (t^2 + 1) \sin\pi x \sin\pi y$. We here select thin plate spline (TPS) function $\varphi(r) = r^{2\beta} \log(r)$ as the RBF for spatial approximation. Let time step $\Delta t = 0.01$ and space step $\Delta x = \Delta y = 0.1$. Through numerical trials, to guarantee acceptable approximations, the integer parameter $\beta$ should exceed one.

**Table 6.4** Error analysis of FDM-MQ ($c = 0.5$) method for solving the one-dimensional fractional diffusion equation

| $\Delta t$ | Maximum absolute error ($t = 0.6$) | Ratio of successive errors (Error rate) | Maximum absolute error ($t = 1.6$) | Ratio of successive errors (Error rate) |
|---|---|---|---|---|
| 0.1 | 9.2e-003 | – | 2.4 e-003 | – |
| 0.05 | 4.4e-003 | $2.09 \approx 0.1/0.05$ | 1.3 e-003 | $1.85 \approx 0.1/0.05$ |
| 0.01 | 1.1e-003 | $4.00 < 0.05/0.01$ | 4.9548 e-004 | $2.62 < 0.05/0.01$ |
| 0.005 | 7.3236e-004 | $1.50 < 0.01/0.005$ | 4.1350 e-004 | $1.20 < 0.01/0.005$ |

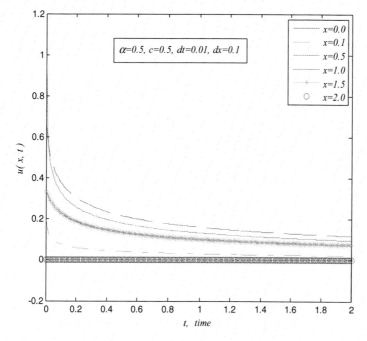

**Fig. 6.11**  Variation of concentrations at different space points with time

Good agreement between analytical and numerical solutions for $t = 0.2$ and $\beta = 4$ is shown in Fig. 6.13.

Figure 6.14 further displays the relative errors related to Fig. 6.13. It is obviously noted that the closer the evaluated points get to the boundary, the larger the errors become. To see the error more clearly, consider a fixed space point $(0.3, 0.2)$. Now we compare the predictions of using MQ ($c = 0.5$) and TPS ($\beta = 4$) functions. The relative errors for this given point at $t = 0.2, 0.4, 0.6$ and $1.0$ are tabulated in Table 6.5. The acceptable accuracy shows the validity of the combination of FDM and RBF methods for solving two-dimensional time-fractional diffusion equation.

## 6.1.2  Summaries and Comparisons of Solution Techniques for TFDEs

The FDMs and IMEs aforementioned are briefly summarized and compared with some iteration methods.

### 1.  Finite difference methods (FDMs)

Since the selection of finite difference schemes for time-fractional diffusion equation is intimately tied to the definitions of the fractional derivatives, we have introduced

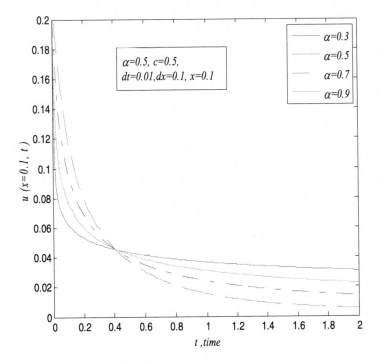

**Fig. 6.12** Variation of concentrations at $x = 0.1$ with time for different $\alpha$

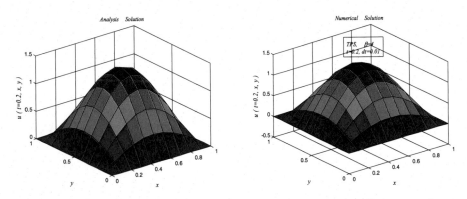

**Fig. 6.13** FDM-TPS ($\beta = 4$) method for two-dimensional fractional diffusion equation: **a** Numerical solution **b** analytical solution

different schemes for different operators. The Grünwald–Letnikov-based scheme is commonly used to construct the explicit finite difference scheme which is condition-ally convergent but at low accuracy [3]. An improved scheme has been presented in [4]. The Riemann–Liouville-based scheme is frequently considered due to its applicability in solving both time and space fractional differential equations [41, 6]. The Caputo-based scheme is realized as a highly accurate and stable approach [16],

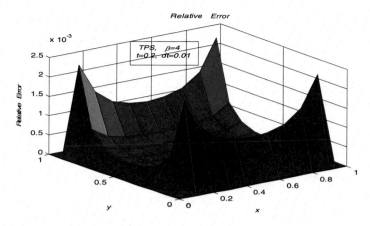

**Fig. 6.14** Relative errors of FDM-TPS ($\beta = 4$) method for solving two-dimensional fractional diffusion equation ($t = 0.2$, $\Delta t = 0.01$, $\Delta x = \Delta y = 0.1$)

**Table 6.5** Error comparison between solutions of FDM-MQ ($c = 0.5$) and FDM-TPS ($\beta = 4$) methods for solving two-dimensional fractional diffusion equation ($x = 0.3$, $y = 0.2$)

| $t$ | Exact values | TPS values | MQ values | Relative errors (TPS) | Relative errors (MQ) |
|-----|--------------|------------|-----------|-----------------------|----------------------|
| 0.2 | 0.49455 | 0.49460 | 0.49461 | 1.1179e-004 | 1.2680e-004 |
| 0.4 | 0.55161 | 0.55163 | 0.55173 | 2.6275e-004 | 2.1111e-004 |
| 0.6 | 0.64672 | 0.64662 | 0.64692 | 1.5851e-004 | 3.0592e-004 |
| 1.0 | 0.95106 | 0.95055 | 0.95150 | 5.3598e-004 | 4.6318e-004 |

thus being constantly utilized in solving TFDEs. The popularity of the Caputo-type schemes in numerical solution follows from the fact that the Grünwald–Letnikov fractional derivative is less considered in recent research and the Riemann–Liouville fractional derivative is often taken into account in theoretical analysis.

## 2. Integral equation methods (IMEs)

IMEs transform the differential equations to the equivalent integral equations, and therefore they only solve the integral equations, which reduces the singularity and simplifies the solution process. IMEs have been widely employed in the numerical solution, of which the predictor–corrector (PC) method is most commonly used.

PC methods amount to the extensions of the Adams–Bashforth methods and can achieve high accuracy despite somewhat high computational cost [20, 21]. The requirement of some mathematical deduction simplifies the solution process and reduces the singularity of the integral kernel. This type of method can also be applied to other types of equations [16, 42].

## 3. Iteration methods

Typical iteration methods for solving TFDEs include Adomian's decomposition method (ADM), differential transform method (DTM), variational iteration method (VIM) and homotopy perturbation method (HPM).

ADM forms the iteration formula by the inverse operation of the differential operator of the largest order (namely, integration) on both sides of the target equation. In this fashion, the initial conditions are naturally involved in the initial estimations, and the resulting iteration formula takes a form that is easy to program [43]. DTM writes the unknown function in form of a series with unknown coefficients and obtains the iteration formula of these coefficients by substituting the series into the target equation [7]. The initial conditions are changed into the initial coefficients as the initial estimations. The iteration formula of VIM is complicated and seems not easy to program [44, 43]. Generally speaking, different iteration formulas can be derived from HPM, yet most of which are complex and hard to program. Adjustments can also be made to establish the equivalence relation between HPM and ADM [45].

Because these iteration methods all take their approximants as the truncation of series solutions, one must increase the number of iterations in order to achieve high accuracy. Moreover, since these methods fail to consider the memory attributes of fractional derivative, that is, the evaluation of a given instant is irrespective of the values at the "past time" (instants prior to that given instant), the convergence of the error curves may thus be destroyed. When the approximate solution be a polynomial series, and if the time interval evaluated is $t \in [0,1]$, then several iterations can recover the exact solution; however, for each $t > 1$, one must increase the number of iterations so as to guarantee the solution accuracy. Hence, for a large $t$, the solution accuracy may drop and the computational cost may rise. To improve the convergence, HPM is enhanced by replacing the polynomials in the initial iteration with exponent functions [46].

We compare the solutions of ADM and DTM for

$$\,_0^C D_t^{0.5} u(t) + u(t) = 0 \text{ with } u(0) = 1. \tag{6.1.61}$$

The numerical solutions and the error curves are illustrated in Figs. 6.15, 6.16, 6.17, 6.18, 6.19 and 6.20. It can be observed that the elevation of the error curve of

**Fig. 6.15** Numerical solutions of fractional relaxation equation of the order 0.5 using ADM (Num.— numerical solution, Ana.—analytical solution)

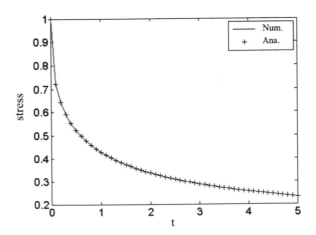

**Fig. 6.16** Absolute errors of fractional relaxation equation of the order 0.5 using ADM

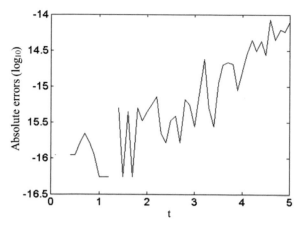

**Fig. 6.17.** Relative errors of fractional relaxation equation of the order 0.5 using ADM

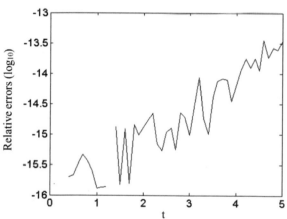

**Fig. 6.18** Numerical solutions of fractional relaxation equation of the order 0.5 using DTM (Num.— numerical solution, Ana.—analytical solution)

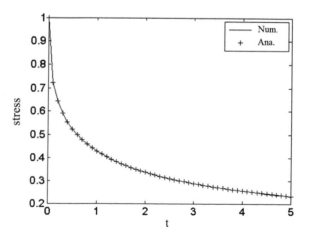

**Fig. 6.19** Absolute errors of
fractional relaxation equation
of the order 0.5 using DTM

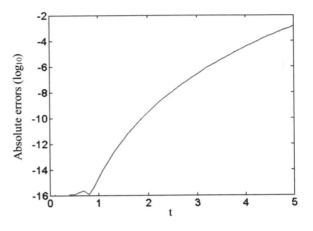

**Fig. 6.20** Relative errors of
fractional relaxation equation
of the order 0.5 using DTM

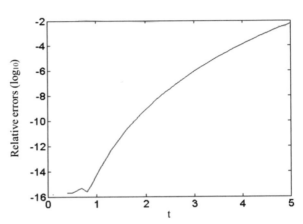

ADM is much slower than that of DTM's curve, which indicates a higher stability
of ADM. Nevertheless, it would still be fair to say that the advantage of DTM over
ADM is to construct simpler and easier-to-program iteration formulas, especially for
solving complicated differential equations.

Iteration methods compare favorably with FDMs and IMEs in achieving
higher accuracy though in the absence of an effective convergence. For long-
duration computation, the computational overhead ascends and the solution accuracy
descends. It is advisable to consider the iteration methods when the time interval to
be evaluated is not very large.

Next, we compare the Grünwald–Letnikov-based FDM, modified PC method,
ADM and DTM for solving

$$\begin{cases} \frac{d^p u(t)}{dt^p} + Bu(t) = f(t) \ (B = \omega^p, 0 < p \leq 2), \\ u(0) = C, \end{cases} \qquad (6.1.62)$$

**Table 6.6** Comparison of absolute errors for four different solution techniques

| Time | Explicit FDM | Modified PC method | ADM | DTM |
|---|---|---|---|---|
| 0 | 0 | 0 | 0 | 0 |
| 0.1 | −0.05751927137794 | 0.05374660166533 | 0.00000000974680 | 0.00419158325687 |
| 0.2 | −0.03714928406556 | 0.01148623726416 | 0.00000042132164 | 0.01569831045673 |
| 0.3 | −0.02661140536844 | 0.00451043674513 | 0.00000378790178 | 0.03366487919926 |
| 0.4 | −0.02044420411395 | 0.00229817886543 | 0.00001792714069 | 0.05756247263562 |
| 0.5 | −0.01645165166670 | 0.00133729997249 | 0.00005972494677 | 0.08700266480073 |
| 0.6 | −0.01367416446148 | 0.00084972695187 | 0.00015939848551 | 0.12167881083381 |
| 0.7 | −0.01163844568144 | 0.00057733984948 | 0.00036510683400 | 0.16133835565525 |
| 0.8 | −0.01008686050512 | 0.00041458449178 | 0.00074784127267 | 0.20576730296459 |
| 0.9 | −0.00886792777195 | 0.00031231107002 | 0.00140654886758 | 0.25478056302483 |
| 1.0 | −0.00788700447651 | 0.00024544413651 | 0.00247345560302 | 0.30821552131499 |

where $d^p/dt^p$ can be the Grünwald–Letnikov or Caputo operator. When $0 < p < 1$, (6.1.62) describes the stress relaxation of concrete, colloid or composite materials, and it can also depict the material creep if the stress is prescribed. In this case, $u(t)$, $f(t)$ and $C$ represent stress, strain and initial stress, respectively. When $1 < p < 2$, (6.1.62) controls the damped vibration of an mass element in viscoelastic media, where $u(t)$ is the displacement, $f(t)$ the external force and $C$ the initial displacement. Here, we take $p = 0.5$, $B, C = 1$, $f = 0$, $\Delta t = 0.1$ and iterations number $= 10$.

It can be observed from Table 6.6 that FDM and modified PC method have large initial errors as discussed in previous sections, while the errors of ADM or DTM gradually turn larger for a larger $t$ as shown in Figs. 6.21 and 6.24. Among these methods, ADM enjoys the highest accuracy whereas DTM possesses the worst stability. Also, since ADM and DTM are both iteration methods, increasing the iteration number will correspondingly improve the accuracy, which has been proved by the numerical trials.

## 6.2 Space Fractional Differential Equations (SFDEs)

Fractional Laplacian is a new type of operator being studied in recent decades, which is defined by the inverse Fourier transform. The explicit expressions of the operator, especially in high space dimensions, are not unique. One of the expressions is based on finite difference operator that encounters super-singularity and is cumbersome to involve the boundary data [47–49]. Up to now, most of researches on fractional Laplacian center on one-dimensional cases. Chen and Holm [48] recently presented a weakly singular, high-dimensional, finite domain definition, which easily takes in boundary data and provides the possibility for further numerical discretization.

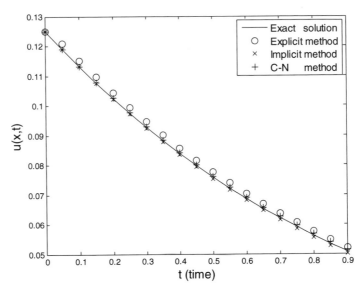

**Fig. 6.21** Exact and numerical solutions for three finite difference schemes at x = 0.50 (Explicit scheme: $N_x = 10, N_t = 20$; implicit and C-N schemes: $N_x = 20, N_t = 20$)

In recent two decades, space fractional differential operators are employed to model a wide range of problems in physics and mechanics. In the absence of closed-form solutions in most cases, numerical solutions are usually required. It would be fair to say the solution techniques for SFDEs are less developed than those for TFDEs. Finite element method, finite volume method and finite difference method are essentially local methods and therefore may not be very suitable for handling SFDEs having non-locality. At present, finite difference methods still dominate the numerical solution to SFDEs but mostly in one space dimension. The prescriptions for two- and three-dimensional SFDEs are rarely reported [50]. On the other hand, since the fractional Laplacian is in nature an integral operator, the operator discretization matrix turns out to be dense, which may lead to $O(N^3)$ operations and $O(N^2)$ memory consumption where $N$ is the dof.

### 6.2.1   Numerical Methods for SFDEs

The subsection focuses on the discretization schemes for space fractional derivatives in the sense of Grünwald–Letnikov, Riemann–Liouville and Riesz–Feller.

1.   **Grünwald–Letnikov-based finite different scheme**

The scheme is more frequently used to solve SFDEs than to solve TFDEs. Since the standard scheme often has a low stability, the shifted scheme is developed and generally preferred. The combination of the shifted scheme in space discretization with the conventional finite difference schemes in time discretization will produce different finite difference methods such as explicit, implicit and the Crank–Nicholson methods. These methods are easy to implement and possess relatively high accuracy. Meerschaert et al. modified the Crank–Nicholson FDM and thus finally achieve a solution accuracy of $O(\Delta t^2) + O(\Delta x^2)$ [50]. For different types of FDMs, the proofs of stability and convergence have been given by Meerschaert et al. [39] and Liu et al.

As discussed in Sect. 6.1.1, the standard Grünwald–Letnikov fractional derivative can be approximated by finite truncation as

$$
{}_{a}^{GL}D_x^\alpha f(i\,\Delta x) \approx \frac{(\Delta x)^{-\alpha}}{\Gamma(-\alpha)} \sum_{k=0}^{i} \frac{\Gamma(k-\alpha)}{\Gamma(k+1)} f_{i-k} \text{ for } i = 1, 2, \dots. \tag{6.2.1}
$$

The right-shifted version of (6.2.1) reads

$$
{}_{a}^{GL}D_x^\alpha f(i\,\Delta x) \approx \frac{(\Delta x)^{-\alpha}}{\Gamma(-\alpha)} \sum_{k=0}^{i} \frac{\Gamma(k-\alpha)}{\Gamma(k+1)} f_{i-k+1} \text{ for } i = 1, 2, \dots. \tag{6.2.2}
$$

2. **Riemann–Liouville-based finite difference scheme** [41]

Space fractional derivative in the Riemann–Liouville sense is given by

$$
D^\alpha(f) = \frac{1}{\Gamma(m-\alpha)} \frac{d^m}{dx^m} \int_0^x \frac{f(y)}{(x-y)^{\alpha-m+1}} dy, \alpha \in (m-1, m). \tag{6.2.3}
$$

If $1 < \alpha < 2$, then

$$
D^\alpha(f) = \frac{f(0)}{\Gamma(1-\alpha)} x^{-\alpha} + \frac{f'(0)}{\Gamma(2-\alpha)} x^{1-\alpha} + \frac{1}{\Gamma(2-\alpha)} \int_0^x \frac{f''(y)dy}{(x-y)^{\alpha-1}}. \tag{6.2.4}
$$

If $0 < \alpha < 1$, then

$$
D^\alpha(f) = \frac{f(0)}{\Gamma(1-\alpha)} x^{-\alpha} + \frac{1}{\Gamma(1-\alpha)} \int_0^x \frac{f'(y)dy}{(x-y)^\alpha}. \tag{6.2.5}
$$

Consider the case of $\alpha \in (1,2)$ and discretize $\int_0^x \frac{f''(y)dy}{(x-y)^{\alpha-1}}$ as $\sum_{j=1}^{i-1} \int_{x_j}^{x_{j+1}} \frac{f''(x-y)dy}{y^{\alpha-1}}$. Use the central difference to further approximate it as

$$
\int_{x_j}^{x_{j+1}} \frac{f''(x-y)dy}{y^{\alpha-1}} \approx \frac{f(x-x_{j+1}) + f(x-x_{j-1}) + 2f(x-x_j)}{(\Delta x)^2} \int_{x_j}^{x_{j+1}} \frac{dy}{y^{\alpha-1}},
$$

$$
\tag{6.2.6}
$$

which can be simplified to

$$D^\alpha(i\Delta x) \approx \frac{f(0)}{\Gamma(1-\alpha)} x_i^{-\alpha} + \frac{f'(0)}{\Gamma(2-\alpha)} x_i^{1-\alpha}$$
$$+ \frac{1}{\Gamma(3-\alpha)(\Delta x)^2} \sum_{j=-1}^{i-1} [(j+1)^{2-\alpha} - j^{2-\alpha}](f_{i-j+1}+f_{i-j+1}-2f_{i-j}).$$

$$(6.2.7)$$

Rearranging (6.2.7) yields

$$D^\alpha(i\Delta x) \approx \frac{f(0)}{\Gamma(1-\alpha)} x_i^{-\alpha} + \frac{f'(0)}{\Gamma(2-\alpha)} x_i^{1-\alpha} + \frac{1}{\Gamma(3-\alpha)(\Delta x)^2} \sum_{j=-1}^{i} w_j(\alpha) f_{i-j},$$

$$(6.2.8)$$

where

$$\begin{cases} w_j = \frac{(j-2)^{2-\alpha}-3(j+1)^{2-\alpha}+3j^{2-\alpha}-(j-1)^{2-\alpha}}{(\Delta x)^\alpha \Gamma(3-\alpha)}, 1 \le j \le i-2, \\ w_{-1} = 1/((\Delta x)^\alpha \Gamma(3-\alpha)), \\ w_0 = (2^{2-\alpha} - 3)/((\Delta x)^\alpha \Gamma(3-\alpha)), \\ w_{i-1} = \frac{-2i^{2-\alpha}+3(i-1)^{2-\alpha}-(i-2)^{2-\alpha}}{(\Delta x)^\alpha \Gamma(3-\alpha)}, \\ w_i = \frac{i^{2-\alpha}-(i-1)^{2-\alpha}}{(\Delta x)^\alpha \Gamma(3-\alpha)}. \end{cases} \quad (6.2.9)$$

The accuracy of (6.2.8) is $O(\Delta x)$ for $\alpha = 1$ and the order $O((\Delta x)^2)$ for $\alpha = 2$. The accuracy continuously changes when $\alpha$ varies from 1 to 2. If four-point difference approximation is taken instead of central difference approximation on the second derivative in the integrand of (6.2.4), one can derive

$$\int_{x_j}^{x_{j+1}} \frac{f''(x-y)}{y^{\alpha-1}} dy \approx \frac{f(x-x_{j+2}) - f(x-x_{j+1}) + f(x-x_{j-1}) - f(x-x_j)}{2(\Delta x)^2}$$

$$\int_{x_j}^{x_{j+1}} \frac{dy}{y^{\alpha-1}}. \quad (6.2.10)$$

Rearrangement leads to

$$D^\alpha(f)_i = \frac{f(0)}{\Gamma(1-\alpha)} x_i^{-\alpha} + \frac{f'(0)}{\Gamma(2-\alpha)} x_i^{1-\alpha} + \frac{1}{\Gamma(3-\alpha)(\Delta x)^2} \sum_{j=-1}^{i+1} \widetilde{w}_j(\alpha) f_{i-j},$$

$$(6.2.11)$$

where

$$
\begin{cases}
\tilde{w}_{-1}(\alpha) = \frac{1}{2(\Delta x)^\alpha \Gamma(3-\alpha)}, \\
\tilde{w}_0(\alpha) = \frac{2^{2-\alpha}-2}{2(\Delta x)^\alpha \Gamma(3-\alpha)}, \\
\tilde{w}_1(\alpha) = \frac{3^{2-\alpha}-2^{3-\alpha}}{2(\Delta x)^\alpha \Gamma(3-\alpha)}, \\
\tilde{w}_j(\alpha) = \frac{(j+2)^{2-\alpha}-2(j+1)^{2-\alpha}+2(j-1)^{2-\alpha}-(j-2)^{2-\alpha}}{2(\Delta x)^\alpha \Gamma(3-\alpha)}, \quad 1 < j \le i-2, \\
\tilde{w}_{i-1}(\alpha) = \frac{-i^{2-\alpha}-(i-3)^{2-\alpha}+2(i-2)^{2-\alpha}}{2(\Delta x)^\alpha \Gamma(3-\alpha)}, \\
\tilde{w}_i(\alpha) = \frac{-i^{2-\alpha}+2(i-1)^{2-\alpha}-(i-2)^{2-\alpha}}{2(\Delta x)^\alpha \Gamma(3-\alpha)}, \\
\tilde{w}_{i+1}(\alpha) = \frac{i^{2-\alpha}-(i-1)^{2-\alpha}}{2(\Delta x)^\alpha \Gamma(3-\alpha)},
\end{cases}
\tag{6.2.12}
$$

and $f_{-1} = f_1$. When $\alpha = 2$, (6.2.11) recovers to conventional four-point difference formula:

$$
\frac{\mathrm{d}^2 f(i\Delta x)}{\mathrm{d}x^2} = \frac{f_{i+1}+f_{i-2}-f_i-f_{i-1}}{2(\Delta x)^2}.
\tag{6.2.13}
$$

### 3. Riesz–Feller-based finite difference scheme

The Riesz–Feller fractional derivative is most commonly considered to be the space fractional derivative, and it characterizes two-sided non-locality and can describe the asymmetric non-local influence on space. All of these determine the crucial role the derivative plays in analyzing anomalous diffusion, fractional Brownian motion and seepage in fractal media. The derivative defined on $[a, b]$ can be written as

$$
D_\theta^\alpha := -\{C_+(\alpha,\theta)\,_a D_x^\alpha + C_-(\alpha,\theta)\,_x D_b^\alpha\}, \quad \alpha \in (0,1) \cup (1,2),
\tag{6.2.14}
$$

where $\theta$ is an asymmetry parameter, and $_a D_x^\alpha$ and $_x D_b^\alpha$ are, respectively, left- and right-sided Riemann–Liouville operators, which are defined as

$$
\begin{cases}
_a D_x^\alpha \phi(x) = \frac{1}{\Gamma(m-\alpha)} \frac{\mathrm{d}^m}{\mathrm{d}x^m} \int_a^x \frac{\phi(\xi)}{(x-\xi)^{\alpha-m+1}}\,\mathrm{d}\xi, \quad x > a, \\
_x D_b^\alpha \phi(x) = \frac{(-1)^m}{\Gamma(m-\alpha)} \frac{\mathrm{d}^m}{\mathrm{d}x^m} \int_x^b \frac{\phi(\xi)}{(\xi-x)^{\alpha-m+1}}\,\mathrm{d}\xi, \quad x < b.
\end{cases}
\tag{6.2.15}
$$

(6.2.15) can also be defined in the shifted Grünwald–Letnikov sense:

$$
\begin{cases}
_a D_x^\alpha \phi(x) = \lim_{N\to\infty} \frac{1}{h^\alpha} \sum_{k=0}^{l+1} g_k \phi[x-(k-1)h], \\
_x D_b^\alpha \phi(x) = \lim_{N\to\infty} \frac{1}{h^\alpha} \sum_{k=0}^{N-l+1} g_k \phi[x-(k-1)h],
\end{cases}
\tag{6.2.16}
$$

where $h = (b-a)/N$, $x = a + lh$ and $g_0 = 1$, $g_k = (-1)^k \alpha(\alpha-1)\ldots(\alpha-k+1)/k!$ for $k = 1, 2,\ldots$. The coefficients in (6.2.14) are given by

$$C_+(\alpha, \theta) = \frac{\sin(\pi(\alpha - \theta)/2)}{\sin(\alpha\pi)}, \, C_-(\alpha, \theta) = \frac{\sin(\pi(\alpha + \theta)/2)}{\sin(\alpha\pi)},$$

$$|\theta| \leq \begin{cases} \alpha, & 0 < \alpha < 1 \\ 2 - \alpha, & 1 < \alpha \leq 2 \end{cases}. \tag{6.2.17}$$

It should be noted that (6.2.15) and (6.2.16) can be approximated by (6.2.8) and (6.2.2), respectively. Ciesielski and Meerschaert et al. [51, 52] used the Riesz–Feller-based finite difference scheme to solve the fractional diffusion equation having initial-boundary conditions. Liu et al. [53, 54] analyzed the Lévy–Feller diffusion equation and its extensions and provided the statistics interpretation. Korabel et al. [55] have given the numerical solutions of the fractional sine–Gordon equation using the four-order Runge–Kutta time approximation.

4.    **Iteration methods**

Iteration methods have been, though not widely, utilized in solving SFDEs. Using the variational iteration method mentioned in Sect. 6.1.2, He [56] solved the space fractional diffusion equation [56] and Mustafa [57] resolved the time–space fractional Burgers equation. The Fractional KdV–Burgers equation and its modified version have been resolved with the homotopy perturbation method by Wang [58] and Abdulaziz et al. [59], respectively.

Nowadays, the researches on the solution techniques for (especially high-dimensional) SFDEs have not been well developed, which manifests itself in two aspects: (1) the developments of numerical methods fall behind those of space fractional derivative models; (2) less is known about the numerical methods for high-dimensional SFDEs with fractional Laplacian.

Owing to the internal relations between SFDEs and the Fredholm integral equations, one can make use of the solution techniques of the latter to solve the former. These techniques are intimately tied to matrix preconditioning and fast solution techniques, such as the fast multi-pole method, H-matrix method, panel clustering method and wavelet preconditioning method, which can relieve the adverse effect of the dense discretization matrix. The key issue to be solved is that how to reduce the computational cost as much as possible when the accuracy and the stability are still acceptable.

## 6.2.2 Comparison of Three Finite Difference Schemes for Space Fractional Diffusion Equations [60]

The analytical solutions of space fractional diffusion equations can only be found for simple geometry and initial-boundary conditions. Thus, numerical methods are usually considered in tackling problems having complex geometry and initial-boundary value conditions. Three key issues regarding numerical methods for SFDEs are: (1) high computation cost for large-scale problems, (2) treatments for the singularity of fractional derivatives and (3) inadequate correlative literature. Aiming at the third item, we make a comparison of three finite difference methods with explicit, implicit and C-N schemes.

1.  **Schemes**

Consider the following one-dimensional space fractional diffusion equation with variable coefficients:

$$\begin{cases} \frac{\partial u(x,t)}{\partial t} = d(x,t)_{x_L}^{GL}D_x^\alpha u(x,t) + q(x,t), \\ u(x_L, t) = a(t), \\ u(x_R, t) = b(t), \\ u(x, 0) = c(x), \end{cases} \qquad (6.2.18)$$

with $\alpha \in (1,2)$, $x \in [x_L, x_R]$ and $t \in [0,T]$. The equation reduces to the conventional diffusion equation for $\alpha = 2$, is anomalous diffusion equation for $\alpha \in (1,2)$ and degenerates to advection equation for $\alpha = 1$. If the medium is homogeneous, the equation can be extended to three space dimensions with a simple modification. Apart from the diffusion equation mentioned here, advection–diffusion and diffusion–reaction equations can also be numerically solved in a similar manner to solving (6.2.18). This is why we only exemplify the diffusion equation. We approximate the time derivative by forward difference and the space derivative by the right-shifted Grünwald–Letnikov formula (6.2.2).

(1)  Explicit scheme

Denote $u_i^j \approx u(x_i, t_j)$ and the explicit finite difference scheme of (6.2.18) reads

$$\frac{u_i^{j+1} - u_i^j}{\Delta t} = \frac{d_i^j}{\Gamma(-\alpha)(\Delta x)^\alpha} \sum_{k=0}^{i+1} \frac{\Gamma(k-\alpha)}{\Gamma(k+1)} u_{i-k+1}^j + q_i^j, \qquad (6.2.19)$$

where $N_x$ is the space node number, $N_t$ the time node number, $\Delta x = (x_R\text{-}x_L)/N_x$ the space step and $\Delta t = T/N_t$ the time step. Letting $g_{\alpha,k}$ denote the operation results of all the gamma functions, one has the following recursive relation:

$$g_{\alpha,0} = 1, g_{\alpha,1} = -\alpha, \quad g_{\alpha,k} - g_{\alpha,k-1}\frac{\alpha-k+1}{k} \quad for \ k = 2, 3, \ldots . \qquad (6.2.20)$$

It follows after arrangement that

$$u_i^{j+1} = u_i^j + \left( \frac{d_i^j}{(\Delta x)^\alpha} \sum_{k=0}^{i+1} g_{\alpha,k} u_{i-k+1}^j + q_i^j \right) \Delta t, \qquad (6.2.21)$$

for $i = 1, 2,\ldots, N_x-1$ and $j = 0, 1, \ldots, N_t-1$.

(2)    Implicit scheme

Replacing the superscript $j$ with $j + 1$ in (6.2.19) and using the notation of $g_{\alpha,k}$ lead to the implicit scheme

$$\frac{u_i^{j+1} - u_i^j}{\Delta t} = d_i^j \delta_{\alpha,x} u_i^{j+1} + q_i^j, \tag{6.2.22}$$

where

$$\delta_{\alpha,x} u_i^{j+1} = \frac{1}{(\Delta x)^\alpha} \sum_{k=0}^{i+1} g_{\alpha,k} u_{i-k+1}^{j+1}. \tag{6.2.23}$$

When $\alpha = 2$, (6.2.23) recovers to central difference:

$$\delta_{2,x} u_i^j = \frac{u_{i+1}^j - 2u_i^j + u_{i-1}^j}{(\Delta x)^2}. \tag{6.2.24}$$

A transposition of (6.2.22) yields

$$\left(1 - d_i^j \Delta t \delta_{\alpha,x}\right) u_i^{j+1} = u_i^j + q_i^j \Delta t, \tag{6.2.25}$$

of which the matrix–vector form reads

$$(\mathbf{I} - \mathbf{A})\mathbf{U}^{j+1} = \mathbf{I}\mathbf{U}^j + \mathbf{Q}^j, \tag{6.2.26}$$

where $\mathbf{I}$ is $N_x$-dimension identity matrix; $\mathbf{U}^j = \{u_0^j, u_1^j, \ldots, d_{N_x-1}^j\}^T$; $\mathbf{D}^j = \{1, d_1^j, d_2^j, \ldots, d_{N_x-1}^j\}^T$; $\mathbf{Q}^j = \left\{0, q_1^j \Delta t, \ldots, q_{N_x-1}^j \Delta t + \eta_{N_x-1}^j b_R^{j+1}\right\}^T$ and $b_R^j = b_R(t_j)$.

The matrix $\mathbf{A}$ is represented as

$$a_{ik} = \begin{cases} 0 & i = 0 \\ \eta_i^j g_{\alpha,i+1-k} & k \le i, i \ne 0 \\ \eta_i^j g_{\alpha,0} & k = i+1, i \ne 0 \\ 0 & k > i+1, i \ne 0 \end{cases} \quad \text{where } \eta_i^j = \frac{d_i^j \Delta t}{(\Delta x)^\alpha} \tag{6.2.27}$$

(3)    Crank–Nicholson scheme (C-N scheme)

The conventional C-N scheme uses the values of six time–space nodes in two successive time layers to approximate the time derivative. Unlike the conventional C-N scheme, due to the non-locality of space fractional derivative, the present C-N scheme involves values of > 6 nodes in two successive time layers. Using the symbol notations above, we write the C-N approximation of (6.2.18) [39] as

$$\frac{u_i^{j+1} - u_i^j}{\Delta t} = \frac{d_i^j}{2}\left(\delta_{\alpha,x}u_i^{j+1} + \delta_{\alpha,x}u_i^j\right) + q_i^j,\tag{6.2.28}$$

of which the matrix–vector form is

$$(\mathbf{I} - \widetilde{\mathbf{A}})\mathbf{U}^{j+1} = (\mathbf{I} + \widetilde{\mathbf{A}})\mathbf{U}^j + \widetilde{\mathbf{Q}}^j,\tag{6.2.29}$$

where $\widetilde{\mathbf{A}}$ has the same form as $\mathbf{A}$ but with $\eta_i^j = \frac{d_i^j \Delta t}{2(\Delta x)^\alpha}$, and $\widetilde{\mathbf{Q}}^j$ is derived by replacing the last entry of $\mathbf{Q}^j$ with $q_{N_x-1}^j \Delta t + \eta_{N_x-1}^j\left(b_R^{j+1} + b_R^j\right)$.

It should be noted that because the right-shifted Grünwald–Letnikov scheme (6.2.2) is only fit for describing the non-locality in one direction (i.e. describing the influence on the left-hand side of the evaluated space point), using the above three finite difference schemes will not lead to dense difference matrices. If replacing the Grünwald–Letnikov operator with the Riesz–Feller operator, two-sided non-locality makes the difference matrices fully dense, which may increase the computation cost. In this connection, an effective approach to reduce the matrix denseness will be of great importance, in particular, when one handles the high-dimensional problems [48].

## 2. Comparison

The local truncated error, stability and time complexity of the above three schemes are listed in Table 6.7.

Explicit scheme enjoys the lowest computation cost in the absence of solving a linear system, but it is conditionally stable. For implicit scheme, stability improves but complexity turns higher. This scheme is preferred at a moderate level. The C-N scheme performances best either in accuracy or in stability, and the accuracy can be further improved to $O((\Delta x)^2 + (\Delta t)^2)$ after some modifications [39]. Nevertheless, C-N scheme possesses the highest complexity and would fail to solve large-scale problems, even though its difference matrix is diagonal-dominant. We refer the readers to Refs. [39, 50, 61] to see how the data in Table 6.7 are derived.

Space fractional diffusion model, described by (6.2.18), is temporally a Markov process while exhibits spatially non-locality. Consider the diffusion of, say, the contamination in groundwater. The concentration at a given space point depends not only on the history concentration at this point, but also on the present whole concentration field and the concentration gradient field. This always produces a dense difference matrix. Accordingly, the computation cost increases at an exponential rate with the enlarging computing domain. It is of great importance to seek approaches to reduce the denseness of the difference matrices of implicit and C-N schemes.

**Table 6.7** Comparison of the three finite difference schemes

| Scheme type | Truncation error | Stability | Complexity |
|---|---|---|---|
| Explicit | $O(\Delta x + \Delta t)$ | Conditionally | $O(N_t N_x^2)$ |
| Implicit | $O(\Delta x + \Delta t)$ | Unconditionally | $O(N_t N_x^3)$ |
| C-N | $O(\Delta x + (\Delta t)^2)$ | Unconditionally | $O(N_t N_x^3)$ |

### 3.   Example

To confirm the analyses for Table 6.7 in the numerical sense, we solve the following equation by using the three finite difference schemes mentioned above:

$$\begin{cases} \frac{\partial u(x,t)}{\partial t} = d(x,t)_{x_L}^{GL} D_x^{1.8} u(x,t) + q(x,t), \ x \in (0,1), \\ u(0,t) = 0, \\ u(1,t) = e^{-t}, \\ u(x,0) = x^3, \end{cases} \quad (6.2.30)$$

where $d(x,t) = \Gamma(2.2)x^{2.8}/6$, $q(x,t) = -(1+x)e^{-t}x^3$, and exact solution is $u(x,t) = e^{-t}x^3$ [39].

Figure 6.21 Exact and numerical solutions for three finite difference schemes at $x = 0.50$ (Explicit scheme: $N_x = 10$, $N_t = 20$; implicit and C-N schemes: $N_x = 20$, $N_t = 20$).

Figure 6.21 gives the comparison of the numerical solutions predicted by three finite difference schemes. The conditional stability requests the space step of the explicit scheme to be larger. The absolute errors are tabulated in Table 6.8, from which it can be seen that the error of explicit scheme at the initial stage is much obvious than those of two other schemes. This phenomenon, which cannot be remedied by decreasing the time step, is due to the absence of the solution of the linear system. The accuracy comparison agrees well with the second column of Table 6.7.

### 4.   Summaries and discussions

Explicit, implicit and C-N finite difference schemes for the solution of space fractional diffusion equation are described and compared. Numerical results show that the explicit scheme enjoys the lowest complexity but possesses the lowest stability and accuracy; the implicit scheme is absolutely convergent, and the accuracy together with the complexity is moderate and C-N scheme keeps absolute convergence and features the highest accuracy yet also the highest complexity.

**Table 6.8** Comparison of absolute errors of three finite difference schemes for space fractional diffusion equation ($x = 0.5$)

| $t$ | Explicit scheme | Implicit scheme | C-N scheme |
|---|---|---|---|
| $t = 0.1$ | 0.00203074277660 | −8.3328735289e-005 | 7.7658939900e-005 |
| $t = 0.2$ | 0.00198975148263 | −0.00017074964197 | 0.00015214898892 |
| $t = 0.3$ | 0.00195890620456 | −0.00026150002085 | 0.00022381649063 |
| $t = 0.4$ | 0.00190813521468 | −0.00035332334556 | 0.00029220706131 |
| $t = 0.5$ | 0.00182279772837 | −0.00044328178702 | 0.00035618915513 |
| $t = 0.6$ | 0.00170661830361 | −0.00052857571027 | 0.00041456196870 |
| $t = 0.7$ | 0.00156999665019 | −0.00060701894027 | 0.00046648316571 |
| $t = 0.8$ | 0.00142350290275 | −0.00067718053692 | 0.00051156685326 |
| $t = 0.9$ | 0.00127554594125 | −0.00073832548072 | 0.00054981111729 |

Aside from the finite difference method, some other methods such as the random walk method [61] and the Monte Carlo method [62] have also been developed. But these methods are simply confined to specific equations. It is worth noting that the explicit expression of an easy-to-implement, high-dimensional space fractional derivative has not yet been given, which would hinder the numerical solution of high-dimensional SFDEs.

## 6.3   Open Issues of Numerical Methods for FDEs

Although researchers have extensively applied the fractional derivative models to fields of mechanics, engineering, physics, biology and even the social sciences, yet the numerical solution techniques are not well developed as they should have been. This is mainly due to the mathematical complexity of the operations regarding fractional derivative and to the high computational overhead arising from the history dependence in time and the non-locality in space. These characteristics confine the applications of fractional derivatives to an academic level and hinder the applications for practical problems.

Finite difference methods are, up to now, in a dominant position among all the solution techniques. Special attention should be paid to the computational complexity, stability analysis and singularity handling of FDMs for FDEs. A key issue is how to effectively apply FDMs to solve SFDEs. FDM solution of SFDEs is much less being seen than that of TFDEs, yet many practical problems are described by SFDEs. This is why we need to develop more FDMs for diverse SFDEs.

Most of the present researches directly apply the solution techniques which are valid for integer-order derivative equations to solve fractional derivative equations. The solution techniques that are based on the characteristics of FDEs themselves are rare. Moreover, the studies on approaches that can reduce the computation cost and memory requirement are far from systemic and adequate. Also, less attention is paid to theoretical studies on the method efficiency, error estimation, stability analysis and adaptive computation. Fractional derivative equations actually belong to a special type of integro-differential equation. This makes the theoretical analysis and algorithm design much different from those of integer-order derivative equations. In recent decades, despite a fast development of computer simulation techniques, the high accurate numerical simulations of fractional derivative models depicting mechanics behaviors in the complex medium are still absent in both numerical and theoretical aspects, which is a principle factor that impedes the improvement of simulation ability. At present, there are very few truly meaningful reports on this research field.

Four open issues on solution techniques for FDEs are briefly discussed in the following.

(1)  Large computational cost and memory requirement

The history dependency in TFDEs leads to a slow variation of unknown function with time and therefore requires a long-duration evaluation. However, owing to the storage of all the values preceding the present value, a long-duration evaluation will bring an obvious increase in the computational cost and memory requirement. It is desirable to develop some preconditioning techniques to reduce the computation effort.

(2)  Large initial-stage error for initial-value problem

A large error arises from the difference between the transient and stable solutions. After a rapid attenuation of the transient solution, the resulting solution recovers to the stable solution. This implies that the stability of the numerical methods is determined by the stable solution while the initial-stage error is influenced by the transient solution. In general, there exists a contradiction between the initial-stage accuracy and the method stability. That is, if we contrive to decrease the initial-stage error, the stability of the method will be lowered. This can be further illustrated by using the PC method to solve the following fractional relaxation equation:

$$\begin{cases} {}^{C}_{0}D^{\alpha}_{t}u(t) = u(t), & 0 < \alpha \leq 1, \\ u(0) = 10. \end{cases} \tag{6.3.1}$$

The comparison of numerical and exact solutions is given in Fig. 6.22.

(3)  Stiffness of the fractional ordinary differential equations

When using the fractional ordinary differential system to describe complex physical or mechanics problems, the system solution is often comprised of fast-varying (transient) and slow-varying (stable) components. Like the solution of a single fractional ordinary differential equation, the system solution also tends

**Fig. 6.22** Comparison of numerical and exact solutions for fractional relaxation equation of the order 0.5 (time step is taken 0.5, Num—numerical solution, Ana. —analytical solution)

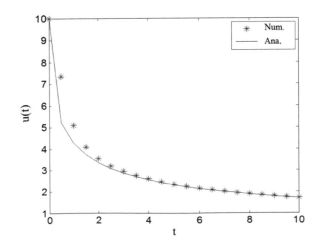

to a stable solution after the transient solution is fast dissipated. Hence, the contradiction of initial-stage accuracy and stability of the numerical method should also be considered as is done in solving a single fractional ordinary differential equation. Since the number of components in the system solution is larger and mutual stiffness is obviously different, the numerical solution of the fractional ordinary differential system will be more complicated.

(4) Discretization of high-dimensional fractional Laplacian

A formalism definition of fractional Laplacian, i.e. $(-\Delta)^{\alpha/2}$, in arbitrary space dimensions, is given by the inverse Fourier transform [49]. This type of space fractional differential operator emerges recently in diverse physical and mechanics models, such as in the anomalous diffusion model aforementioned. Nevertheless, compared to the time-fractional differential operators, the theoretical and numerical analyses of fractional Laplacian are far from sound. An equivalent definition of fractional Laplacian, which can be directly discretized, is based on a super-singular integral whose integrand is a finite difference expression. This definition fails to involve the boundary conditions [49, 63] and is thus hardly applied to engineering modeling and computation. Up to now, the researches on the discretization of the fractional Laplacian are mainly focused on one-dimensional cases [52]. Of special care are two issues pertaining to the discretization of fractional Laplacian. Since finite element method, finite volume method and finite difference method are essentially local methods, they seem not suitable for solving equations with non-local fractional Laplacian. It would be advisable to develop some global methods such as global radial basis function methods for space discretization. On the other hand, the non-locality of fractional Laplacian yields a dense discretization matrix, even though local methods, such as the finite element method, are employed as solvers. Handling the dense matrix always consumes a large overhead of computation, which will be a bottleneck for the solution of a high-dimensional problem having large geometry.

Based on the introduction and discussions above, we may obtain that despite some achievements in studying the numerical methods for FDEs, there still exist some open issues to be settled. To our knowledge, special attention should be taken on the following facts of numerical computation of fractional derivative equations [64]:

(1) Fast algorithm for time and/or space fractional derivative equations;
(2) Basics of computation theory for factional derivative equations;
(3) Mesh-free methods for space fractional derivative equations with fractional Laplacian;
(4) Development of the corresponding computation mechanics software packages;
(5) Internal relation between algorithms of fractional calculus and those of statistic mechanics.

## 6.4  Numerical Methods for Fractal Derivative Equations

The fractal derivative can be approximated by a backward difference:

$$\frac{du(t_n)}{dt^\alpha} \approx \frac{u(t_n) - u(t_{n-1})}{t_n^\alpha - t_{n-1}^\alpha}. \tag{6.4.1}$$

Consider the following two fractal derivative equations.

1. **Fractal relaxation–oscillation equation**

$$\frac{du(t)}{dt^\alpha} + Bu(t) = f(t). \tag{6.4.2}$$

Substitute (6.4.1) into (6.4.2) to obtain the finite difference scheme:

$$\frac{u_n - u_{n-1}}{t_n^\alpha - t_{n-1}^\alpha} + Bu_n = f_n \text{ for } n = 1, 2, \ldots, \tag{6.4.3}$$

which can be rearranged to

$$\left(B(t_n^\alpha - t_{n-1}^\alpha) + 1\right)u_n - u_{n-1} = (t_n^\alpha - t_{n-1}^\alpha)f_n, \tag{6.4.4}$$

with $B = 1$, $u_0 = 1$ and $f \equiv 0$. The numerical solutions are shown in Fig. 6.23.

2. **Fractal damped vibration equation**

$$m\ddot{u}(t) + c\frac{du(t)}{dt^\alpha} + ku(t) = f(t). \tag{6.4.5}$$

Substituting (6.4.1) in (6.4.5) gives the finite difference scheme, namely

**Fig. 6.23** Numerical solution of fractal relaxation–oscillation equation ("standard": standard oscillation)

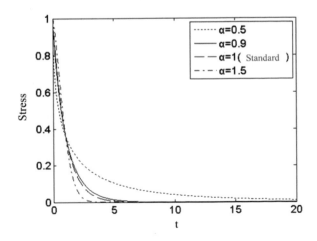

**Fig. 6.24** Numerical solutions of fractal damped vibration equation

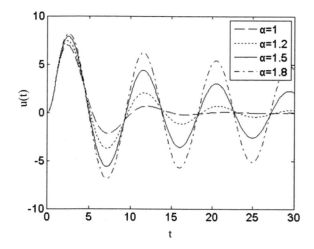

$$m\frac{u_{n+1} - 2u_n + u_{n-1}}{(\Delta t)^2} + c\frac{u_n - u_{n-1}}{t_n^\alpha - t_{n-1}^\alpha} + ku_n = f_{n-1}. \tag{6.4.6}$$

Given $m = 1, c=0.5, k = 0.5$ and $f = 8$ for $t \in [0,1]$ (zero otherwise), the numerical solutions are given in Fig. 6.24.

## 6.5 Numerical Methods for Positive Fractional Derivative Equations

The present section introduces how the predictor–corrector method and finite difference method solve the positive fractional relaxation–oscillation equation as well as the damped vibration equation, respectively.

1. **PC method for positive fractional relaxation–oscillation equation**

$$(-1)^{[\alpha]-1}\frac{d^{|\alpha|}u(t)}{dt^{|\alpha|}} + Bu(t) = f(t) \; (B = \omega^\alpha), \tag{6.5.1}$$

where $[\alpha]$ is the smallest integer equal to or larger than $\alpha$, and the positive fractional derivative is in the Caputo sense, i.e.

$$\frac{d^{|\alpha|}f(t)}{dt^{|\alpha|}} = \begin{cases} \frac{-1}{\alpha q(\alpha)}\int_0^t \frac{f'(\tau)}{(t-\tau)^\alpha}d\tau, & 0 < \alpha \le 1 \\ \frac{1}{\alpha(\alpha-1)q(\alpha)}\int_0^t \frac{f''(\tau)}{(t-\tau)^{\alpha-1}}d\tau, & 1 < \alpha < 2 \end{cases}. \tag{6.5.2}$$

From (6.5.2), one can derive the relation of the Caputo derivative and positive derivative:

$$\phi(\alpha)_a^C D_t^\alpha u(t) = \frac{d^{|\alpha|}u(t)}{dt^{|\alpha|}}, \qquad (6.5.3)$$

where

$$\phi(\alpha) = \begin{cases} -\frac{\Gamma(1-\alpha)}{\alpha q(\alpha)}, & 0 < \alpha \le 1, \\ \frac{\Gamma(2-\alpha)}{\alpha(\alpha-1)q(\alpha)}, & 1 < \alpha < 2, \end{cases}$$

with $q(\alpha) = \frac{\pi}{2\Gamma(\alpha+1)\cos[(\alpha+1)\pi/2]}$.

Using (6.5.3) to rearrange (6.5.1) into the fractional relaxation–oscillation equation

$$_a^C D_t^\alpha u(t) = f(t)/\phi(\alpha) - Bu(t)/\phi(\alpha). \qquad (6.5.4)$$

As is shown in Sects. 2.1 and 2.2, (modified) PC method can be utilized to solve the above equation [20]. Given $B = 1$, $u(0) = 1$, $u'(0) = 0$ (for the case of $\alpha > 1$) and $f(t) = 0$, Figs. 6.25 and 6.26 graph the numerical solutions for different $\alpha$.

2.   **FDM for positive fractional damped vibration equation**

$$mu'' + (-1)^{[\alpha]-1} c \frac{d^{|\alpha|}u}{dt^{|\alpha|}} + ku = f(t), \qquad (6.5.5)$$

where

$$\frac{d^{|\alpha|}u(t)}{dt^{|\alpha|}} = \frac{1}{q(\alpha)} \int_a^t \frac{u(\tau)}{(t-\tau)^{\alpha+1}} d\tau = \frac{\Gamma(-\alpha)}{q(\alpha)} \frac{1}{\Gamma(-\alpha)} \int_a^t \frac{u(\tau)}{(t-\tau)^{\alpha+1}} d\tau, \quad (6.5.6)$$

with $q(\alpha) = \frac{\pi}{2\Gamma(\alpha+1)\cos[(\alpha+1)\pi/2]}$.

**Fig. 6.25** Numerical solutions for positive fractional relaxation equation

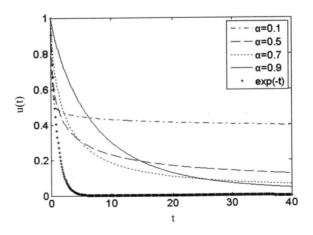

**Fig. 6.26** Numerical solutions for positive fractional oscillation equation

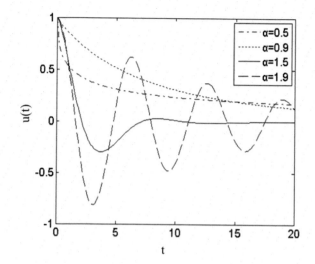

Based on the relation between the Grünwald–Letnikov integral and Riemann–Liouville integral [65].

$$\frac{1}{\Gamma(-\alpha)} \int_a^t \frac{u(\tau)}{(t-\tau)^{\alpha+1}} d\tau = \lim_{\Delta t \to 0} (\Delta t)^\alpha \sum_{r=0}^{(t-a)/\Delta t} (-1)^r \binom{-\alpha}{r} u(t - r\Delta t), \quad (6.5.7)$$

one can obtain the finite difference scheme of positive fractional derivative:

$$\frac{d^{|\alpha|} u(n\Delta t)}{dt^{|\alpha|}} \approx \frac{2\Gamma(-\alpha)\Gamma(\alpha+1)\cos((\alpha+1)\pi/2)(\Delta t)^\alpha}{\pi}$$

$$\sum_{r=0}^n (-1)^r \binom{-\alpha}{r} u(t - r\Delta t). \quad (6.5.8)$$

Substitute (6.5.8) into (6.5.5) and use the backward difference for the second derivative to obtain the finite difference scheme [17, 55]:

$$m \frac{u_n - 2u_{n-1} + u_{n-2}}{(\Delta t)^2} - a(\Delta t)^\alpha \sum_{r=0}^n \omega_r^{(\alpha)} u_{n-r} + u_{n-1} = f_n, \quad (6.5.9)$$

where

$$a = c \frac{2\Gamma(-\alpha)\Gamma(\alpha+1)\cos((\alpha+1)\pi/2)}{\pi}, \ \omega_r^{(\alpha)} = (-1)^r \binom{-\alpha}{r} = \frac{\alpha(\alpha+1)\cdots(\alpha+r-1)}{r!}.$$

The numerical solution can be further written as

**Fig. 6.27** Numerical
solutions for positive
fractional damped vibration
equation

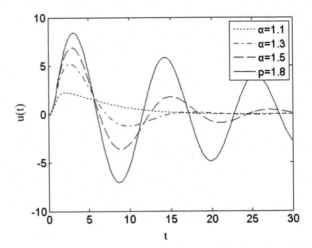

$$u_n = \frac{f_n - ku_{n-1} + a(\Delta t)^\alpha \sum_{r=1}^{n} \omega_r^{(\alpha)} u_{n-r} + m(dt)^{-2}(2u_{n-1} - u_{n-2})}{m(\Delta t)^{-2} - a(\Delta t)^\alpha}. \quad (6.5.10)$$

Given $m = 1, c = 0.5, k = 0.5, f(t) = 8$ for $t \in [0,1]$ (zero otherwise), $u(0) = 0$ and $u'(0) = 0$, Fig. 6.27 plots the numerical solutions for different $\alpha$.

# References

1. S.C. Lim, C.H. Eab, Riemann-Liouville and Weyl fractional oscillator processes [J]. Phys. Lett. A **355**, 87–93 (2006)
2. K. Yao, Y.S. Liang, J.X. Fang, The fractal dimensions of graphs of the Weyl-Marchaud fractional derivative of the Weierstrass-type function [J]. Chaos Solitons Fractals **35**, 106–115 (2008)
3. R. Gorenflo, E.A. Abdel-Rehim, Convergence of the Grünwald-Letnikov scheme for time-fractional diffusion [J]. J. Comput. Appl. Math. **205**, 871–881 (2007)
4. R. Lin, F. Liu, Fractional high order methods for the nonlinear fractional ordinary differential equation [J]. Nonlinear Anal. **66**, 856–869 (2007)
5. D.A. Murio, Stable numerical solution of a fractional-diffusion inverse heat conduction problem [J]. Comput. Math. Appl. **53**, 1492–1501 (2007)
6. C.-M. Chen, F. Liu, I. Turner, V. Anh, A Fourier method for the fractional diffusion equation describing sub-diffusion [J]. J. Comput. Phys. **227**, 886–897 (2007)
7. N.J. Ford, A. Charles Simpson, The approximate solution of fractional differential equations of order greater than 1. Numerical Analysis Report No.286. http://www.ma.man.ac.uk/MCCM/MCCM.html
8. K. Subbaraj, M.A. Dokainish, A survey of direct time-integration methods in computational structural dynamics-II: Implicit methods [J]. Comput. Struct. **32**(6), 1387–1401 (1989)
9. K.J. Bathe, E.L. Wilson, *Numerical Methods in Finite Element Analysis [M]* (Prentice-Hall, Inc., Englewood Cliffs, New Jersey, 1976), pp. 309–333
10. Y. Qiang, F.W. Liu, Implicit difference approximation for the time-fractional order reaction-diffusion equation [J]. J. Xiamen Univ. (Nat. Sci.) **45**(3), 315–319 (2006) (in Chinese)

11. Z.D. Xu, Z. Zhou, H.T. Zhao et al., A new model on viscoelastic dampers [J]. Eng. Mech. **18**(6), 88–93 (2001). ((in Chinese))
12. W. Zhang, N. Shimizu, Numerical algorithm for dynamic problems involving fractional operators [J]. JSME Int. J. (Ser. C) **41**(3), 364–370 (1998)
13. Y. Lin, C. Xu, Finite difference/spectral approximations for the time-fractional diffusion equation [J]. J. Comput. Phys. **225**, 1533–1552 (2007)
14. T.A.M. Langlands, B.I. Henry, The accuracy and stability of an implicit solution method for the fractional diffusion equation [J]. J. Comput. Phys. **205**, 719–736 (2005)
15. G.J. Fix, J.P. Roop, Least squares finite-element solution of a fractional order two-point boundary value problem [J]. Comput. Math. Appl. **48**, 1017–1033 (2004)
16. S. Momani, An algorithm for solving the fractional convection–diffusion equation with nonlinear source term [J]. Commun. Nonlinear Sci. Numer. Simul. **12**, 1283–1290 (2007)
17. S.J. Shen, F.W. Liu, A computationally effective numerical method for the fractional-order Bagley-Torvik equation [J]. J. Xiamen Univ. (Nat. Sci.) **43**(3), 306–311 (2004) (in Chinese)
18. R. Gorenflo, S. Vessella, *Abel Integral Equations [M]* (Springer, Berlin/Heidelberg, 1991)
19. J. Leszczynski, M. Ciesielski, A numerical method for solution of ordinary differential equations of fractional order [J]. ArXiv:math.NA/0202276. v1 26 Feb 2002
20. K. Diethelm, N.J. Ford, A.D. Freed, A Predictor-Corrector approach for the numerical solution of fractional differential equations [J]. Nonlinear Dyn. **29**, 3–22 (2002)
21. K. Diethelm, N.J. Ford, A.D. Freed, Yu. Luchko, Algorithms for the fractional calculus: a selection of numerical methods [J]. Comput. Methods Appl. Mech. Eng. 743–773 (2005)
22. C.H. Yang, F.W. Liu, A new predictor-corrector method for fractional relaxation-oscillation equation [J]. J.Xiamen Univ. (Nat. Sci.) **44**(6), 761–765 (2005) (in Chinese)
23. W. Deng, Numerical algorithm for the time fractional Fokker-Planck equation [J]. J. Comput. Phys. **227**, 1510–1522 (2007)
24. K. Adolfsson, M. Enelund, S. Larsson, Adaptive discretization of fractional order viscoelasticity using sparse time history [J]. Comput. Methods Appl. Mech. Eng. **193**, 4567–4590 (2004)
25. K. Diethelm, N.J. Ford, A.D. Freed, Detailed error analysis for a fractional Adams method [J]. Numer. Algorithms **36**, 31–51 (2004)
26. C.P. Li, C.X. Tao, On the fractional Adams method [J]. Comput. Math. Appl. **58**, 1573–1588 (2009)
27. P. Kumar, O.P. Agrawal, An approximate method for numerical solution of fractional differential equations [J]. Signal Process. **86**, 2602–2610 (2006)
28. Y.-P. Liu, T. Lü, Mechanical quadrature methods and their extrapolation for solving first kind Abel integral equations [J]. J. Comput. Appl. Math. **201**, 300–313 (2007)
29. F. Liu, S. Shen, V. Anh, I. Turner, Analysis of a discrete non-Markovian random walk approximation for the time fractional diffusion equation [J]. Anziam J. **46**(E), C488–C504 (2004)
30. D.A. Murio, Implicit finite difference approximation for time fractional diffusion equations [J]. Comput. Math. Appl. **56**, 1138–1145 (2008)
31. Z.M. Wu, Radial basis function scattered data interpolation and the meshless method of numerical solution of PDEs [J]. Chin. J. Eng. Math. **19**(2), 1–12 (2002). ((in Chinese))
32. P. Zhuang, F. Liu, Finite difference approximation for two-dimensional time fractional diffusion [J]. J. Algorithms Comput. Technol. **1**(1), 1–15 (2007)
33. B.T. Jin, Mesh-free method for the inverse problems of a kind of elliptical partial differential equations [D], in *Hangzhou: Science College of Zhejiang University* (2005), pp. 14–17 (in Chinese)
34. E.J. Kansa, Multiquadrics-A scattered data approximation scheme with application to computation fluid dynamics, I. Surface approximations and partial derivative estimates [J]. Comput. Math. Appl. **19**, 127–145 (1990)
35. X. Zhang, Y. Liu, *Mesh-Free Methods [M]* (Tsinghua University Press, Beijing, 2004). ((in Chinese))

36. M. Zerroukat, H. Power, C.S. Chen, A numerical method for heat transfer problems using collocation and radial basis functions [J]. Int. J. Numer. Meth. Eng. **42**, 1263–1278 (1998)

37. M. Zerroukat, K. Djidjeli, A. Charafi, Explicit and implicit meshless methods for linear advection-diffusion-type partial differential equations [J]. Int. J. Numer. Meth. Eng. **48**, 19–35 (2000)

38. O.P. Agrawal, Solution for a fractional diffusion-wave equation defined in a bounded domain [J]. Nonlinear Dyn. **29**, 145–155 (2002)

39. C. Tadjeran, M.M. Meerschaert, H.P. Scheffler, A second-order accurate numerical approximation for the fractional diffusion equation [J]. J. Comput. Phys. **213**, 205–213 (2006)

40. B. Baeumer et al., Advection and dispersion in time and space [J]. Phys. A **350**, 245–262 (2005)

41. V.E. Lynch, B.A. Carreras, D. del-Castillo-Negrete, K.M. Ferreira-Mejias, H.R. Hicks, Numerical methods for the solution of partial differential equations of fractional order [J]. J. Comput. Phys. **192**, 406–421 (2003)

42. L.J. Sheu et al., Chaos in the Newton-Leipnik system with fractional order [J]. Chaos Solitons Fractals **36**, 98–103 (2008)

43. S. Momani, Z. Odibat, Numerical approach to differential equations of fractional order [J]. J. Comput. Appl. Math. **207**, 96–110 (2007)

44. N.H. Sweilam, M.M. Khader, R.F. Al-Bar, Numerical studies for a multi-order fractional differential equation [J]. Phys. Lett. A **371**, 26–33 (2007)

45. S. Momani, Z. Odibat, Homotopy perturbation method for nonlinear partial differential equations of fractional order [J]. Phys. Lett. A **365**, 345–350 (2007)

46. S.H. Hosein Nia et al., Maintaining the stability of nonlinear differential equations by the enhancement of HPM [J]. Phys. Lett. A **372**(6), 2855–2861 (2008)

47. T. Bojdecki, L.G. Gorostiza, Fractional Brownian motion via fractional Laplacian [J]. Statist. Probab. Lett. **44**, 107–108 (1999)

48. W. Chen, S. Holm, Fractional Laplacian time-space models for linear and nonlinear lossy media exhibiting arbitrary frequency power-law dependency [J]. J. Acoust. Soc. Am. **115**(4), 1424–1430 (2004)

49. S.G. Samko, A.A. Kilbas, O.I. Marichev, Fractional integrals and derivatives: theory and applications [M]. Gordon and Breach Science Publishers (1993)

50. M.M. Meerschaert, C. Tadjeran, Finite difference approximations for fractional advection–dispersion flow equations [J]. J. Comput. Appl. Math. **172**, 65–77 (2004)

51. M. Ciesielski, J. Leszczynski, Numerical treatment of an initial-boundary value problem for fractional partial differential equations [J]. Signal Process. **86**, 2619–2631 (2006)

52. M.M. Meerschaert, C. Tadjeran, Finite difference approximations for two-sided space-fractional partial differential equations [J]. Appl. Numer. Math. **56**, 80–90 (2006)

53. Q. Liu, F. Liu, I. Turner, V. Anh, Approximation of the Lévy-Feller advection–dispersion process by random walk and finite difference method [J]. J. Comput. Phys. **222**, 57–70 (2007)

54. H. Zhang, F. Liu, V. Anh, Numerical approximation of Lévy-Feller diffusion equation and its probability interpretation [J]. J. Comput. Appl. Math. **206**, 1098–1115 (2007)

55. N. Korabel, G.M. Zaslavsky, V.E. Tarasov, Coupled oscillators with power-law interaction and their fractional dynamics analogues [J]. Commun. Nonlinear Sci. Numer. Simul. **12**, 1405–1417 (2007)

56. J.H. He, Approximate analytical solution for seepage flow with fractional derivatives in porous media [J]. Comput. Methods Appl. Mech. Eng. **167**, 57–68 (1998)

57. M. Inc, The approximate and exact solutions of the space- and time-fractional Burgers equations with initial conditions by variational iteration method [J]. J. Math. Anal. Appl. **345**, 476–484 (2008)

58. Q. Wang, Homotopy perturbation method for fractional KdV-Burgers equation [J]. Chaos Solitons Fractals **35**, 843–850 (2008)

59. O. Abdulaziz et al., Approximate analytical solution to fractional modified KdV equations [J]. Math. Comput. Model. **49**(1–2), 136–145 (2009)

60. H.G. Sun, W. Chen, X. Cai, Comparative study of numerical algorithms for "anomalous" diffusion equation with spatial fractional derivatives [J]. Chin. J. Comput. Phys. **26**(5), 719–724 (2009). ((in Chinese))

61. E.A. Abdel-Rehim, R. Gorenflo, Simulation of the continuous time random walk of the space-fractional diffusion equations [J]. J. Comput. Appl. Math. **222**(2), 274–283 (2008)
62. M. Marseguerra, A. Zoia, Monte Carlo evaluation of FADE approach to anomalous kinetics [J]. Math. Comput. Simul. **77**, 345–357 (2008)
63. Q.Y. Guan, Z.M. Ma, Reflected symmetric α-stable processes and regional fractional Laplacian [J]. Probab. Theory Relat. Fields **134**, 649–694 (2006)
64. W. Chen, H.G. Sun, Status and problems of numerical algorithm on fractional differential equations [J]. Comput. Aided Eng. **19**(2), 1–3 (2010). ((in Chinese))
65. I. Podlubny, *Fractional Differential Equations [M]* (Academic Press, San Diego, 1999)
66. R. Gorenflo, F. Mainardi, Random walk models for space-fractional diffusion processes [J]. Fract. Calc. Appl. Anal. **1**, 167–191 (1998)
67. J.W. Hanneken et al., A random walk simulation of fractional diffusion [J]. J. Mol. Liq. **114**, 153–157 (2004)

# Chapter 7
# Current Development and Perspectives of Fractional Calculus

## 7.1 Summary and Discussion

### 7.1.1 Current Research and Application

1. **Theoretical Study of the Fractional Calculus**

   The main study includes.

   (1) Improvement of the fractional calculus definition: there are dozens of definitions about fractional calculus, and there is a close link between these definitions. However, because the conditions include application background, initial conditions are different, and there are selection uncertainties in the applications. Therefore, the classification and unity of fractional calculus will be a very meaningful groundbreaking work.
   (2) The numerical solution of fractional calculus and extension of the definition of fractional calculus (such as fractal derivative and positive fractional calculus).
   (3) The properties and integral transform of fractional calculus. The properties which are different from that of the integer-order calculus should be explored. The integral transform such as Fourier transform, Laplace transform and Z-transform are also important theoretical research directions of fractional calculus.

2. **Numerical algorithm of fractional calculus**

   In view of the potential applications of fractional calculus, and the great difficulties on the analytical solution for fractional calculus, the development of numerical algorithm is a hotspot. The algorithms include finite difference

© Science Press 2022

W. Chen et al., *Fractional Derivative Modeling in Mechanics and Engineering*,
https://doi.org/10.1007/978-981-16-8802-7_7

method, finite element method, Lagrange multiplier method used in integer-order calculus, and also include the random walk method, Monte Carlo algorithm; variational algorithm, perturbation method and other methods. In addition, space calculus will encounter great difficulties, not only because of the global correlation of fractional calculus but also because of the difference scheme in the calculation of complex shapes; thus, it needs a big effort to solve the problem. The numerical algorithm is in a stage of rapid development.

3.   **Application research of fractional calculus**

The research history of fractional calculus is as long as the integer-order calculus. The concept of fractional calculus was first proposed by Leibniz in the seventeenth century. The basic theory of fractional calculus is established under the efforts of Liouville, Grünwald, Letnikov, Riemann, etc. until the end of the nineteenth century. However, due to the absence of a physics and mechanics background, fractional calculus has merely been a purely theoretical mathematics research issue for three centuries. The main reason is that the fractional operators are contradicted with the classical theory of physics and Newtonian force in mechanics system established on the basis of the integer-order calculus system. Until the 1970s, the fractional calculus theory has been rapidly applied to various fields of natural science and engineering, such as turbulence, anomalous diffusion, seepage analysis and control, signal processing, medical imaging, fluctuations, damping analysis, friction control, viscous elastic material, and rheological, optical and mechanical control. Because it can well describe the history dependency and non-locality of the system or physical processes, fractional calculus has become an important mathematical tool of differential equations modeling complex mechanics phenomena.

4.   **The combination of fractional calculus and mechanics principles to study some basic issues**

It includes fractional Brownian motion, the random walk theory, fractional-order controller, soft material constitutive model, fractional quantum mechanics, power-law phenomenon, etc. Although fractional calculus combined with some of the basic principles of mechanics has extended the existing mechanics system, the complete theoretical system needs to be established, and it will also be an important direction of the applied research of the calculus.

## 7.1.2   Key Issues

1.   **Improve the mathematical theory of fractional calculus system**

Nowadays, although the definition of fractional calculus has been proposed, the theoretical system of fractional calculus has yet to be further improved, such as the unification of the time-fractional derivative definition. There are even more serious problems in the definition of the space fractional derivative. The definition of the space fractional calculus widely used in numerical computations is the definition of Grünwald–Letnikov and Riesz–Feller, followed by the definition of Riemann–Liouville. The multi-dimensional space fractional definition, such as the fractional Laplacian definition, is complicated and has not been applied to numerical solution research on differential equations. It can be said that fractional calculus will be more popular in the natural scientific fields only when the perfect definition of fractional calculus is proposed. It needs to be noted that the nature of fractional calculus has not been fully explored, such as Fourier transform, Laplace transform, the relationship and distinction between fractional calculus and integer-order calculus and so on.

## 2. Application in the field of engineering, physics and mechanics

As a novel mathematical tool, academic articles on the applications of fractional calculus are showing a "blowout" growth trend, but how to apply it to solve the real-world physical, mechanical, biological, signal processing and materials discipline problems is still a long way to go.

In the physical field, the concept of fractional quantum mechanics has been proposed, but theoretical research is still in its infancy. As now fractional Schrödinger equation, fractional derivative model-related power-law phenomenon has been widely concerned, but in-depth studies still need more efforts. In addition, fractional Brownian motion has very in-depth research, and its associated anomalous diffusion phenomena are also showing a flourishing trend. However, how to combine these theories with practical experiments or how to apply theory to guide the engineering practices encountered still needs more effort.

In the mechanics fields, fractional calculus theory has been widely used in the modeling of anomalous mechanics behaviors, for example, the constitutive relationship of viscoelasticity, seepage in complex media and turbulence transport. But most theories are phenomenological models. How to establish a new theoretical framework based on fractional theory is a challenging task.

In the area of signal processing, the idea of the fractional-order derivative has been introduced into the field of digital image processing and data processing but is not mature. As for the relationship between fractal and fractional calculus, although some scholars have proposed some qualitative links between them, these conclusions have yet to be fully proofed. How to combine fractional calculus with fractal theory to solve some engineering problems or explain specific natural phenomena is still the problem that needs to be worked on.

Obviously, there are still a lot of obstacles when fractional calculus theories are introduced to various engineering and physical, mechanical research areas at this stage.

## 3. Numerical problems for long-time process and large computational domain in space

The main numerical method for fractional calculus is the finite difference algorithm. However, there are still some key difficulties confronted with the numerical computation for the long-time process or large computational domain.

Firstly, for finite difference algorithms, the computation cost increases exponentially with time, due to the integral definition of fractional calculus. Although some scholars have proposed a number of short-term memory algorithms, it has been proven that these algorithms are not universal, and only work for some special cases. Therefore, it is still an unsolved issue to solve the time-fractional equations by finite difference algorithm for a long time range. Meanwhile, the numerical computation of the space fractional derivative equations with a large computational domain also encountered the same and even more serious problems. Because of the global correlation property of the space fractional calculus, there are fewer numerical algorithms for space fractional derivative equations. In addition, the two-dimensional and three-dimensional fractional derivative definitions have not been well introduced in papers. This is also a serious obstacle to developing the numerical schemes for space fractional differential equations.

4.  **The intrinsic links among fractional calculus, statistical mechanics and physical random phenomena**

Obviously, there is a close relationship among fractional calculus, statistical mechanics and physical random phenomena. Some scholars have found that a lot of statistical mechanics problems can be dealt with using statistical methods or fractional stochastic equations. And the results of the two different methods should be consistent. It requires further investigations for links between the fractional calculus theory and statistical distribution, to find the bonding point of the quantitative relationship between the differential equations and the statistical distribution. Especially in the aspects of quantum mechanics, fractional calculus has been applied to explain some random phenomena, but in-depth study is still a daunting task.

5.  **The development of application software to solve practical engineering problems**

The numerical computation of fractional calculus is still in the exploratory stage, and there is a great distance to mature applications. Now, some mature numerical discretization programs can be downloaded from the network, such as from the personal websites by I. Podlubny, Y. Q. Chen, M. M. Meerschaert, etc. Numerical programs on computing some special functions, signal processing and anomalous diffusion equation are released on MATLAB Central website; refer to Appendix II. It can be said that although the emergence of the application software used for numerical computation still requires a lot of research work to do, now you can hear its footsteps.

## 7.2 Perspectives

In the roundtable discussion on the international conference on fractional calculus
and its applications held in Turkey in 2008, some scholars have proposed several
unresolved issues on fractional calculus and its application. This book will include
them as supplementary to the previous chapters and discussions.

**O. P. Agrawal:**

(1) The analytical and numerical methods for the equations with the presence of
left- and right-side fractional calculus operators at the same time;
(2) The programming and application of fractional controller;
(3) Fractional optimal control;
(4) The new fractional variational method and its applications.

**T. Machado:**

(1) The simple applications of fractional calculus;
(2) The development of industrial products with respect to fractional calculus.

**R. Nigmatullin:**

(1) Whether it is possible to derive the Newtonian equation with memory

$$
m\frac{\mathrm{d}^2\overline{r}}{\mathrm{d}t^2} = \int_0^t \kappa(t-\tau)\overline{F}(\overline{r}, \overline{v}, \tau)\mathrm{d}\tau.
$$

(2) How to link the power law and thermal dynamics $v(p,T,v)$?

**CRONE research team:**

(1) To extend the concept of fractional calculus in the field of nonlinear systems;
(2) To establish the baseline of the test;
(3) The development of the industrial applications of fractional calculus.

**D. Baleanu:**

(1) To establish a consistent fractional quantum mechanics, quantum field theory
and fractional differential geometry;
(2) Delayed fractional variational principle;
(3) Experimental evidence and performance of fractional dynamics in complex
systems.

**T. Kaczorek:**

(1) Positive fractional system (1-D and 2-D systems);
(2) Realization of positive system:

Known: transform matrix $T(s)$ and $T(z)$.

Solving: morphological space description.

Definition: if $x_0 \in R_+^n$ and $u \in R_+^m$, then $x \in R_+^n$ and $y \in R_+^p$, so we call the system

$$\begin{cases} {}_0D_t^\alpha x(t) = Ax(t) + Bu(t) \\ y = Cx(t) + Du_k \end{cases}$$

is positive, where $x \in R_+^n$, if and only if $x_i \in R_+$, $i = 1, \cdots, n$.

(3)   Predictive control of the fractional system (chemical industry);

(4)   Adaptive control of the fractional system.

# Appendix A
# Special Functions

Because some special functions are often involved in the fractional calculus research, to fully understand the knowledge of fractional calculus, here we introduce six related special functions. For more details, refer to Refs. [1, 2].

## *Gamma function*

The Gamma function $\Gamma(z)$ is generally defined as

$$\Gamma(z) = \int_0^\infty e^{-t} t^{z-1} dt \qquad (A1.1)$$

which is applied in the right half of the complex plane $Re(z) > 0$ and guarantees the integral convergent at $t = 0$. This definition is also known as the second category Euler integration, which is often used in practical applications and can be further extended to the whole complex plane. Other forms of the definition (e.g. Euler's infinite series expressions, Weierstrass' infinite series, etc.) will be briefly presented in the last part of this section.

The sign of Gamma function $\Gamma(z)$ is used in most cases; in addition, there are two other signs $\Pi(z)$ and $z!$, which are both equal to $\Gamma(z+1)$, $z! = \Pi(z) = \Gamma(z+1)$. The sign $z!$ is normally used only in the case of a positive integer $z$, but is not restricted in this book. Thus, Eq. (A1.1) can be understood as the promotion of any real number $z$, non-integer and even complex. Figure A.1 shows the Gamma function graphics. It is easy to observe from Fig. A.1 that there are singularities of Gamma function when $z = 0, -1, -2, \ldots, -n, \ldots$.

1. **Basic Properties**

$\Gamma(z)$ satisfies the following recurrence relations:

© Science Press 2022
W. Chen et al., *Fractional Derivative Modeling in Mechanics and Engineering*,
https://doi.org/10.1007/978-981-16-8802-7

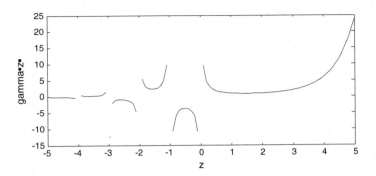

**Fig. A.1** Gamma function

$$\Gamma(z+1) = z\Gamma(z) \tag{A1.2}$$

which can be easily proved by integrating by parts:

$$\Gamma(z+1) = \int_0^\infty e^{-t} t^z dt = (-e^{-t} t^z)_{t=0}^{t=\infty} + z\int_0^\infty e^{-t} t^{z-1} dt = z\Gamma(z). \tag{A1.3}$$

If $z$ is assumed to be a positive integer, formula (A1.1) can be generalized as

$$\Gamma(z+n) = (z+n-1)(z+n-2)\ldots(z+1)z\Gamma(z), \tag{A1.4}$$

or

$$\Gamma(z) = \frac{\Gamma(z+n)}{z(z+1)\ldots(z+n-1)} = \frac{1}{(z)_n}\int_0^\infty e^{-t} t^{z+n-1} dt, \tag{A1.5}$$

where $(z)_n = z(z+1)\ldots(z+n-1)$.

Obviously, Eq. (A1.5) extends the definition of $\Gamma(z)$ to $Re(z) > -n$, where $n$ is an arbitrary positive integer.

In Equation (A1.1), let $z = 1$, we have

$$\Gamma(1) = 0! = \int_0^\infty e^{-t} dt = 1. \tag{A1.6}$$

In Eq. (A1.4), let $z = 1$, we have $\Gamma(n+1) = n! = n(n-1)\ldots2\cdot1$. This shows when $z$ is a positive integer, $\Gamma(n+1)$ is the factorial $n!$.

## 2. Euler's Infinite Product Formula

According to the limit relation $e^{-t} = \lim\limits_{n \to \infty} (1 - t/n)^n$, the Gamma function $\Gamma(z)$ can be expressed as the limit of the following integration:

$$P_n(z) = \int_0^n (1 - t/n)^n t^{z-1} dt, \tag{A1.7}$$

where the proof is omitted here. Let $t = n\tau$, integrating $P_n(z)$ by parts for n times, thus,

$$P_n(z) = n^z \int_0^1 (1 - \tau)^n \tau^{z-1} d\tau$$

$$= n^z \left[ \frac{\tau^z}{z} (1 - \tau)^n \right]_0^1 + \frac{n^z \cdot n}{z} \int_0^1 (1 - \tau)^{n-1} \tau^z d\tau$$

$$= \cdots = \frac{n^z n(n - 1) \cdots 2 \cdot 1}{z(z + 1) \cdots (z + n - 1)} \int_0^1 \tau^{z+n-1} d\tau$$

$$= \frac{1 \cdot 2 \cdots n}{z(z + 1) \cdots (z + n)} n^z, \tag{A1.8}$$

that is, $\Gamma(z) = \lim\limits_{n \to \infty} \frac{n! n^z}{z(z+1)\ldots(z+n)}$. Because $\lim\limits_{n \to \infty} n/(z + n) = 1$, this formula can be rewritten as

$$\Gamma(z) = \lim_{n \to \infty} \frac{(n - 1)! n^z}{z(z + 1) \ldots (z + n - 1)}. \tag{A1.9}$$

And because $n^z$ can be written as $n^z = \prod\limits_{m=1}^{n-1} \left(1 + \frac{1}{m}\right)^z$, moreover,

$$\frac{(n - 1)!}{z(z + 1) \ldots (z + n - 1)} = \frac{1}{z} \prod_{m=1}^{n-1} \left(1 + \frac{z}{m}\right)^{-1}. \tag{A1.10}$$

In the end, we obtain another form of expression of the Gamma function, which is Euler's infinite series formula.

$$\Gamma(z) = \frac{1}{z} \prod_{n=1}^{\infty} \left\{ \left(1 + \frac{z}{n}\right)^{-1} \left(1 + \frac{1}{n}\right)^z \right\}. \tag{A1.11}$$

The Weierstrass infinite series can be expressed as

$$\frac{1}{\Gamma(z)} = ze^{\gamma z} \prod_{n=1}^{\infty} \left\{ \left(1 + \frac{z}{n}\right)e^{-z/n} \right\}, \tag{A1.12}$$

where $\gamma$ is the Euler constant

$$\gamma = \lim_{n \to \infty} \left\{ \sum_{m=1}^{n} \frac{1}{m} - \ln n \right\} = 0.577\,215\ldots. \tag{A1.13}$$

For the specific derivation, refer to [1].

3. **Important Properties**

   (1)  When $z \to 0^+$, $\Gamma(z) \to +\infty$.
   (2)  Euler's reflection formula: $\Gamma(z)\Gamma(1-z) = \frac{\pi}{\sin(\pi z)}$.
   (3)  $\Gamma(n+1/2) = \frac{\sqrt{\pi}\,(2n)!}{2^{2n}n!}$.
   (4)  Multiplication theorem:

$$\Gamma(z)\Gamma(z + 1/2) = 2^{1-2z}\sqrt{\pi}\,\Gamma(2z),$$

$$\Gamma(z)\Gamma\left(z + \frac{1}{m}\right)\Gamma\left(z + \frac{2}{m}\right)\ldots\Gamma\left(z + \frac{m-1}{m}\right)$$
$$= (2\pi)^{(m-1)/2}m^{1/2-mz}\Gamma(mz).$$

4. **Special value**

$$\Gamma\left(\frac{-3}{2}\right) = \frac{4\sqrt{\pi}}{3},\; \Gamma\left(\frac{-1}{2}\right) = -2\sqrt{\pi},$$

$$\Gamma\left(\frac{1}{2}\right) = \sqrt{\pi},\, \Gamma(1) = 0! = 1,$$

$$\Gamma\left(\frac{3}{2}\right) = \frac{\sqrt{\pi}}{2},\, \Gamma(2) = 1,$$

$$\Gamma\left(\frac{5}{2}\right) = \frac{3\sqrt{\pi}}{4},\, \Gamma(3) = 2!,$$

$$\Gamma\left(\frac{7}{2}\right) = \frac{15\sqrt{\pi}}{8},\, \Gamma(4) = 3!.$$

## Beta Function

Beta function, also known as the first-class Euler integration, is another special function defined as

$$B(z, w) = \int_0^1 \tau^{z-1}(1 - \tau)^{w-1}d\tau, \tag{A2.1}$$

where the above equation needs to satisfy the condition $Re(z) > 0$, $Re(w) > 0$.

### Basic Properties

(1) Beta function is symmetrical, which can be proved by the variable transformation

$$B(z, w) = B(w, z). \tag{A2.2}$$

Beta function has many other forms, including

$$B(z, w) = \frac{\Gamma(z)\Gamma(w)}{\Gamma(z + w)}, \tag{A2.3}$$

$$B(z, w) = 2 \int_0^{\pi/2} (\sin \theta)^{2z-1}(\cos \theta)^{2w-1}d\theta, \quad Re(z) > 0, Re(w) > 0, \tag{A2.4}$$

$$B(z, w) = 2 \int_0^\infty \frac{t^{z-1}}{(1 + t)^{z+w}}dt, \quad Re(z) > 0, Re(w) > 0, \tag{A2.5}$$

$$B(z, w) = \sum_{n=0}^\infty \frac{\binom{n - w}{n}}{z + n}, \tag{A2.6}$$

$$B(z, w) = \prod_{n=0}^\infty \left(1 + \frac{zw}{n(z + w + n)}\right)^{-1}, \tag{A2.7}$$

$$B(z, w) \cdot B(z + w, 1 - w) = \frac{\pi}{z \sin(\pi z)}, \tag{A2.8}$$

$$B(z, w) = \frac{1}{w}\sum_{n=0}^\infty (-1)^n \frac{y^{n+1}}{n!(z + n)}. \tag{A2.9}$$

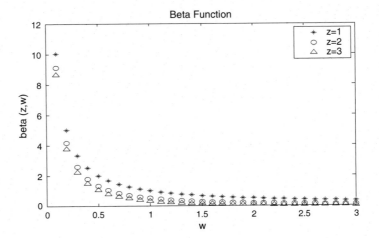

**Fig. A.2** Beta function

(2)   From the relation between the Beta function and Gamma function, it is easy to obtain the following equation:

$$B(z, 1 - z) = \Gamma(z)\Gamma(1 - z), \qquad (A2.10)$$

$$B(z, z) = 2^{1-2z} B(z, 1/2). \qquad (A2.11)$$

Figure A.2 shows the Beta function changes with variable $w$ under three selected different constants $z$

Next, the specific proof of the relationship between the Beta function and Gamma function will be given in the following part.

Considering the following equation:

$$\Gamma(z)\Gamma(w) = \int_0^\infty e^{-u} u^{z-1} \mathrm{d}u \int_0^\infty e^{-v} v^{w-1} \mathrm{d}v, \qquad (A2.12)$$

let $u = x^2$, $v = y^2$,

thus,

$$\Gamma(z)\Gamma(w) = 4 \int_0^\infty e^{-x^2} x^{2z-1} \mathrm{d}x \int_0^\infty e^{-y^2} y^{2w-1} \mathrm{d}y,$$

$$= 4 \int_0^\infty \int_0^\infty e^{-(x^2+y^2)} x^{2z-1} y^{2w-1} \mathrm{d}x\mathrm{d}y. \qquad (A2.13)$$

Introducing plane polar coordinates: $x = r \cos \theta$, $y = r \sin \theta$, the above equation changes into the following form:

$$\Gamma(z)\Gamma(w) = 4 \int_0^\infty e^{-r^2} r^{2(z+w)-1} dr \int_0^{\pi/2} (\cos \theta)^{2z-1} (\sin \theta)^{2w-1} d\theta. \qquad \text{(A2.14)}$$

In the first integration, let $r^2 = t$, then

$$\int_0^\infty e^{-r^2} r^{2(z+w)-1} dr = \frac{1}{2} \int_0^\infty e^{-t} t^{z+w-1} dt = \frac{1}{2} \Gamma(z+w); \qquad \text{(A2.15)}$$

in the second integration, let $\cos^2 \theta = x$, then

$$\int_0^{\pi/2} (\cos \theta)^{2z-1} (\sin \theta)^{2w-1} d\theta = \frac{1}{2} \int_0^1 x^{z-1} (1-x)^{w-1} dx = \frac{1}{2} B(z, w). \qquad \text{(A2.16)}$$

Substituting the above two equations into Eq. (A2.14), the relationship between the Gamma function and Beta function is obtained as $\Gamma(z)\Gamma(w) = \Gamma(z+w)B(z, w)$.

## Dirac Delta Function

Dirac delta function is a special function widely used in the physical realm. It is a great help for the analysis of physical problems. In particular, it is often used in the analysis of problems, such as diffusion, seepage and wave. A brief description of the function will be given in the following part.

1. The expression of the Dirac delta function

$$\delta(x) = \begin{cases} \infty, x = 0 \\ 0, x \neq 0 \end{cases}. \qquad \text{(A3.1)}$$

2. The properties of the Dirac delta function:

   (1) Integral property: $\int_{-\infty}^{+\infty} \delta(x) dx = 1$
   (2) Fourier transform properties: $F(1) = \delta(x)$; $F^{-1}(\delta(x)) = 1$.
   (3) Limit Properties:

$$\lim_{\sigma \to 0} \frac{1}{\sqrt{2\pi}\sigma} \exp\left(-\frac{x^2}{2\sigma}\right) = \delta(x), \qquad \text{(A3.2)}$$

$$\lim_{\alpha \to \infty} \sqrt{\frac{\alpha}{\pi}} \exp(-\alpha x^2) = \delta(x), \tag{A3.3}$$

$$\lim_{\alpha \to \infty} \sqrt{\frac{\alpha}{\pi}} e^{i\pi/4} e^{-i\alpha x^2} = \delta(x), \tag{A3.4}$$

$$\lim_{\alpha \to \infty} \frac{\sin(\alpha x)}{\pi x} = \delta(x), \quad \lim_{\alpha \to \infty} \frac{\sin^2(\alpha x)}{\pi \alpha x^2} = \delta(x), \tag{A3.5}$$

$$\lim_{\varepsilon \to 0} \frac{1}{2\varepsilon} e^{-|x|/\varepsilon} = \delta(x), \tag{A3.6}$$

$$\lim_{\varepsilon \to 0} \frac{\varepsilon}{x^2 + \varepsilon^2} = \pi \delta(x), \tag{A3.7}$$

$$\sqrt{i} = \exp\left(\frac{i\pi}{4}\right), \tag{A3.8}$$

$$\frac{1}{2\pi} \int_{-\infty}^{+\infty} \exp(ikx)dk = \delta(x). \tag{A3.9}$$

**Proof** $\frac{1}{2\pi} \int_{-\infty}^{+\infty} \exp(ikx)dk = \frac{\sin(\alpha x)}{\pi x} = \lim_{\alpha \to \infty} \frac{\sin(\alpha x)}{\pi x} = \delta(x).$

(4)    The properties of the derivative:

Step function: $\theta(x) = \begin{cases} 1 \ x > 0 \\ 0 \ x < 0 \end{cases} \quad \theta'(x) = \delta(x).$

$$\delta'(-x) = -\delta'(x), \quad \delta^{(n)}(-x) = (-1)^n \delta^{(n)}(x),$$

$$\frac{d(\ln x)}{dx} = \frac{1}{x} - i\pi\delta(x).$$

(5)    Because $\delta(x)$ is an even function, there exists the following result:

$$\int_0^\infty \delta(x)dx = \frac{1}{2}.$$

(6)    Convolution Properties:

$$\delta(ax) = \frac{1}{|a|}\delta(x), \tag{A3.10}$$

$$\int\limits_{-\infty}^{+\infty} f(x)\delta(x-a)\mathrm{d}x = f(a), \tag{A3.11}$$

$$\delta(x-x') = \sum_{n=0}^{\infty} \frac{1}{\sqrt{\pi}2^n n!} e^{-(x^2+x'^2)/2} H_m(x') H_m(x). \tag{A3.12}$$

## *Mittag-Leffler Function*

For a long time, the Mittag-Leffler (M-L) function, especially the generalized (two-parameter) M-L function is not familiar to the public. In fact, Mathematics Subject Classification in 1991 even didn't include the introduction of the M-L function and related content, and the American Mathematical Society (AMS) classification forecast doesn't have its new entry (33E12) until 2000. In recent years, the generalized M-L function has been widely applied in the study of fractal dynamics, fractional anomalous diffusion and fractal random field [1, 3, 4] and coherent states in quantum field theory [5]. The application in these areas, in turn, promotes the development of the study of the function. For example, in the study of the theory of generalized fractional calculus, a recently developed multi-index Mittag-Leffler function has obtained a full use [6].

Exponential function $e^z$ plays an important role in the integer-order differential equation; it can be written in the form of a series: $e^z = \sum_{k=0}^{\infty} \frac{z^k}{\Gamma(k+1)}$; it is a special case of the single-parameter Mittag-Leffler function. The function $E_\alpha(z)$ was proposed by G. M. Mittag-Leffler, and A. Wiman also did some research on this function. The generalized Mittag-Leffler function has a very important role in the fractional calculus, and it is derived from solving fractional differential equations using the Laplace transform by Humbert and Agarwal.

1. **The definition of the Mittag-Leffler function**

   (1) Single-parameter Mittag-Leffler function:

   $$E_\alpha(x) = \sum_{k=0}^{\infty} \frac{x^k}{\Gamma(\alpha k + 1)}, \quad \alpha > 0. \tag{A4.1}$$

Figure A.3 shows the image of the Mittag-Leffler function of several special cases. The value of the parameter $\alpha$ is 0, 1, 2, 3, 4 and 5.

When $\alpha = 0$, $E_0(x) = \sum_{k=0}^{\infty} x^k = \frac{1}{1-x}.x = 1$ is the singular point of the function. When $x \in [-50, 1)$, the function value is a positive number, and the function value increases with $x$. When $x \to 1^-$, function value tends to $+\infty$. When $x \in (1, 10]$, the function value is negative, and when $x \to 1^+$, the function value tends to $-\infty$. When $\alpha = 1$, M-L function is the exponential function $e^x$, so we can say that the exponential function is a special case of the M-L function. Taking other different

**Fig. A.3** Single-parameter
Mittag-Leffler function [29]

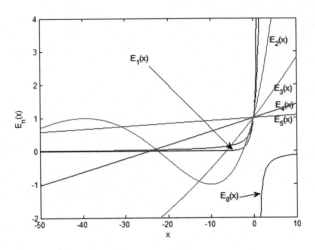

values of $\alpha$, the function represented by the M-L function will be listed below:

$$\alpha = 2,\ E_2(x) = \sum_{k=0}^{\infty} \frac{x^k}{\Gamma(2k+1)} = \cosh(\sqrt{x}),$$

$$\alpha = 3,\ E_3(x) = \sum_{k=0}^{\infty} \frac{x^k}{\Gamma(3k+1)} = \frac{1}{3}\left[e^{x^{1/3}} + 2e^{-x^{1/3}/2}\cos\left(\frac{1}{2}\sqrt{3}z^{1/3}\right)\right],$$

$$\alpha = 4,\ E_4(x) = \sum_{k=0}^{\infty} \frac{x^k}{\Gamma(4k+1)} = \frac{1}{2}\left[\cos(x^{1/4}) + \cosh(z^{1/4})\right],$$

For single-parameter Mittag-Leffler function, let $z = -t^\beta$, thus,

$$E_\beta(-t^\beta) = \sum_{k=0}^{\infty} \frac{(-t^\beta)^k}{\Gamma(\beta k + 1)}.$$

The function has the following limiting form:

When $\beta = 1$, the Mittag-Leffler function degrades to an exponential decay function $e^{-t}$.

When $0 < \beta < 1$, if $t \to 0$, the Mittag-Leffler function can be approximated by the extended exponential decay function: $\exp(-t^\beta/a)$, $a = \Gamma(\beta + 1)$; if $t \to \infty$, the Mittag–Leffler function can be approximated by a power function: $bt^{-\beta}$, $b = \Gamma(\beta)\sin(\beta\pi)/\pi$ (see Fig. A.4).

(2)  Two-parameter (generalized) Mittag-Leffler function [1]:

$$E_{\alpha,\beta}(z) = \sum_{k=0}^{\infty} \frac{z^k}{\Gamma(\alpha k + \beta)}, \quad \alpha > 0,\ \beta > 0. \tag{A4.2}$$

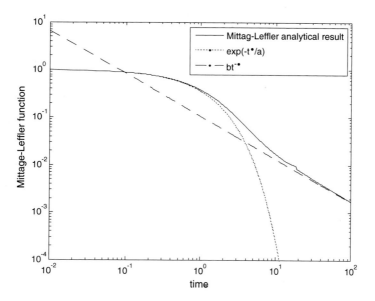

**Fig. A.4** The approximation of the Mittag-Leffler function

From the definition of the generalized M-L function, it is not difficult to find that single-parameter M-L function is its special case (when $\beta = 1$); therefore, Fig A.3 can be regarded as a graphic of the generalized M-L function when $\beta = 1$.

Considering several special cases:

(1)  $\alpha = 1, \beta = 1$,

$$E_{1,1}(z) = e^z.$$

(2)  If $\alpha = \frac{1}{2}, \beta = 1$,
By definition, it can be obtained that

$$E_{1/2,1}(z) = \sum_{k=0}^{\infty} \frac{z^k}{\Gamma(k/2 + 1)} = e^{z^2} erfc(-z),$$

$$erfc(z) = \frac{2}{\sqrt{\pi}} \int_{z}^{\infty} e^{-t^2} dt,$$

where $erfc()$ is the error function.

(3)  When $\beta \neq 1$,

$$E_{1,2}(z) = \frac{e^z - 1}{z}.$$

(3)    Generalized M-L function with changed parameters

Let $z = 1$ and $\beta = 1$, studying the function changes with parameter $\alpha$. It is shown in Fig A.5, with the increase of parameter $\alpha$, the value of the function at the same point reduces, i.e. the smaller $\alpha$, the greater function value.

Let $z = 1$ and $\alpha = 1$, considering that the function changes with parameter $\beta$. It is shown in Fig A.6, for a fixed $z$, M-L function decreases with the increase in $\beta$. The decreasing rate is smaller than that in Fig A.5, and it decreases slowly at first, then increases with the increase in $\beta$.

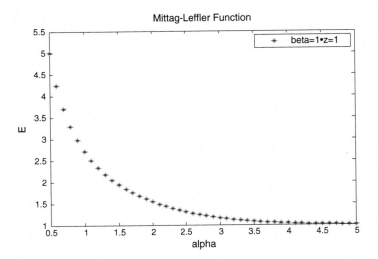

**Fig A.5**  When $\beta = 1$ and $z = 1$, M-L function changes with $\alpha$

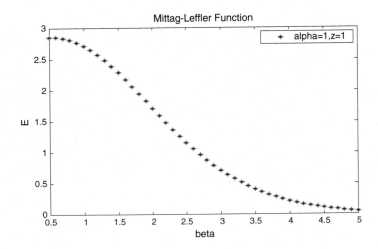

**Fig A.6**  When $\alpha = 1$ and $z = 1$, M-L function changes with $\beta$

(4)   Some special functions

Some special functions are introduced below, which can all be expressed as the form containing the M-L function [1].

(1)   The Miller–Ross function

$$\varepsilon_t(v, a) = t^v \sum_{k=0}^{\infty} \frac{(at)^k}{\Gamma(v + k + 1)} = t^v E_{1, v+1}(at),$$

(2)   The Rabotnov function

$$\Im_\alpha(\beta, t) = t^\alpha \sum_{k=0}^{\infty} \frac{\beta^k t^{k(\alpha+1)}}{\Gamma((k + 1)(1 + \alpha))} = t^\alpha E_{\alpha+1, \alpha+1}(\beta t^{\alpha+1}),$$

(3)   A class of fractional sine and cosine functions

$$Sc_\alpha(z) = \sum_{n=0}^{\infty} \frac{(-1)^n z^{(2-\alpha)n+1}}{\Gamma((2 - \alpha)n + 2)} = z E_{2-\alpha, 2}(-z^{2-\alpha}),$$

$$Cs_\alpha(z) = \sum_{n=0}^{\infty} \frac{(-1)^n z^{(2-\alpha)n}}{\Gamma((2 - \alpha)n + 1)} = E_{2-\alpha, 1}(-z^{2-\alpha}),$$

These two functions proposed by Plotnikov and Tseytlin are called "fractional sine and cosine functions".

(4)   Another class of fractional sine and cosine functions

$$\sin_{\lambda, \mu}(z) = \sum_{n=0}^{\infty} \frac{(-1)^k z^{2k+1}}{\Gamma(2\mu k + 2\mu - \lambda + 1)} = z E_{2\mu, 2\mu-\lambda+1}(-z^2),$$

$$\cos_{\lambda, \mu}(z) = \sum_{n=0}^{\infty} \frac{(-1)^k z^{2k}}{\Gamma(2\mu k + \mu - \lambda + 1)} = E_{2\mu, \mu-\lambda+1}(-z^2),$$

These two functions are proposed by Luchko and Srivastava; they also can be expressed in the form of M-L function.

(5)   Double M-L function

$$\xi_{\alpha, \beta, \lambda, \mu}^{v, \sigma}(x, y) = \sum_{m=0}^{\infty} \sum_{n=0}^{\infty} \frac{x^{m + \frac{\beta(vn+1)-1}{\alpha}} y^{n + \frac{\mu(\sigma m+1)-1}{\lambda}}}{\Gamma(m\alpha + (vn + 1)\beta)\Gamma(n\lambda + (\sigma m + 1)\mu)},$$

This function is proposed by P. Hubert, P. Delerue and A. M Chak and further expanded by H. M Srivastava.

(5)    M-L functions with multi-parameter

$$E_{(\alpha_1,\ldots,\alpha_m),\beta}(z_1,\ldots,z_m) = \sum_{k=0}^{\infty} \sum_{\substack{l_1+\cdots+l_m=k \\ l_1>0,\ldots,l_m>0}} \frac{(k;l_1,\ldots,l_m)\prod_{i=1}^{m} z_i^{l_i}}{\Gamma\left(\beta + \sum_{i=1}^{m} \alpha_i l_i\right)},$$

where $(k;l_1,\ldots,l_m)$ denotes polynomial coefficients. This function was originally proposed by Hadid and Luchko for solving fractional differential equations with linear constant coefficient.

## 2.    The Laplace transform of the generalized Mittag-Leffler function

The Laplace transform of the generalized Mittag-Leffler function plays an important role in solving fractional differential equations. And the inverse Laplace transforms of this function are always applied to get the analytical solution of some simple fractional-order equations.

The derivation of the Laplace transforms of the generalized M-L function will be given in the following section.

Firstly, substituting $e^{\pm zt}$ in the integration $\int_0^{\infty} e^{-t}e^{\pm zt}dt$ with the Mittag-Leffler function, then performing integral with respect to $t$, lastly, we have

$$\int_0^{\infty} e^{-t}e^{\pm zt}dt = \int_0^{\infty} e^{-t}\left(\sum_{k=0}^{\infty} \frac{(\pm zt)^k}{k!}\right)dt,$$

$$= \sum_{k=0}^{\infty} \frac{(\pm z)^k}{k!}\int_0^{\infty} e^{-t}t^k dt,$$

According to the definition of Gamma function $\int_0^{\infty} e^{-t}t^k dt = \Gamma(k+1) = k!$, thus

$$\int_0^{\infty} e^{-t}e^{\pm zt}dt = \sum_{k=0}^{\infty} (\pm z)^k = \frac{1}{1\mp z}. \tag{A4.3}$$

Adding $t^k$ into the above integral term and performing the above integration again, thus

$$\int_0^{\infty} e^{-t}t^k e^{\pm zt}dt = \frac{k!}{(1\mp z)^{k+1}}, \quad (|z|<1). \tag{A4.4}$$

Making appropriate substitution to the equation, the Laplace transform of function $t^k e^{\pm at}$ is obtained as follows:

$$\int_0^\infty e^{-pt} t^k e^{\pm at}\,dt = \frac{k!}{(p \mp a)^{k+1}}, \quad (\mathrm{Re}(p) > |a|). \tag{A4.5}$$

According to Eqs. (A4.2) and (A1.1), the Laplace transform of generalized M-L function is considered as follows:

$$\int_0^\infty e^{-t} t^{\beta-1} E_{\alpha,\beta}(zt^\alpha)\,dt = \sum_{k=0}^\infty \frac{z^k}{\Gamma(\alpha k + \beta)} \int_0^\infty e^{-t} t^{\alpha k + \beta - 1}\,dt$$

$$= \frac{1}{1-z}, \quad (|z| < 1). \tag{A4.6}$$

Then, making the same transform to Eq. (A4.5), the Laplace transform of function $t^{\alpha t + \beta - 1} E_{\alpha,\beta}^{(k)}(\pm at^\alpha)$ is obtained as follows:

$$\int_0^\infty e^{-pt} t^{\alpha k + \beta - 1} E_{\alpha,\beta}^{(k)}(\pm at^\alpha)\,dt = \frac{k! p^{\alpha-\beta}}{(p^\alpha \mp a)^{k+1}}, \quad \left(\mathrm{Re}(p) > |a|^{1/\alpha}\right), \tag{A4.7}$$

where $E_{\alpha,\beta}^{(k)}(y) = \frac{d^k}{dy^k} E_{\alpha,\beta}(y)$.
Simplified as

$$L\left\{t^{k\alpha + \beta - 1} E_{\alpha,\beta}^{(k)}(\mp at^\alpha), p\right\} = \frac{k! p^{\alpha-\beta}}{\left(p^\alpha \pm a^{k+1}\right)}, \quad \mathrm{Re}(p) > |a|^{1/a}, \tag{A4.8}$$

where $E_{\alpha,\beta}^{(k)} = \frac{d^k}{dz^k} E_{\alpha,\beta}(z)$, the Ref. [7] has given a rigorous and simple proof of this formula.

In particular, if let $\alpha = \beta = \frac{1}{2}$, we can get

$$\int_0^\infty e^{-pt} t^{\frac{k-1}{2}} E_{\frac{1}{2},\frac{1}{2}}^{(k)}(\pm a\sqrt{t})\,dt = \frac{k!}{(\sqrt{p} \mp a)^{k+1}}, \quad \left(\mathrm{Re}(p) > |a|^2\right). \tag{A4.9}$$

It is noted here that Eq. (A4.9) is extremely useful for solving fractional derivative equation when the order of derivative equals 1/2.

3. **Derivative and integral of the M-L function**

(1) Derivative of the M-L function

For equations given in this section, the derivation process is no longer given, and only the conclusion is given, and the interested reader can try to derive it.

Because individually differentiating the M-L function is relatively complicated, the differential of the product of the M-L function and the power function of $t$ is generally considered. Choosing the definition of the Riemann–Liouville fractional differential, then

$$\frac{d^\gamma}{dt^\gamma}\left(t^{\alpha k+\beta-1}E_{\alpha,\beta}^{(k)}(\lambda t^\alpha)\right) = t^{\alpha k+\beta-\gamma-1}E_{\alpha,\beta-\gamma}^{(k)}(\lambda t^\alpha), \tag{A4.10}$$

Let $k = 0, \lambda = 1, \gamma$ is an integer and $m = \gamma$, thus Eq. (A4.10) can be rewritten as

$$\left(\frac{d}{dt}\right)^m\left(t^{\beta-1}E_{\alpha,\beta}(t^\alpha)\right) = t^{\beta-m-1}E_{\alpha,\beta-m}(t^\alpha), \ (m = 1, 2, 3, \ldots). \tag{A4.11}$$

Considering the following two cases of Eq. (A4.11):

(1)   When $\alpha = \frac{m}{n}$, and $m, n$ are natural numbers, thus

$$\left(\frac{d}{dt}\right)^m\left(t^{\beta-1}E_{m/n,\beta}(t^{m/n})\right) = t^{\beta-1}E_{m/n,\beta}(t^{m/n}) + t^{\beta-1}\sum_{k=1}^n \frac{t^{-\frac{m}{n}k}}{\Gamma(\beta-\frac{m}{n}k)}. \tag{A4.12}$$

If let $n = 1$, then

$$\left(\frac{d}{dt}\right)^m\left(t^{\beta-1}E_{m,\beta}(t^m)\right) = t^{\beta-1}E_{m,\beta}(t^m) + t^{\beta-1}\frac{t^{-m}}{\Gamma(\beta-m)}. \tag{A4.13}$$

According to the property of Gamma function,

$$\frac{1}{\Gamma(-v)} = 0, \ (v = 0, 1, 2, \ldots).$$

Let $m = 1, 2, 3, \ldots; \quad \beta = 0, 1, 2, \ldots, m$, we can obtain

$$\left(\frac{d}{dt}\right)^m\left(t^{\beta-1}E_{m,\beta}(t^m)\right) = t^{\beta-1}E_{m,\beta}(t^m). \tag{A4.14}$$

Substituting $t = z^{n/m}$ into Eq. (A4.11), thus

$$\begin{aligned}&\left(\frac{m}{n}z^{1-\frac{n}{m}}\frac{d}{dz}\right)^m\left(z^{(\beta-1)n/m}E_{m/n,\beta}(z)\right)\\&= z^{(\beta-1)n/m}E_{m/n,\beta}(z) + t^{(\beta-1)n/m}\sum_{k=1}^n \frac{z^{-k}}{\Gamma(\beta-\frac{m}{n}k)}\end{aligned}, \ (m, n = 1, 2, 3, \ldots), \tag{A4.15}$$

and let $m = 1$, we obtain

$$\frac{1}{n}\frac{d}{dz}\left(z^{(\beta-1)n}E_{1/n,\beta}(z)\right)$$
$$= z^{(\beta n-1)}E_{1/n,\beta}(z) + z^{\beta n-1}\sum_{k=1}^{n}\frac{z^{-k}}{\Gamma(\beta-\frac{k}{n})}, \quad (n = 1, 2, 3, \ldots). \qquad (A4.16)$$

(2) The integral of M-L function

Performing integral itemized to the left side of the following equation, thus we obtain

$$\int_0^z E_{\alpha,\beta}(\lambda t^\alpha)t^{\beta-1}dt = z^\beta E_{\alpha,\beta+1}(\lambda z^\alpha), \, \beta > 0, \qquad (A4.17)$$

then considering the integral

$$\frac{1}{\Gamma(v)}\int_0^z (z-t)^{v-1}E_{\alpha,\beta}(\lambda t^\alpha)t^{\beta-1}dt,$$

this integration is also relatively easy, and the integral can be solved as follows:
The above integral

$$\frac{1}{\Gamma(v)}\int_0^z (z-t)^{v-1}E_{\alpha,\beta}(\lambda t^\alpha)t^{\beta-1}dt$$

$$= \sum_{k=0}^{\infty}\frac{\lambda^k}{\Gamma(\alpha k+\beta)}\frac{1}{\Gamma(v)}\int_0^z (z-t)^{v-1}t^{\alpha k+\beta-1}dt$$

$$= \sum_{k=0}^{\infty}\frac{\lambda^k}{\Gamma(\alpha k+\beta)}\frac{1}{\Gamma(v)}\int_0^{\frac{\pi}{2}}\left(z-z\sin^2\theta\right)^{v-1}z\sin^2\theta^{\alpha k+\beta-1}d\left(z\sin^2\theta\right)$$

$$= z^{\beta+v-1}E_{\alpha,\beta+v}(\lambda z^\alpha). \quad \beta > 0, v > 0. \qquad (A4.18)$$

As a result, some special integral equations are obtained:

$$\frac{1}{\Gamma(\alpha)}\int_0^z (z-t)^{\alpha-1}e^{\lambda t}dt = z^\alpha E_{1,\alpha+1}(\lambda z), \quad (\alpha > 0)). \qquad (A4.19)$$

$$\frac{1}{\Gamma(\alpha)}\int_0^z (z-t)^{\alpha-1}\cosh(\sqrt{\lambda}t)dt = z^\alpha E_{2,\alpha+1}(\lambda z^2), \quad (\alpha > 0). \qquad (A4.20)$$

$$\frac{1}{\Gamma(\alpha)} \int_0^z (z-t)^{\alpha-1} \frac{\sinh(\sqrt{\lambda t})}{\sqrt{\lambda t}} dt = z^{\alpha+1} E_{2,\alpha+2}(\lambda z^2), \quad (\alpha > 0). \tag{A4.21}$$

The analytical solutions of fractional differential equations involve M-L function many times. Therefore, in order to compare the error relationship between the numerical solution and analytical solution, it is necessary to calculate the function values of M-L function. For this reason, many MATLAB Programs for calculating this function have been written by a lot of scholars until now, and readers can download them on the website [27, 28].

## Wright Function

The Wright function plays an important role in solving linear fractional partial differential equations, such as the wave equation. There are some connections between this function and the generalized M-L function. This function was first proposed by British mathematician Wright [8], and a large number of useful equality relations are derived from the Laplace transform of fractional differential equations summarized by Humbert and Agarwal [9].

Series form definition of the Wright function:

$$W(z; \alpha, \beta) = \sum_{k=0}^{\infty} \frac{z^k}{k! \Gamma(\alpha k + \beta)}, \tag{A5.1}$$

and Eq. (A5.1) can be written as the following integral form:

$$W(z; \alpha, \beta) = \frac{1}{2\pi i} \int_{Ha} \tau^{-\beta} e^{\tau + z t^{-\alpha}} d\tau, \tag{A5.2}$$

The Properties of the Wright function [10, 11].

**Property 1** *If* $\arg(-z) = \zeta, |\zeta| \leq \pi$, *and*

$$Z_1 = (\alpha |z|)^{1/(\alpha+1)} e^{i(\zeta+\pi)/(\alpha+1)}, \quad Z_2 = (\alpha |z|)^{1/(\alpha+1)} e^{i(\zeta-\pi)/(\alpha+1)},$$

*thus*

$$W(z; \alpha, \beta) = H(Z_1) + H(Z_2), \tag{A5.3}$$

*in which*

$$H(Z) = Z^{1/2-\beta} e^{\{1+(1/\alpha)\}z} \left\{ \sum_{m=0}^{M} \frac{(-1)^m a_m}{Z^m} + O\left(\frac{1}{|Z|^{M+1}}\right) \right\}, \quad Z \to \infty, \quad (A5.4)$$

where if the value m in Eq. (A5.4) is fixed, the value am can be calculated directly, e.g. $a_0 = (2\pi(\rho+1))^{-1/2}$.

**Property 2** *The relationship between the Wright Function and the Bessel function:*

$$J_v(z) = \left(\frac{z}{2}\right)^v W\left(-\frac{z^2}{4}; 1, v+1\right);$$
(A5.5)

$$I_v(z) = \left(\frac{z}{2}\right)^v W\left(\frac{z^2}{4}; 1, v+1\right).$$
(A5.6)

**Property 3** *The relationship between the Wright function and the Mittag–Leffler function [1]:*

$$L\{W(t; \alpha, \beta); s\} = s^{-1} E_{\alpha,\beta}(s^{-1}).$$
(A5.7)

**Property 4** *The relationship between the Wright function and the Meijer G-function.*
*When $\alpha$ is a rational number, and $\alpha = p/q$, the Wright function can be expressed with the Meijer G-function as follows:*

$$W(-z; \alpha, \beta) = (2\pi)^{(p-q)/2} q^{1/2} p^{-\beta+1/2}$$

$$\times G_{0,p+q}^{q,0} \left[ \frac{z^q}{q^q p^p} \middle| \begin{array}{c} - \\ 0, \frac{1}{q}, \frac{2}{q}, \dots, \frac{q-1}{q}, 1-\frac{\beta}{p}, 1-\frac{1+\beta}{p}, \dots, 1-\frac{p-1+\beta}{p} \end{array} \right].$$
(A5.8)

**Property 5** *The relationship between the Wright function and the Fox H-function.*
*When $\rho$ is an arbitrary positive number, the Wright function is a special case of the Fox H-function [12–14]:*

$$W(-z; \alpha, \beta) = H_{0,2}^{1,0} \left[ z \middle| \begin{array}{c} - \\ (0, 1), (1-\beta, \alpha) \end{array} \right].$$
(A5.9)

In addition, the Generalized Wright function generally can be expressed as

$$W(z; (\mu, a), (v, b)) = \sum_{k=0}^{\infty} \frac{z^k}{\Gamma(a+\mu k)\Gamma(b+vk)} . \mu, v \in R, a, b \in C. \quad (A5.10)$$

## *H-Fox Function*

H-Fox function is also called as the Fox function, H-function, the generalized Mellin–Barnes function or generalized Meijer's G-function in different papers. In order to unify and extend the existing results of the symmetric Fourier kernel, Fox has defined the H-function using the general Mellin–Barnes-type integral. It is widely used in the problems of statistics, physics and engineering to get the solution of fractional linear differential equations. It is necessary to note that almost all the special functions applied in the mathematical and statistical area are the special cases of H-Fox function. Even the complex functions such as the Mittag-Leffler function, Meijer's G-function [18], the Maitland generalized hypergeometric function and the Wright generalized Bessel functions are included. This section is compiled based primarily on the literature [2, 19–22].

H-Fox function based on the Mellin–Barnes-type integral is [14, 17, 23]

$$
H_{p,q}^{m,n}(z) = H_{p,q}^{m,n}\left[z\Big|_{(b_q,\beta_q)}^{(a_p,\alpha_p)}\right] = H_{p,q}^{m,n}\left[z\Big|_{(b_1,\beta_1),(b_2,\beta_2),\ldots,(b_q,\beta_q)}^{(a_1,\alpha_1),(a_2,\alpha_2),\ldots,(a_p,\alpha_p)}\right]
$$
$$
= \frac{1}{2\pi i}\int_L \chi(s)z^s \mathrm{d}s, z \neq 0, \tag{A6.1}
$$

in which integral density

$$
\chi(s) = \frac{A(s)B(s)}{C(s)D(s)},
$$

and

$$
A(s) = \prod_{j=1}^{m}\Gamma(b_j - \beta_j s), \quad B(s) = \prod_{j=1}^{n}\Gamma(1 - a_j + \alpha_j s),
$$
$$
C(s) = \prod_{j=m+1}^{q}\Gamma(1 - b_j + \beta_j s), D(s) = \prod_{j=n+1}^{p}\Gamma(a_j - \alpha_j s), \tag{A6.2}
$$

where $m, n, p$ and $q$ are non-negative integers, which satisfy $0 \leq n \leq p, 1 \leq m \leq q$. When $n = 0$, $B(s) = 1$; $q = m$, $C(s) = 1$; $p = n$, $D(s) = 1$. The parameters $a_j(j = 1, 2, \ldots, p)$ and $b_j(j = 1, 2, \ldots, q)$ are complex numbers, and $\alpha_j(j = 1, 2, \ldots, p)$ and $\beta_j(j = 1, 2, \ldots, q)$ are positive numbers. These parameters satisfy the conditions:

$$
P(A) \cap P(B) = \varnothing, \tag{A6.3}
$$

where

$$P(A) = \left\{ s = \frac{b_j + k}{\beta_j} \mid j = 1, \ldots, m; k = 0, 1, 2, \ldots \right\},$$

$$P(B) = \left\{ s = \frac{a_j - 1 - k}{\alpha_j} \mid j = 1, \ldots, n; k = 0, 1, 2, \ldots \right\},$$

are the sets of poles $A(s)$ and $B(s)$, respectively. The integral path $L$ is from $s = c - i\infty$ to $s = c + i\infty$ and makes the poles set separate. Then the points in $A(s)$ locate in the right of $L$, and the points in $B(s)$ locate in the left of $L$. Equation (A6.3) can also be written as $\alpha_j(b_h + \upsilon) \neq \beta_h(a_j - \lambda - 1)$, $(\upsilon, \lambda = 0, 1, 2, \ldots; h = 1, \ldots, m; j = 1, \ldots, n)$. Note that the path integral (A6.1) is the inverse Mellin transform of $\chi(s)$.

H-Fox function has the following important properties [15, 17, 24]:

**Property 1** H-Fox function has the property of permutation symmetry with respect to $(a_1, \alpha_1), \ldots, (a_n, \alpha_n), (a_{n+1}, \alpha_{n+1}), \ldots, (a_p, \alpha_p), (b_1, \beta_1), \ldots, (b_m, \beta_m)$ and $(b_{m+1}, \beta_{m+1}), \ldots, (b_q, \beta_q)$.

**Property 2** *If an element of the array in* $(a_j, \alpha_j)(j = 1, 2, \ldots, n)$ *equals an element of the array in* $(b_j, \beta_j)(j = m + 1, m + 2, \ldots, q)$ *[or an element of the array in* $(b_j, \beta_j)(j = 1, 2, \ldots, m)$ *equals an element of the array in* $(a_j, \alpha_j)(j = n + 1, n+2, \ldots, p)$*], the H-Fox function can be simplified to a low-level H-Fox function, namely subtract 1 from p, q and n (or m), respectively.*

Therefore, there is the simplification formula:

$$H_{p,q}^{m,n}\left[ z \Big|_{(b_1,\beta_1),(b_2,\beta_2),\ldots,(b_{q-1},\beta_{q-1}),(a_1,\alpha_1)}^{(a_1,\alpha_1),(a_2,\alpha_2),\ldots,(a_p,\alpha_p)} \right] = H_{p-1,q-1}^{m,n-1}\left[ z \Big|_{(b_1,\beta_1),\ldots,(b_{q-1},\beta_{q-1})}^{(a_2,\alpha_2),\ldots,(a_p,\alpha_p)} \right],$$

$$\text{(A6.4)}$$

where $n \geq 1, q > m$.

**Property 3**

$$H_{p,q}^{m,n}\left[ z \Big|_{(b_j,\beta_j)}^{(a_j,\alpha_j)} \right] = H_{q,p}^{n,m}\left[ \frac{1}{Z} \Big|_{(1-a_j,\alpha_j)}^{(1-b_j,\beta_j)} \right] \qquad \text{(A6.5)}$$

According to this nature, we can rewrite the Fox function under the condition of $\mu = \sum_{j=1}^{g} \beta_j - \sum_{j=1}^{p} \alpha_j < 0$ into another Fox function satisfying $\mu > 0$.

**Property 4**

$$H_{p,q}^{m,n}\left[ z \Big|_{(b_q,\beta_q)}^{(a_p,\alpha_p)} \right] = k H_{q,p}^{n,m}\left[ Z^k \Big|_{(b_q,k\beta_q)}^{(a_p,k\alpha_p)} \right], \qquad \text{(A6.6)}$$

where $k > 0$.

**Property 5**

$$z^\sigma H^{m,n}_{p,q}\left[z\Big|^{(a_p,\alpha_p)}_{(b_q,\beta_q)}\right] = H^{n,m}_{q,p}\left[z\Big|^{(a_{p+\sigma\alpha_p},\alpha_p)}_{(b_q+\sigma\beta_q,\beta_q)}\right].$$  (A6.7)

In order to discuss the analytic properties of the H-Fox function and asymptotic expansion, define the following symbols:

$$\mu = \sum_{j=1}^{q}\beta_j - \sum_{j=1}^{p}\alpha_j;$$  (A6.8)

$$\alpha = \sum_{j=1}^{n}\alpha_j - \sum_{j=n+1}^{p}\alpha_j + \sum_{j=1}^{m}\beta_j - \sum_{j=m+1}^{q}\beta_j;$$  (A6.9)

$$\beta = \prod_{j=1}^{p}\alpha_j^{\alpha_j}\prod_{j=1}^{q}\beta_j^{-\beta_j};$$  (A6.10)

$$\gamma = \sum_{j=1}^{q}b_j - \sum_{j=1}^{p}a_j + \frac{p-q}{2};$$  (A6.11)

$$\lambda = \sum_{j=1}^{m}\beta_j - \sum_{j=m+1}^{q}\beta_j - \sum_{j=1}^{p}\alpha_j;$$  (A6.12)

$$\delta = \left(\sum_{j=1}^{m}\beta_j - \sum_{j=n+1}^{p}\alpha_j\right)\pi.$$  (A6.13)

H-Fox function is the analytic function of $z$ and meaningful if the following existence conditions are met [15, 17, 24, 25]:

Situation 1 If $\mu > 0$, $z \neq 0$.
Situation 2 If $\mu = 0$, $0 < |z| < \beta^{-1}$.

Then generally, the H-Fox function is multivalued; however, it is single-valued in the Riemann surface of $\log z$ and can be obtained as follows:

$$H^{m,n}_{p,q}(z) = -\sum_{s\in P(A)}\text{Res}\left(\frac{A(s)B(s)}{C(s)D(s)}z^s\right).$$  (A6.14)

When the pole of the function $\prod_{j=1}^{m}\Gamma(b_j - \beta_j s)$ is a single pole, i.e. when $j \neq h$; $j, h = 1, 2, \ldots, m$; $\lambda, \upsilon = 0, 1, 2, \ldots$, $\beta_h(b_j + \lambda) \neq \beta_j(b_h + \upsilon)$, we get the H-Fox function as follows:

$$H_{p,q}^{m,n}(z) = \sum_{h=1}^{m} \sum_{v=0}^{\infty} \frac{\prod_{j=1, j \neq h}^{m} \Gamma\big(b_j - \beta_j(b_h + v)/\beta_h\big)}{\prod_{j=m+1}^{q} \Gamma\big(1 - b_j + \beta_j(b_h + v)/\beta_h\big)}$$
$$\times \frac{\prod_{j=1}^{n} \Gamma\big(1 - a_j + \alpha_j(b_h + v)/\beta_h\big)}{\prod_{j=n+1}^{p} \Gamma\big(a_j - \alpha_j(b_h + v)/\beta_h\big)} \times \frac{(-1)^v z^{(b_h+v)/\beta_i}}{v!\beta_h}. \qquad (A6.15)$$

Braaksma pointed out [17, 24, 26]

$$H_{p,q}^{m,n}(z) = O\big(|z|^c\big), z \leq 1, \qquad (A6.16)$$

where $\mu \geq 0$, $c = \min \Re\big(b_j/B_j\big)(j = 1, 2, \ldots, m)$;

$$H_{p,q}^{m,n}(z) = O\big(|z|^d\big), z \geq 1, \qquad (A6.17)$$

in which $\mu \geq 0$, $\alpha > 0$, $|\arg z| < \alpha\pi/2$, $d = \max \Re\left(\frac{a_j-1}{\alpha_j}\right)(j = 1, 2, \ldots, n)$.

Especially, if $\lambda > 0$, $|\arg z| < \lambda\pi/2$ and $\mu > 0$, then when $n = 0$, for the bigger $z$, H-Fox function tends to 0 exponentially, thus

$$H_{p,q}^{m,0}(z) \to O\big\{\exp\big(-\mu z^{1/\mu}\beta^{1/\mu}\big)z^{(\gamma+1/2)/\mu}\big\}. \qquad (A6.18)$$

If $n > 0$, $\delta > \mu\pi/2$, when $|z| \to \infty$, we get the H-Fox function at every closed subspace of $|\arg z| < \delta - \pi\mu/2$ as follows:

$$H_{p,q}^{m,n}(z) \to \sum_{s \in P(-1)} \text{Re } s\left(\frac{A(s)B(s)}{C(s)D(s)}z^s\right). \qquad (A6.19)$$

In addition, if $\omega$ and $\eta$ are both complex numbers, $\omega \neq 0$ and $\eta \neq 0$, $\mu > 0$ ($\mu$ is defined in Eq. (A6.8)), thus we get

$$H_{p,q}^{m,n}\left(\eta\omega\big|_{(b_q,\beta_q)}^{a_p,\alpha_p}\right) = \eta^{b_q/\beta_q} \sum_{r=0}^{\infty} \frac{\big(\eta^{1/\beta_q} - 1\big)^r}{r!}$$
$$\times H_{p,q}^{m,n}\left(\omega\big|_{(b_1,\beta_1),\ldots,(b_{q-1},\beta_{q-1}),(r+b_q,\beta_q)}^{(a_1,\alpha_1),\ldots,(a_p,\alpha_p)}\right), \qquad (A6.20)$$

where $q > m$, $\big|\eta^{1/\beta_q-1}\big| < 1$, $\arg(\eta\omega) = \beta_q \arg\big(\eta^{1/\beta_q}\big) + \arg(\omega)$, $\big|\arg\big(\eta^{1/\beta_q}\big)\big| < \pi/2$.

Some special cases of the Fox functions are discussed below. When $\alpha_j = 1(j = 1, 2, \ldots, p)$, $\beta_j = 1(j = 1, 2, \ldots, q)$, H-Fox function reduces to a Meijer G-function [2,18]:

$$H_{p,q}^{m,n}\left[z\big|_{(b_1,1),\ldots,(b_q,1)}^{(a_1,1),\ldots,(a_p,1)}\right] = G_{p,q}^{m,n}\left[z\big|_{b_1,\ldots,b_q}^{a_q,\ldots,a_p}\right]. \qquad (A6.21)$$

If adding other conditions $m = 1$ and $p \leq q$, the Fox function can be expressed as a generalized hypergeometric function $_pF_q$ as follows:

$$
H_{p,q}^{m,n}\left[ Z\big|_{(b_1,1),...,(b_q,1)}^{(a_1,1),...,(a_p,1)} \right] = \frac{\prod_{j=1}^{n} \Gamma\left(1 + b_1 - a_j\right) z^{b_1}}{\prod_{j=2}^{q} \Gamma\left(1 + b_1 - a_j\right) z^{b_1}}
$$
$$
\times_p F_{q-1}\left( \begin{array}{c} 1 + b_1 - a_1, \ldots, 1 + b_1 - a_p \\ 1 + b_1 - b_2, \ldots, 1 + b_1 - b_q \end{array}; (-1)^{p-n-1} z \right).
$$

$$(A6.22)$$

Many of the so-called special functions, such as the error function, the Bessel functions, the Whittaker functions, the Jacobi polynomials and elliptic integrals, are special cases of the generalized hypergeometric function.

An important H-Fox function not included in the G-function class is shown as follows:

$$
H_{p,q+1}^{1,p}\left[ Z\big|_{(0,1),(1-b_1,\beta_1)...,(1-b_q,\beta_q)}^{(1-a_1,a_1),\quad ...,(1-a_p,a_p)} \right] = \sum_{r=0}^{\infty} \frac{\prod_{j=1}^{p} \Gamma\left(a_j + \alpha_j r\right)}{\prod_{j=2}^{g} \Gamma\left(b_j + \beta_j r\right)} \times \frac{(-z)^r}{r!}
$$
$$
=_p \Psi_q\left( \begin{array}{c} (a_1,\alpha_1), \ldots, (a_p,\alpha_p) \\ (b_1,\beta_1), \ldots, (b_q,\beta_q) \end{array}; -z \right),
$$

$$(A6.23)$$

where $_p\Psi_q(z)$ is called the Maitland generalized hypergeometric function. A special case in Eq. (A6.23) gives the relationship between the H-fox function and the generalized Mittag-Leffler function $E_{\alpha,\beta}(z)$ as follows:

$$
H_{1,2}^{1,1}\left( z\big|_{(0,1),(1-\beta,\alpha)}^{(0,1)} \right) = E_{\alpha,\beta}(-z).
$$

$$(A6.24)$$

### References

1. I. Podlubny, *Fractional Differential Equations* [M] ( Academic Press, San Diego, 1999)
2. C.X. Wang, D.R. Guo, *Special Functions* [M] (Peking University Press, Beijing, 2000)
3. V.V. Anh, N.N. Leomenko. Scaling laws for fractional diffusion-wave equations with singular data [J]. Stat. Prob. Lett. **48**(3), 239–252 (2000)
4. Y.Q. Liu, *The Solution of Fractional Anomalous Diffusion Equation* [D] (Shandong University, 2006)
5. J.M. Sixdeniers, K.A. Penson, A.I. Solomon, Mittag-Leffler coherent states [J]. J. Phys. A Math. General **32**, 7543–7563 (1999)
6. V.S. Kiryakova, Multiple Mittag-Leffler functions and relations to generalized fractional calculus [J]. J. Comput. Appl. Math. **118**(1–2), 241–259 (2000)

7. M.Y. Xu, W.C Tan, Theoretical analysis of the generalized second-order fluid fractional anomalous diffusion velocity field, stress field and vortex layer [J]. Sci. China (Ser. A) **31**(7), 626–638 (2001)

8. E.M. Wright, On the coefficients of power series having exponential singularities [J]. J. Lond. Math. Soc. **8**, 71–79 (1933)

9. P. Humbert, R.P. Agarwal, Sur la function de Mittag-Leffler et quelques-unes de ses generalizations [J]. Bulletin des Sciences Mathématiques **77**(10), 180–185 (1953)

10. R. Gorenflo, Y. Luchko, F. Mainardi, Analytical properties and applications of the Wright function [J]. arXiv preprint math-ph/0701069 (2007)

11. R. Gorenflo, Y. Luchko, F. Mainardi, Wright function as scale-invariant solutions of the diffusion-wave equation [J]. J. Comput. Appl. Math. **118**, 175–191 (2000)

12. R. Gorenflo, F. Mainardi, H.M. Srivastava, Special functions in fractional relaxation–oscillation and fractional diffusion-wave phenomena [C], in: Proceedings VIII International Colloquium on Differential Equations, ed. By D. Bainov, Plovdiv 1997, VSP, Utrecht, 1998, pp. 195–202

13. V. Kiryakova, *Generalized Fractional Calculus and Applications* [M] (Longman, Harlow, 1994)

14. H.M. Srivastava, K.C. Gupta, S.P. Goyal, *The H-Functions of One and Two Variables with Applications* [M] (South Asian Publishers, New Delhi, 1982)

15. R. Hilfer, *Applications of Fractional Calculus in Physics* [M] (World Scientific Press, Singapore, 2000)

16. B.D. Carter, M.D. Springer, The distribution of products quotients and powers of independent H-function variates [J]. SIAM J. Appl. Math. **33**(4), 542–558 (1977)

17. A.M. Mathai, R.K. Saxena, *The H-Function with Applications in Statistics and Other Disciplines* [M] (Wiley Eastern Limited, New Delhi, 1978)

18. E.W. Weisstein, *CRC Concise Encyclopedia of Mathematics CD-ROM* [M]. (CRC Press, Boca Raton, FL, 1999)

19. F. Mainardi, G. Pagnini, R.K. Saxena, Fox H functions in fractional diffusion [J]. J. Comput. Appl. Math. **178**, 321–331 (2005)

20. J.G. Liu, *Some Applications of Fractional Calculus to the Constitutive Equations of Viscoelastic Materials* [D] (Shandong University, 2006), pp. 19–24

21. X.Y. Guo, *Some Applications of Fractional Calculus to the Researches on Quantum Mechanics and Non-Newtonian Fluid Mechanics* [D] (Shandong University, 2007) pp. 18–21

22. S.W. Wang, *Some Applications of the Theory of Fractional Calculus in the Viscoelastic Fluid Mechanics and Quantum Mechanics* [D] (Shandong University, 2007), pp. 17–21

23. W.G. Glöckle, T.F. Nonnenmacher, Fractional integral operators and Fox functions in the theory of viscoelasticity [J]. Macromolecules **24**, 6426–6434 (1991)

24. W. Wyss, The fractional diffusion equation [J]. J. Math. Phys. **27**(11), 2782–2785 (1986)

25. W.R. Schneider, W. Wyss, Fractional diffusion and wave equations [J]. J. Math. Phys. **30**(1), 134–144 (1989)

26. B.L.J. Braaksma, Asymtotic expansions and analytic continuations for a class of Barnes-integrals [J]. Compos. Math. **15**, 239–341 (1964)

27. http://www.mathworks.com/matlabcentral/fileexchange/8738-mittag-leffler-function

28. http://www.mathworks.com/matlabcentral/fileexchange/21454-generalized-generalized-mittag-leffler-function

29. http://mathworld.wolfram.com/Mittag-LefflerFunction.html25.

# Appendix B
# Related Electronic Resources of Fractional Dynamics

## *Web Resources*

1. Power-law phenomenon and Fractional dynamic system

   (http://www.ismm.ac.cn/ismmlink/PLFD/index.html).
2. Center for Self-Organizing and Intelligent Systems

   (http://www.csois.usu.edu).
3. Fractional calculus in Utah State University

   (http://www.mechatronics.ece.usu.edu/foc).
4. Institute of Soft Matter Mechanics

   (http://www.ismm.ac.cn/).
5. Group of Robotics and Intelligent Systems

   (http://www.ave.dee.isep.ipp.pt/~gris/index.htm).
6. Fractional calculus modeling

   (http://www.fracalmo.org/).
7. Jordan Research Group in Applied Mathematics (JRGAM)

   (http://www.mutah.edu.jo/jrgam/index.html)
8. Equipe CRONE

   (http://www.ims-bordeaux.fr/IMS//pages/accueilEquipe.php?guidPage=les_
   equipes).

## *Professional Journals*

1. Fractional Calculus & Applied Analysis (Fract. Calc. Appl. Anal.), ISSN 1311–
   0454 Website: http://www.math.bas.bg/~fcaa/

© Science Press 2022

W. Chen et al., *Fractional Derivative Modeling in Mechanics and Engineering*,
https://doi.org/10.1007/978-981-16-8802-7

2. Journal of Fractional Calculus, ISSN 0918-5402.
3. Fractional Dynamic Systems Website: http://www.fds.ele-math.com/.


## *Open Source Codes*

1. Program package on the Adams method and finite difference method by Kai. Diethelm

   (http://www-public.tu-bs.de:8080/~diethelm/software/software.html).
2. Predictor corrector method for solving the relaxation equation

   (http://www.mathworks.com/matlabcentral/fileexchange/26407-predictor-corrector-method-for-variable-order-random-order-fractional-relaxation-equation).
3. Matrix method for solving fractional partial differential equations

   (http://www.mathworks.com/matlabcentral/fx_files/22071/14/content/html/Matrix_Approach.html).
4. CRONE Toolbox (http://wWW.ims-bordeaux.fr/IMS//pages/pageAccueilPerso.php?email=alain.outloup)
5. Mittag-Leffler function curve

   (http://www.mathworks.com/matlabcentral/fileexchange/8738-mittag-leffler-function).
     (http://www.mathworks.com/matlabcentral/fileexchange/21454-generalized-generalized-mittag-leffler-function )
6. The random number generator of Mittag–Leffler distribution

   (http://www.mathworks.com/matlabcentral/fileexchange/19392-mittag-leffler-random-number-generator).
7. Fractional chaotic system

   (http://www.mathworks.com/matlabcentral/fileexchange/27336-fractional-order-chaotic-systems).
8. Impulse response invariant discretization of distributed-order low-pass filter

   (http://www.mathworks.com/matlabcentral/fileexchange/26868-impulse-response-invariant-discretization-of-distributed-order-low-pass-filter).
9. Digital Fractional-Order Differentiator/integrator—FIR type

   (http://www.mathworks.com/matlabcentral/fileexchange/3673-digital-fractional-order-differentiatorintegrator-fir-type).
10. A New IIR-type Digital Fractional-order differentiator

    (http://www.mathworks.com/matlabcentral/fileexchange/3518-a-new-iir-type-digital-fractional-order-differentiator).
11. Variable-order derivative

(http://www.mathworks.com/matlabcentral/fileexchange/24444-variable-order-derivatives).

12.  Fractional-order–differential-order equation solver

(http://www.mathworks.com/matlabcentral/fileexchange/13866-fractional-order-differential-order-equation-solver).

13.  Fractional-order control

(http://www.mathworks.com/matlabcentral/fileexchange/8312-ninteger).

14.  Part of the program code of Professor Mark M. Meerschaert

(http://www.stt.msu.edu/~mcubed/).

## *Key Words*

Fractional calculus

Fractional derivative

Fractional differential equation

Anomalous diffusion

Power law

Frequency-dependent dissipation

Softer matter

Path dependency

Stable distribution

Fractional Brownian motion

Fractal

Fractal derivative

Variable-order derivative

Random-order derivative

Distributed-order derivative

Fractional Fourier transform

Stretched Gaussian distribution

Fractional variational principle

Time-fractional derivative

Spatial/space fractional derivative

Continuous-time random walk

## *The Relevant Pages of This Book*

Owing to the limitation of our knowledge, although the book has been modified several times, surely there are many errors or improprieties. We urge readers of this book if you find any error or irregularity, please tell us your opinion by email, and we will further improve the book.

    Wen chen: chenwen@hhu.edu.cn.

    HongGuang Sun: shg@hhu.edu.cn.

    Xicheng Li: xichengli@hhu.edu.cn.

### References

1. J.T. Machado, V. Kiryakova, F. Mainardi, Recent history of fractional calculus [J]. Commun. Nonlinear Sci. Numer. Simul. **16**(3), 1140–1153 (2011).
2. http://www.ismm.ac.cn/ismmlink/PLFD/index.html.

Printed in the United States
by Baker & Taylor Publisher Services